DIGITAL EDUCATION FOR THE 21ST CENTURY

Technologies and Protocols

DIGITAL EDUCATION FOR THE 21ST CENTURY

Technologies and Protocols

Edited by

Souvik Pal, PhD

Ton Quang Cuong, PhD

R. S. S. Nehru, PhD

AAP | APPLE ACADEMIC PRESS

First edition published 2022

Apple Academic Press Inc.
1265 Goldenrod Circle, NE,
Palm Bay, FL 32905 USA

4164 Lakeshore Road, Burlington,
ON, L7L 1A4 Canada

CRC Press
6000 Broken Sound Parkway NW,
Suite 300, Boca Raton, FL 33487-2742 USA

2 Park Square, Milton Park,
Abingdon, Oxon, OX14 4RN UK

Library and Archives Canada Cataloguing in Publication

Title: Digital education for the 21st century : technologies and protocols / edited by Souvik Pal, PhD, Ton Quang Cuong, PhD, R.S.S. Nehru, PhD.

Names: Pal, Souvik, editor. | Ton, Quang Cuong, 1970- editor. | Nehru, R. S. S., 1974- editor.

Description: First edition. | Includes bibliographical references and index.

Identifiers: Canadiana (print) 20210169915 | Canadiana (ebook) 20210169974 | ISBN 9781774630075 (hardcover) | ISBN 9781774638064 (softcover) | ISBN 9781003180517 (ebook)

Subjects: LCSH: Educational technology. | LCSH: Educational innovations. | LCSH: Education—Effect of technological innovations on. | LCSH: Internet of things.

Classification: LCC LB1028.3 .D56 2022 | DDC 371.33—dc23

Library of Congress Cataloging-in-Publication Data

Names: Pal, Souvik, editor. | Ton, Quang Cuong, 1970- editor. | Nehru, R. S. S., 1974- editor.

Title: Digital education for the 21st century : technologies and protocols / Edited by Souvik Pal, PhD, Ton Quang Cuong, PhD, R.S.S. Nehru, PhD.

Description: First edition. | Palm Bay, FL : Apple Academic Press, 2021. | Includes bibliographical references and index. | Summary: "Digital Education for the 21st Century: Technologies and Protocols highlights the evolution of digital education related issues and by reporting on effective IoT-based technologies for the teaching-learning process. The volume brings together a selection of leading academic policymakers, researchers, educationalists, and education scholars to share their experiences and research results on many aspects of digital pedagogy in the Education of Things. The volume discusses recent innovations, trends, and concerns as well as the practical challenges encountered and solutions adopted in the fields of digital pedagogies and educational design. The chapters cover the concepts of IoT-based digital technologies regarding teacher and teaching education, IoT-based education, flipped learning, assessment process, and more. Key features: Introduces the integration of technology with digital education Explains the functional framework workflow in the Education of Things and and networked learning Explores basic and high-level concepts of teaching-learning pedagogy in IoT-based education Covers the major challenges, issues, and advances in flipped and blended learning based on IoT technologies Looks at digital education pedagogy collaborations with organizations outside the academy Explores teaching education and the process of assessment, testing, and evaluation Digital Education for the 21st Century: Technologies and Protocols provides a rich resource for academic and administrative policymakers, academicians, researchers, educationalists and experts who are concerned with educational research"-- Provided by publisher.

Identifiers: LCCN 2021012714 (print) | LCCN 2021012715 (ebook) | ISBN 9781774630075 (hardcover) | ISBN 9781774638064 (paperback) | ISBN 9781003180517 (ebook)

Subjects: LCSH: Computer-assisted instruction. | Educational technology. | Internet in education. | Digital communications. | Web-based instruction.

Classification: LCC LB1028.5 .D5534 2021 (print) | LCC LB1028.5 (ebook) | DDC 371.33/4--dc23

LC record available at https://lccn.loc.gov/2021012714

LC ebook record available at https://lccn.loc.gov/2021012715

ISBN: 978-1-77463-007-5 (hbk)
ISBN: 978-1-77463-806-4 (pbk)
ISBN: 978-1-00318-051-7 (ebk)

About the Editors

Souvik Pal, PhD, is an Associate Professor and Head of the Computer Science and Engineering Department at the Global Institute of Management and Technology, West Bengal, India. Prior to that, he was associated with Brainware University, Kolkata, India; JIS College of Engineering, Nadia; Elitte College of Engineering, Kolkata; and Nalanda Institute of Technology, Bhubaneswar, India. Dr. Pal received his BTech, MTech, and PhD degrees in the field of Computer Science and Engineering. He has more than a decade of academic experience. He is author or co-editor of 12 books from reputed publishers, including Elsevier, Springer, CRC Press, and Wiley, and he holds three patents. He is serving as a series editor for "Advances in Learning Analytics for Intelligent Cloud-IoT Systems," published by Scrivener Publishing (Scopus-indexed) and "Internet of Things: Data-Centric Intelligent Computing, Informatics, and Communication," published by CRC Press, Taylor & Francis Group, USA. Dr. Pal has published a number of research papers in Scopus/SCI-indexed international journals and conferences. He is the organizing chair of RICE 2019, Vietnam; RICE 2020 Vietnam; ICICIT 2019, Tunisia. He has been invited as a keynote speaker at ICICCT 2019, Turkey, and ICTIDS 2019, Malaysia. His professional activities include roles as associate editor and editorial board member for more than 100+ international journals and conferences of high repute and impact. His research area includes cloud computing, big data, internet of things, wireless sensor network, and data analytics. He is a member of many professional organizations, including MIEEE; MCSI; MCSTA/ACM, USA; MIAENG, Hong Kong; MIRED, USA; MACEEE, New Delhi; MIACSIT, Singapore; and MAASCIT, USA.

Ton Quang Cuong, PhD, is currently Dean of the Faculty of Educational Technology at the University of Education at Vietnam National University, Hanoi (VNU-UED). The university provides a new model of teacher education in Vietnam; develops research and service in educational technology, management and leadership; and collaborates with other institutions in Vietnam and overseas. Ton Quang Cuong has extensive teaching experience in teacher education, educational technology, and management skills. He has over 18 years of teaching experience in education. As a honor and prestige expert and Head of the Department of Educational Technology, Learning Theories and Instructional Technology, he is responsible for the development of curricula, learning methodology and technology, and leadership and management instruction at VNU-UED.

R. S. S. Nehru, PhD, is an Assistant Professor in the Department of Education at Sikkim University (Central University), Gangtok, Sikkim, India. He was previously a Postdoctoral Fellow and Invited Lecturer at the Vietnam National University and at the University of Education, Hanoi, Vietnam. Dr. Nehru has published 14 books as main and co-author and more than 30 edited book chapters related to education and educational technologies as well as several research papers in Scopus-indexed national and international journals and conferences. He is an editorial board member for two multidisciplinary international journals. Dr. Nehru has served at many conferences as chair, keynote speaker, co-chair, and resource person for national workshops on multidisciplinary research. He has mentored students for MPhil and PhD degrees. Dr. Nehru is an approved faculty member of Indira Gandhi National Open University, Kuvempu University, and other university open and distance learning mode programs on education and teacher education. Dr. Nehru is a member of many professional organizations, including the International Comparative Education Society (ICES), USA; Comparative Education Society of Asia (CESA), Japan; WWF-India; Dolphin Nature Conservation Society (DNCS); Indian Psychiatry Society (IPS); and others. His research area includes ICT in education, comparative education, and teacher education with a multidisciplinary focus.

Dr. Nehru received his MEd degree in education from Osmania University, Hyderabad, and MPhil from Madurai Kama Raj University, Madurai, Tamil Nadu, India. He received his PhD degree in education from Andhra University, Visakhapatnam, India. He has also acquired an MSc degree in Zoology, MSc (Psychology), MSW, PGDHR, PGDSE&IP, Honors Diploma in Computer Science (HDCS), Certificate Program in IoT, VR, IR, and Rabbits.

Contents

Contributors

Sonali Bhattacharyya
Department of Computer Science & Engineering, JIS College of Engineering, West Bengal, India

Nguyen Thi Bich
Hanoi National University of Education

Runjun Borah
Department of Education, Normal School Nagaon, State Council of Educational Research and Training, Assam

Pham Kim Chung
VNU University of Education, Hanoi, Vietnam

Tran Van Cong
University of Education, Vietnam National University, Hanoi, Vietnam

Ton Quang Cuong
University of Education, Vietnam National University, Hanoi, Vietnam

Bui Thi Thanh Huong
University of Education, Vietnam National University, Hanoi, Vietnam

Nguyen Thi Huong
Faculty of Preschool Education, Hanoi Pedagogical University 2, Vinh Phuc, Vietnam

Ton Quang Cuong
VNU University of Education, Hanoi, Vietnam

Aindrila Das
JIS College of Engineering, Kalyani, India

Nguyen Mau Duc
Thai Nguyen University of Education, Vietnam

Ninh Thi Hanh
Hanoi Pedagogical University

Nguyen Thi Hong Hanh
Quang Trung High School, Hanoi, Vietnam

Nguyen Thi Hao
Center for Higher Education Studies, The Vietnam National Institute of Educational Sciences, Hanoi, Vietnam

Tran Thi Hong Hoa
Faculty of Publishing, Academy of Journalism and Communication, Hanoi, Vietnam

Puvvada George Raja Kumar
Department of Education, Central University of Odisha, Odisha, India

Lai Phương Lien
VNU University of Education, Hanoi, Vietnam

Ha Thu Ly
VNU University of Education, Hanoi, Vietnam

Ira Nath
Department of Computer Science & Engineering, JIS College of Engineering, West Bengal, India

R.S.S. Nehru
Department of Education, Sikkim University, Gangtok, Sikkim

Nguyen Quang Ngoc
Cao Ba Quat High School—Quoc Oai District, Hanoi, Vietnam

Nguyen Duc Nguyen
University of Education, Vietnam National University, Hanoi, Vietnam

Do Thi Huong Nhu
School of Business and Management, RMIT University Vietnam, Hanoi, Vietnam

Adam Paul Patteti
Maulana Azad National Urdu University, College of Teacher Education, NUH, (Mewat), Haryana, India

P. K. Paul
Department of CIS, Information Scientist, Raiganj University (RGU), West Bengal, India

Ta Duy Phuong
Hanoi Institute of Mathematics, Vietnam

Pham Thi Thanh Phuong
Faculty of Educational Technology, University of Education, Vietnam National University, Hanoi, Vietnam

Nguyen Thi Thuy Quynh
VNU University of Education, Hanoi, Vietnam

Rukmini S
School of Social Sciences and Languages, VIT University, India

Dwaipayan Saha
JIS College of Engineering, Kalyani, India

Bikramjit Sarkar
JIS College of Engineering, Kalyani, India

Shabreen Sultana Shaik
Department of English, Bapatla Engineering College, Bapatla, Andhra Pradesh, India

A. N. Shankar
Department of Commerce, Sikkim University, Gangtok, Sikkim, India

Dharmpal Singh
Department of Computer Science & Engineering, JIS College of Engineering West Bengal, India

Nguyen Thi Bich Thuy
University of Science, Vietnam National University, Hanoi, Vietnam

Tran Le Thuy
University of Education, Vietnam National University, Hanoi, Vietnam

Pham Quang Tiep
Faculty of Pedagogy, University of Education, Vietnam National University, Hanoi, Vietnam

Vu Minh Trang
University of Education, Vietnam National University, Hanoi, Vietnam

Nguyen Thi Linh Yen
VNU University of Languages and International Studies, Hanoi, Vietnam

Abbreviations

AI	artificial intelligence
ANPA	American Newspaper Publishers Association
AR	augmented reality
BL	blended learning
BYOD	bring your own device
CA	computer application
CBI	content-based instruction
CDU	Charles Darwin University
CIEs	cumulative internal examinations
CO	course outcome
CS	computer science
CSE	computer science and engineering
DEd	Doctor of Education
DU	DEAKIN University
EdD	Doctor of Education in Educational Technology
EdTech	education technology
EGPs	English taught for general purposes
ELT	English language teaching
ESL	English as a second language
ESPs	English for specific purposes
EU	University of Edinburgh
FET	free educational timetable
GC	Google Calendar
GIS	geographic information system
HEIs	higher educational institutes
HES	high school of education sciences
IASLE	International Association of Smart Learning Environments
ICT	Information and Communication Technology
IDT	Instructional Design and Technology
IEQ	improving educational quality
IT	information technology
JHU	John Hopkins University
LMSs	learning management systems
LoS	levels of significance

MEd	Master of Education
MAA	Mathematical Association of America
MBA	Master of Business Administration
MCom	Master of Commerce
MDL	Master's Degree in Digital Learning
MHRD	Ministry of Human Resource Development
MOOCs	Massive Open Online Courses
MP	mobile pedagogy
MSc	Master of Science
NPIU	National Project Implementation Unit
OER	Open educational resource
SEE	Semester-end examination
SPU	Seattle Pacific University
TPACK	Technological Pedagogical Content Knowledge
UGs	undergraduates
VACI	Vietnam Anti-Corruption Initiative Program
VR	virtual reality

Preface

This book, *Digital Education for the 21st Century: Technologies and Protocols,* discusses the evolution of digital-education-related issues using the Internet of Things. This volume brings the IoT-based digital teaching-learning process-related technologies to a single platform, so that postgraduate students, researchers, and above all academicians and teaching staffs can easily understand digital teaching-learning methodology. This edited volume aims to bring together leading academic policymakers, researchers, educationalists, and education scholars to exchange and share their experiences and research results on all aspects of digital pedagogies in the Education of Things. It also provides a premier interdisciplinary platform for researchers, practitioners, and educators to present and discuss the most recent innovations, trends, and concerns as well as practical challenges encountered and solutions adopted in the fields of digital pedagogies and educational design.

This volume discusses the concepts of related IoT-based digital technologies regarding teacher and teaching education, IoT-based education, flipped learning, process of assessments, and novel findings of the researchers through its chapter organization.

It is becoming increasingly important to develop adaptive, robust, scalable, and digital teaching-learning mechanisms in academics. This book covers digital pedagogy and the integration of IoT in the teaching-learning process, MOOCs—the paradigm shift from traditional to digital pedagogy in academics, Flipped Learning and Blended Learning in Education of Things, Technology-based digital Learning and its challenges, IoT semantics and knowledge extraction in Education of Things, Digital pedagogy: Collaboration, Protocols for Educational and administrative policies, Intelligent object identification in smart classroom using Internet of Things, IoT-based teaching-learning methodology and different case-study-based approach in digital teaching-learning methodology. The book serves as a useful guide for academic persons and also helps beginners to learn things from basic to advance of digital education and teaching-learning process.

This book aims to attract academic and administrative policymakers, academicians, researchers, educationalists, and experts who are occupied with educational research. The book serves as a useful guide for academicians and teaching staffs and also helps beginners to learn things from basic advancement in the area of digital education.

Benefits to the Readers:

- Deals with the complete functional framework workflow in Education of Things.
- Explores basic and high level concepts, thus serving as a manual for those in the academics while also helping beginners to understand both basic and advanced aspects in teaching-learning pedagogy in IoT-based education.
- Covers the major challenges, issues, and advances in flipped learning and blended learning.
- Explores digital education pedagogy collaborations with organizations outside the academy
- Explains concepts of Education of Things and networked learning.
- Deals with the integration of technology with digital education.
- Explores teaching education and the process of assessment, testing and evaluation.

CHAPTER 1

The Potential of App Using for Interactive Smart Learning Environments

TON QUANG CUONG[1*], PHAM KIM CHUNG[1], and NGUYEN THI LINH YEN[2]

[1] *VNU-University of Education, Hanoi, Vietnam*

[2] *VNU-University of Languages and International Studies*

[*] *Corresponding author. E-mail: cuongtq@vnu.edu.vn*

ABSTRACT

A "smart learning" environment requires a combination of both formal and informal learning experiences along with the involvement of immersive activities. Accordingly, contextualized learning resources can be formed instantly, including digital content, activities, and connection during the learning process. Applications (apps) developed on smartphones (or mobile devices), with their ultimate purposes, can assist the adaptive, personalized, and ubiquitous learning experience for students. However, creating algorithms for the learning process by apps remains a problem for both teachers and students. Apps can be integrated as both specific subject tools and smart learning activities for digital literacy in various education programs of higher education institutions.

1.1 INTRODUCTION

In recent years, mobile learning (M-learning) has become a popular trend in education, in general, and higher education in Vietnam, in particular. There are some viewpoints that M-learning is the developing step of E-learning. However, from the perspective of intervention and application into the learning process, M-learning can be considered as an emphasis on "mobility/instantly/ubiquitous" in adaptive and personalized learning, web app development priorities in education, and the mobile intersection of cloud computing and online education. Also, it is an open opportunity for bring your own devices (BYOD) perspective in education.

The population of Vietnam was about 97 million (April 2019), 70% of which were working age, 40% were under 25, and 24% were members of the millennial generational bracket, with 145.8 million mobile phone connection, 68.17 million Internet users, and 65 million active social users. In other words, Vietnam is large, young, and increasingly tech-savvy. Nowhere is this more apparent than in their mobile app market, which has already established itself as one of the most productive markets in Asia (https://datareportal.com/reports/digital-2020-vietnam?rq=vietnam).

In the context of education (2020), with 22 million students across all grades and 1.5 million teachers and educators, educational managers make use of the apps in education, which promises to be a thriving application/development field. This enforces higher education institutions to rethink their instructional leadership, management, and learning processes. Toward digital platforms for M-learning, blended and flipped learning prospective educational apps facilitate collaboration among online learning professionals, provide professional development opportunities, and advocate and educate policymakers and researchers in best practices. Educational apps are opening access to courses in higher education, providing a new learning network, and changing the status of qualified participants: teachers and students. The perspective of changes may be:

1. The way of thinking about teaching and learning;
2. The way of practice on teaching and learning (including the way of assessment and evaluation);
3. The way of social interaction in the classroom (pre-duepost-class time) and learning environment; and
4. The way of content and curriculum design and development.

Using apps in class has numerous benefits, including course content and time extensions, supporting customized, and personalized, adaptive learning opportunities, allowing struggling students to master a subject by themselves by pace, interest, and competencies, providing a rigorous, interactive learning model of data-rich and digital communication. Hence, learning using apps in learning is considered to be a new approach to the pedagogical principles, teaching/learning content, which defines the integration of digital tools and resources that now support portable, flexible, and M-learning.

1.2 THEORETICAL FRAMEWORK

When designing and developing education programs in M-learning mode, higher education institutions should take into consideration the technology solutions, digital pedagogy approaches, content knowledge, the constituents (teachers and administrators), as well as the learning environment and learner-specific factors.

There is currently a common question in educational technology (EdTech) studies: What can EdTech do, and how should it be done to advance the 21st-century learning platforms? Recent studies in Vietnam showed that digital transformation in education is a popular trend, taking place at a relatively fast pace. Within the process, there is an emphasis on one key point that it is possible to digitize teaching activities with its elements with current technologies. Educators should not only "digitize" and replace the teachers' roles (digitalization activities in the learning process with all of its components but not replace teachers with their functional role). Besides, learning theories share one thing in common that people learn best when they are actively *involved* (minds-on), *engaged* with the learning materials and undistracted by peripheral elements, have *meaningful experiences* that relate to their lives, and socially *interact* with others in high-quality ways around new material, within a context that provides a clear learning goal (Kathy Hirsh-Pasek et al., 2015). Effective learning is facilitated in a flexible context that supports scaffolded exploration, questioning, and discovery as children work toward well-defined learning goals.

The relationship between some learning theories in light of technology integration have been shown in Table 1.1.

TABLE 1.1 The Relationship between Learning Theories and Technology Integration

Learning Theories	Technology Integration	Example
Behaviorism	Keep track of students' effort, attitude, and behavior modification, immediacy and drill techniques	Immersive, simulation, digital environment, mobile apps
Constructivism	Evidence, data gathering, instant feedback, assessment	Multimedia tools, mobile apps for specific subjects
Experiential	Experiments, rich toolkits	Simulation, AR/VR/XR, mobile apps
Situated cognition	Contextual knowledge	GPS, simulation, 3D, mobile apps
Cognitivism	Linkage of knowledge	AI, Big data, search engine, network

TABLE 1.1 *(Continued)*

Learning Theories	Technology Integration	Example
Connectivism	Communication, portability, ubiquitous	Social media, the Internet, mobile apps
Multimedia learning	Processing and integrating information, digitalization	Digital tools, platform, mobile apps

Notes. AR = augmented reality; VR = virtual reality; XR = extended reality; GPS = Global Positioning System; AI = artificial intelligence.

TPACK: This is a theory that explores technology integration. It provides a framework of three specific areas concerning the nature of knowledge within the learning environment. In the Technological Pedagogical Content Knowledge (TPACK) model, there is a dynamic concept of the learning process, working together to address the concepts of teachers' knowledge and pedagogy, through appropriate technical support and import of the content of the subject matter (Koehler and Mishra, 2005).

TPACK-XL: There is a model developed by Milad M. Saad and others (2012) that makes a transformative view of integration technology in general mode, adding the constituents such as learner-specific (L) and context-specific (X) in the learning process. Following this model, EdTech (T) considers essential roles in learning (i.e., platform, solutions and tools for interaction, sharing content, learning resources, assessment, etc. between teachers and students in the online environment). T in Figure 1.1 must be interconnected to other constituents in the model TPACK-XL (TPCLX in short version).

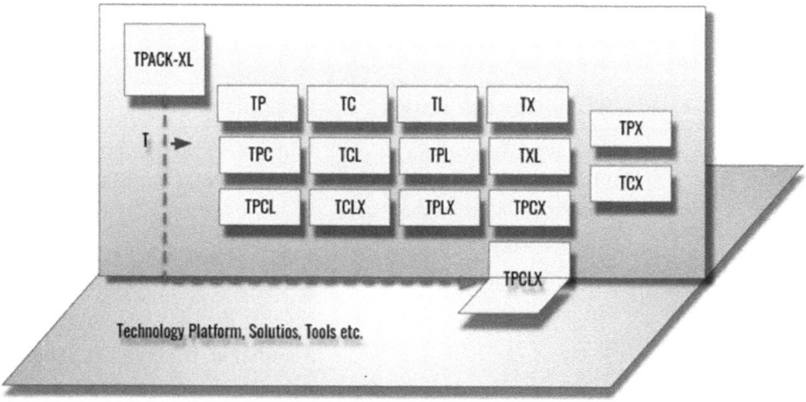

FIGURE 1.1 Educational technology interconnected constructs (adapted from M. Saad et al., 2012).

1.2.1 SMART LEARNING ENVIRONMENT

Smart learning environments, which can be online or physical, or cyber–physical learning environments, are being defined and conceptualized by the members of the International Association of Smart Learning Environments. The concept of the smart learning environment can be implemented at different levels, such as learning activities, classroom management, operational management, and institutional management. It may be a flexible combination. The integration between the when, where, and how of learning that occurs in the "smart" technology support the context. A smart learning environment "is an adaptive system that puts the learner at the forefront; improves learning experiences for the learner based on learning traits, preferences, and progress; features increased degrees of engagement, knowledge access, feedback and guidance; and uses rich media with seamless access to pertinent information, real life, and on-the-go mentoring, with high use of artificial intelligence (AI), neural networks, and smart technologies to continuously enhance the learning environment".

Focusing on the "smart" at learning activities and management level, the smart learning environment may be defined by the following dimensions:

Active involvement (minds on): It is the practice of selecting and using apps that match the content and subject goals. Apps installed on smartphones or devices directly support the students' critical thinking and creativity when conducting learning activities, testing, evaluating, responding, and seeking new knowledge.

Engagement: The use of educational apps creates a connection between various learning activities, encouraging learners to participate in answers, switch from search to practical, creative activities, promote positive and share ideas among learners, minimize distractions, and increase sensory and visualization.

Meaningful experiences: Mobile educational apps create links between the known and new-born knowledge, have search engines, and enable personalization of learning content and activities.

High-quality social interaction: Apps have tools that allow interactive learning, collaboration, and sharing in learning activities, providing instant feedback, connecting with teachers, and other social sources.

Technology-rich support: Apps leverage technology to maximize learning access and opportunities within an environment, resources, and system; and offer students the options of time, place, and pace; and emphasize different learning styles such as synchronous and asynchronouslearning.

Mobile learning: M-learning can be identified as "the use of mobile technology, alone or in combination with other information and communication technology, to enable learning anytime and anywhere". There are a variety of ways to unfold learning using this model; for instance, to access educational resources and materials, to connect with other participants, to create content within and without the physical classrooms. Also, comprehensive educational goals can be achieved via M-learning with an effective system of administrations and communications between schools and families. Four elements of M-learning were categorized as a method of learning, which is across multiple contexts, through social and content interactions, and via personal electronic devices. Besides, Hayman claimed that M-learning encourages learners to use mobile devices to access learning materials enriching his/her knowledge background, given the current limits of a physical classroom and limited interacting time with a course facilitator.

Based on the definition and Crompton's categorization of the four elements of M-learning, M-learning can be described as a process that occurs when a learner accesses higher education knowledge content at his/her convenience via any mobile devices, and it can be delivered through any formats such as games, storytelling, presentations, simulations, role-play, discussions, debate, case studies, and reflection. This process allows learners to achieve a high level of thinking and learning goals toward the 21st century's expectation.

1.3 THE CONSIDERATION OF MODEL TPACK-XL TO SMART LEARNING ENVIRONMENT

1.3.1 MOBILE APPS FOR SMART LEARNING

In terms of mobile technology for smart learning, apps are considered as a convenient and effective technology solution to implement BYOD policies, learning activities, instructional leadership, and management. Furthermore, mobile apps can offer new features for smart learning, such as follows:

1. Mobility: considers learning no longer restricted by the limitations of location, time, physical subjects, and context.
2. Connectivity: It is the consequence of mobility. It allows for a diverse, multiobject, multipurpose connection while showing that

the main relationship in the learning process is a connective network (connectionism).
3. Adaptability: It is the ability to meet the diverse requirements of all subjects participating in activities based on analyzing the characteristics of that subject. From there, it is possible to organize, manage, and provide appropriate services to the audience.
4. Ubiquity: It is a supplement just-in-time for all activities of learners inside and outside the classroom (Grab-learning/Uber-learning).
5. Continuity: It is a continuous connection between constituents of learning that blur the boundaries between informal, nonformal, and formal learning.
6. Contextual: It allows learners to choose the most appropriate conditions for their personal learning, because of their success and development.
7. Social: It allows learners to make effective social interaction during the learning process, turning learning into social interaction with learning goals (social cognitivism).
8. Multimode: It supports learning to be multimode (operational, academic, metacognitive, emotional, social, etc.).

Studies show that smart learning is a remarkable step-forward education, integrating many new technologies such as cloud computing, AI, Big Data, and Blockchain toward learner-specific, context-sensitive social technologies (Huang et al., 2020). Moreover, Huang (2020) states that the smart learning environment offers instant and adaptive support to learners based on their needs from different perspectives in real-world contexts in which they are situated.

In this context, the learners need to have a new user interface, the platform of presenting subject contents to meet the personal factors toward their new learning status. It means that the new learning environment just requires new ways of recognition, acquisition, and representation of learners' information, an effective process of learning styles and preferences development, and specific individual learners' performance in context. Hence, learners can interact with the learning environment (context) via technology, such as mobile apps, wearable devices, or even ubiquitous computing systems embedded in everyday content objects. Therefore, it is a challenging issue to adapt the L-, X-component interface to meet the T-P-C-K constituents in the model of TPACK-XL.

A mobile application or a mobile app is a software program that is designed to run on a mobile device such as a smartphone, tablet, or other types of mobile devices. These are individual software units that are coded to perform a particular function. Currently, apps are mostly designed and installed on the web platform, Android operating system, or IOS, which appear in almost all areas of human's life such as commerce, education, transportation, health, and entertainment. There are two types of apps: native apps installed on mobile operating systems and hybrid apps run on both web and mobile platforms (see Table 1.2). Educational mobile apps are software programs that help individuals to facilitate learning on their terms and their location choices. These apps provide the same access to learning material irrespective of the place you are. E-learning apps help in boosting engagement through knowledge-oriented activities.

TABLE 1.2 The Consideration TPACK-XL to the Smart Learning Environment by Using Mobile Apps

Characteristics	App Using for Smart Learning Environment
TPCL	Apps focusing on:
(including TP and TPC)	• individual knowledge
	• high-level cognitive objectives
	• knowledge construction of community collaboration
	• diversity of learning methods
	• learner sensitivity
TCLX	Apps providing:
(including TC and TCL)	• digital resources based on rich media
	• context sensitivity
	• deliver on-demand resources
	• seamless connection or automatic
	• online 24 × 7 access and synchronization
	• personal independent selection resources
TLPX	Apps integrating:
(including TL and TLP)	• tools with all-in-one functions, systematized, and specialized
	• learners judge the technology, the learning scenarios, and the environment
	• automatically sensing technology

TABLE 1.2 *(Continued)*

Characteristics	App Using for Smart Learning Environment
TPCX	Apps emphasizing learning resource design:
(including TX, TXL, TCX and TPX)	• formative and summative evaluation of learning outcomes (based on learners' cognition, behaviors, attitude, and competencies)
	• adaptive evaluation of learning; personalized learning
	• instant reflection, feedback, and task support
	• intervention on learning activities
TPCLX	Apps engaging:
	• virtual community focusing on online communication
	• automatically sensing technology environment
	• combination with the mobile interconnected real community to communicate anytime and anywhere
	• learning scenarios are automatically, personally recognized

On average, more than 3000 new applications appear on the two most popular platforms of the world every day, Android and IOS, in which the number of educational apps is always at TOP1 (Android) and TOP2 (IOS). According to the statistics, Android currently has 275,267 free and 14,884 paid education apps, and IOS has 151,546 and 26,793, respectively (https://42matters.com/stats).

According to the recent statistics, educational-content mobile apps are the third most frequently downloaded applications on mobile phones, with the current rate at 8.47%. Currently, the development of mobile apps continues to follow the AI trend, voice search engine, virtual assistant, Chatbot, VR/AR/XR, mobile Blockchain, and on-demand apps. Most of the platforms and digital solutions that used to run primarily on the web are now switched to run in parallel on the mobile (mobile application) platform. This allows hybrid apps to be used more widely, increasing mobility and flexibility for users.

Systematic review shows that educational apps can be divided into the following main groups:

1. Platform solutions: They provide flexible support and tools for students with online, real-time synchronous and asynchronous meetings, video-based real-time learning sessions, instant feedback,

forum/chatting. Therefore, learners have support for further practices and reflections on the learning content and material. Based on smart learning systems and mobile cloud computing, these apps allow learners to be embedded in the learning environment with personalized support for individual characteristics (fully flexible and M-learning preferences). It is currently common to use apps such as "portable" Learning Management System, Massive Open Online Courses, Small Private Open Courses, and social and collaboration platforms and spaces.

2. Learning activities tools: By using mobile apps, an individual learner can choose appropriate learning techniques, method, or approaches (lectures, seminar group work, self-study, discussion, games, etc.), as well as time, pace, places, and tutors to work with (individually, in groups, collaboratively, etc.). Some versions of apps also enhance learning experiences activated in cyber–physical environments, social studying by the ways of flexibility and mobility.
3. Learning authoring tools: The apps with Web 4.0 tools can support learners to generate content independently, learning materials explanation, VR/AR/XR objects, video/audio, animation, simulation, multimedia instant messages, and interact with peers and teachers. Additionally, various technology-based presentations, spreadsheet tools, graphics and infographics, screen capture, and screencasting techniques can make learning more interactive, ubiquitous, and smarter for students.
4. Content development tools: Apps allow convenient linkage to open educational resources (OER). Moreover, apps also support students instantly to capture, adapt, or mix fragments of contents given from OER, digital textbooks, or open license libraries by individual needs. In the learning context, students can quickly create reflective content presentations, interactive images/video/audio, 3D, simulation, VR/AR/XR objects, and reading resources in multimedia formats (screen capture, video, graphics, digital storyline, etc.) from the Web social computing, and its tools.
5. Instructional management tools: Based on mobile cloud computing, apps provide and deliver management tools such as time management, schedule of learning task delivery, automatic grading dashboard, e-portfolio, students' enrolment, assignment submission for the courses, including in campus-based learning or web-based learning, independent study or online learning, live virtual classrooms, etc.

6. Authentic assessment and evaluation tools: By using apps, students can submit authentic productions and assignments such as research papers, group presentations, team projects, or standardized tests. Also, the e-portfolio can be offered as a mobile, flexible formative assessment that allows students to continuously update the proof of their learning achievements and progress over time. Online and adaptive tests or paper-based tests can be uploaded via apps instantly. Thus, mobile applications provide dashboards or reports, instant feedback, and reflection by the analytics approach to learners, individually and flexibly.
7. Interactive tools for the learning environment: Mobile applications support time, place, subject role, and context for what students may interact with immersive and seamless learning. Location-based learning and gamification can be offered to learners via mobile devices, VR/AR glasses for enriching their digital literacy, spatial awareness, and immersion with their learning material, etc. in an M-learning environment.

1.3.2 MOBILE APPS IN THE TRANSFORMATIVE MODEL MPCLX

The concept of EdTech is broadly understood. However, in this study, the constituent "T" has been replaced by "M" (in terms of mobility or mobile technology). Hence, the model TPACK-XL can be transformed and represented as MPCLX (i.e., mobile pedagogy, mobile content, mobile learner, and mobile context as MPCLX) in the aspect of the mobile apps solution.

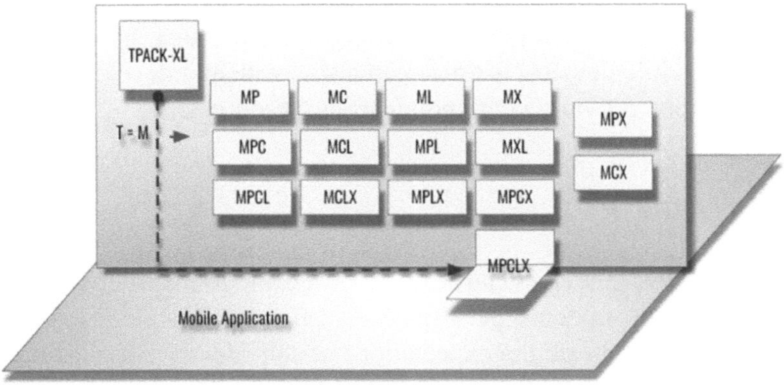

FIGURE 1.2 The transformative model MPACK-XL (M=Mobile).

MOBILE PEDAGOGY

Currently, the concept of mobile pedagogy (MP) is still quite controversial. However, it can be confirmed that all basic issues of MP have been implemented and developed by the known learning theories, and mobile technologies prospects for their use in learning (e.g., cognitivism, behaviorism, constructivism, situated learning, connectivism, transactional distance theory, collaborative learning, conversational learning, activity theory, problem-based learning, context-awareness learning, sociocultural theory, lifelong learning, enquiry-based learning, or informal learning). MP may be focused on personality, mobility, network, ubiquitous, and durability. In short, "learning is mobile, situated, contingent, context-aware, and authentic. Learning happens both formally and informally, in and between classrooms, homes, transportations, and other spaces in communities, extending beyond learners' immediate physical environments and networks" (Kukulska-Hulme et al., 2015).

MOBILE CONTENT

1. Using apps allows learners to access information, services, learning materials, offering courses at any time, anywhere with any interaction, and with anybody according to their needs, attention, or competence level.
2. The use of mobile apps allows for instant access to digitized, informative educational input and educational data sources, capable of "multisensitization" and strong interaction for learners. Developed based on AI technology, Big Data, web, and app, mobile digital content is easy to reuse, accessible, look content up, share, and contribute.... On the other hand, during learning, by using features of mobile devices, learners can also create and develop new mobile content, providing common learning resources.

MOBILE LEARNER

Learner-centered strategies can be transformed into user-centered M-learning toward mastering students' personalized academics. Mobile learner orientation and targets indicate the expectations of self-identifying gaps in knowledge, social connectivity with other users, application knowledge for real purposes in real-world contexts, and developing multimodal digital (mobile) literacies. Mobilities also highlight the learners' role and experience in enacting the process where and when they learn with mobile technologies. Moreover, learners create personalized, authentic artifacts content for their learning, using mobile device features in contexts.

MOBILE CONTEXT

1. Learning contextualization can be personalized at high levels of both learners' and teachers' activities. App users should adapt to technological, metacognitive, academic skills in social-share networks in cyber–physical environments.

   In the learning process, learners who use apps will have to follow strict course-related instructions, self-direction, and choose an appropriate way to handle learning tasks that are relevant to the context, personal conditions, and competencies. Moreover, thanks to technological tools, the learning context can be flexibly adjusted toward a higher level of thinking and competence-based development (i.e., 4Cs: communication, collaboration, critical thinking, and creativity). Moreover, learning coontextualization is able to break the training program into modules/content package/topic unit, etc. It also allows the creation of a user-created mobile context that can stop operating in intermediate stages and continue at more favorable times. For instance, learners can select options according to the optimal plan, which can be repeated several times until the result is achieved.

   In short, in the M-learning context, to increase the effectiveness of the process, learners must develop several competencies concurrently, that is, technology (operational), learning (academic, metacognitive, and affective), and social (relational) skills.
2. In contrast to traditional learning, app users receive content knowledge (factual, conceptual, procedural, and metacognitive) and teachers' instructions through technological tools. So, pedagogically, the M-learning context can be added with phenomena such as *technophobia*, *technophilia*, and maybe *technomania*. Both teachers and learners should be active "players" in the learning process following technical indicators. In this context, the learners may face new ethics, misbehaviors, academic plagiarism issues, and unexpected conflict management to teachers. Hence, apps use both mindset shifts and skillset in a flexible context.

3. On the other hand, the mobile context has a close relationship with the learning environment with mobile and ubiquitous learning strategies (i.e., blended, flipped, and personalized learning approaches). Through gamification, VR/AR/MX, simulation, and 3D environment, digital resources will motivate learners to interact actively and perform multiple academic, metacognitive, and technology competencies. Lectures, narrations, presentations, discussions, debate case studies, assessment, and reflection, etc. via mobile devices encourage learners to transform from the followers, that is, customers to producers, that is, creators at their convenience, flexibility, and portability. The well-organized and structured mobile context also indicates one's personality, characteristics of individual learners' progression, control, mastering, and development throughout the learning process.

1.4 RESEARCH DESIGN

On the whole, a smart learning environment with technology must be designed to support the mechanism that explores and develops the learning capacity of learners by utilizing mobile apps in the learning process with personalization and collaboration.

The survey of students' knowledge of technology is based on a 22-item Likert-type scale survey for teachers and 14-item- Likert-type scale survey for students that include 10 subscale measures for MPCLX domains (MPCL, MCLX, MPLX, MPCX, and MPCXL), including three subscales for different areas of specialization (English and Language Arts, Teacher Education in Math, Physics, Chemistry, Biology, History, and Philology, EdTech Management, School Management, Educational Sciences, Educational Quality Management, and Schooling Counseling) (Tables 1.3–1.5). Participants rated their agreement with each item on a 5-point Likert scale (5 = Strongly Agree, 4 = Agree, 3 = Neither Agree or Disagree, 2 = Disagree, 1 = Strongly Disagree). The measure of MPCXL domains used in this study represents participants' self-assessment of their knowledge and apps using.

The MPCXL focuses on general apps using themes in M-learning and has been used in this study because of its generic nature since the sample consists of students of various areas of specialization education, attending different study programs and courses. The original MPCXL inventory was written in English with Vietnamese translation and was proof read.

TABLE 1.3 Descriptive Statistics of Teachers

Item	N	Mean	Std. Dev.	Variance
Item 1: App on smartphone helps me better perform as an experienced and knowledgeable professional	39	3.026	0.8732	0.762
Item 2: Easy to find interesting, effective and effective teaching content	39	3.179	0.7905	0.625
Item 3: Provide teaching content and support student learning	39	3.205	0.7320	0.536
Item 4: Teach different students and students from different backgrounds	38	2.684	0.8732	0.762
Item 5: Learn about students' interests, habits, and emotions	39	2.615	0.9066	0.822
Item 6: Communicate and chat with students	39	3.205	0.8329	0.694
Item 7: Monitor learners and provide pertinent guidance	39	2.821	0.8847	0.783
Item 8: Design and select appropriate teaching activities	39	2.615	0.7819	0.611
Item 9: Provide timely, informative feedback to students	39	3.205	0.8006	0.641
Item 10: Adapts learning content for individual learners	38	2.816	0.8654	0.749
Item 11: Adapts learning tasks or objectives for individual learners	39	2.744	0.9095	0.827
Item 12: Provide feedback or guidance to individual students or groups of students	39	2.821	0.8847	.783
Item 13: Provide learning instruction or support across contexts (in the classroom, on campus, in the library, in the field)	39	2.795	0.8006	.641
Item 14: Introduce students to learning tools or strategies	39	2.897	0.7879	.621
Item 15: Considers the real-world learning status of learners	39	2.564	0.7538	.568
Item 16: Facilitates both formal and informal learning	39	2.846	0.7448	.555
Item 17: Interact with students through multiple channels	38	3.421	0.6831	.467
Item 18: Assist students with applications that adapt to real and virtual contexts	39	2.615	0.8465	.717

TABLE 1.3 *(Continued)*

Item	N	Mean	Std. Dev.	Variance
Item 19: Easy Wi-Fi connection in the classroom	38	2.658	1.0469	1.096
Item 20: Skills to use the smartphone for learning	39	2.897	0.8206	.673
Item 21: Internet speed is sufficient to use a smartphone in studies	39	2.615	0.8465	.717
Item 22: Satisfied using a smartphone in teaching	39	3.103	0.8521	.726
Valid *N* (listwise)	**35**			

TABLE 1.4 Descriptive Statistics of Students

Item	N	Mean	Std. Dev.	Variance
Item 1: Flexible use to study at any time	186	3.124	0.7284	0.531
Item 2: Learn more using smartphones in university study	186	2.925	0.8086	0.654
Item 3: Helps to do homework faster	186	2.941	0.8132	0.661
Item 4: Save time in the learning process	186	3.070	0.7990	0.638
Item 5: Learn and conduct research at the university	184	2.745	0.8396	0.705
Item 6: Find new ways of learning at the university	186	2.860	0.8066	0.651
Item 7: Interact with other students	184	3.147	0.7997	0.640
Item 8: Interaction with the teacher	186	3.081	0.7773	0.604
Item 9: Effective teamwork	185	2.973	0.8105	0.657
Item 10: Teachers encourage the use app of smartphones in the classroom	185	3.065	0.7194	0.518
Item 11: Easy Wi-Fi connection in the classroom	185	2.735	1.0426	1.087
Item 12: Skills to use the smartphone for learning	183	3.071	0.7706	0.594
Item 13: Internet speed is sufficient to use a smartphone in studies	186	2.597	0.9380	0.880
Item 14: Satisfied using a smartphone in university's studies	185	3.049	0.7752	0.601
Valid N (listwise)	177			

TABLE 1.5 Compare Mean Statistic in Area Specialization

		Frequency	Percentage	Valid Percentage	Cumulative Percentage
Valid	1. English and Language Arts	16	8.6	8.6	8.6
	2. Teacher Education in Math	16	8.6	8.6	41.4
	3. Teacher Education in Physics	16	8.6	8.6	50.0
	4. Teacher Education in Chemistry	16	8.6	8.6	58.6
	5. Teacher Education in Biology	16	8.6	8.6	67.2
	6. Teacher Education in History	16	8.6	8.6	75.8
	7. Teacher Education in Philology	15	8.1	8.1	83.9
	8. Educational Technology Management	15	8.1	8.1	91.9
	9. School Management	15	8.1	8.1	100.0
	10. Educational Sciences	15	8.1	8.1	16.7
	11. Educational Quality Management	15	8.1	8.1	24.7
	12. Schooling Counseling	15	8.1	8.1	32.8
	Total	186	100.0	100.0	

1.5 DISCUSSION

To be able to apply educational apps effectively in teaching, it is necessary to synchronize the relationships and implementation conditions toward the implementation of educational goals and policies. The relationship expressed between entities operating on the psychological, academic, and technological levels (teachers, learners, managers, administrators, service providers, etc.), compatibility and response of public groups tools (tools for content construction and development, learning activities and analytics, instructional management, assessment and feedback, personalization and collaboration, etc.), and the principles set by digital and MP.

Furthermore, the M-learning undisruptive characteristic that responds to lifelong and informal learning also sets requirements for future apps development at policy, system operation, and individual user levels. In the field of education, this requirement is also expressed in cooperation between the government–society–enterprise–school system with essential conditions of what and how smart learning works in a digital smart society.

Using apps is a prospective solution to technology-enhanced M-learning. It is more than technological integration in the smart learning process and environment. Therefore, at the microlevel, the following discussion may be raised to acknowledge the presence of easiness and usefulness of friendly learning tools that make the MPCLX model more popular and adaptable for different learning modes. It is also important to suggest the framework, standards, or indicators of quality and effectiveness for apps using different education levels, M-learning organizations, and social networks of participants. Last but not least, this conception also relates closely to the better understanding of the learners' behaviors (flexibility, mobility, adaptability, etc.) and the challenges of conducting smart and seamless, technology-enriched learning activities.

1.6 CONCLUSION

The smart learning environment is both a characteristic and an indispensable result of digital transformation in education. The process of teaching based on technology (M-learning, ubiquitous learning) with the support of learning apps has restructured all teaching activities, content, methods, and assessment methods. The MPCLX model contributes a new perspective on digital transformation in education through the application of MP, mobile content, M-learning, and sufficient mobile context in M-learning.

Currently, the Vietnam government and Ministry of Education and Training have issued national policies and programs on digital transformation in education with clear national goals. This allows expectations for new developments in smart learning based on digital technology in terms of program development, digital pedagogy transformation, open digital learning resources, and the opportunity to participate in all subjects. In which the connection between government, businesses, and educational institutions is essential. On the other hand, smart learning still leaves many fields for educational experts, educational policymakers, teachers to continue researching interdisciplinary/transdisciplinary in EdTech at the macro/meso and micro levels.

ACKNOWLEDGMENT

The National Education and Science Programs have partly supported this work under the project KHGD/16–20DT.042

KEYWORDS

- **smart learning environment**
- **apps using**
- **mobile learning**

REFERENCES

Hirsh-Pasek, K., Zosh, J. M., Golinkoff, R. M., Gray, J. H., Robb, M. B., & Kaufman, J. Putting education in "Educational" apps: Lessons from the science of learning. *Psychological Science in the Public Interest.* 16(1), 2015: 3–34. https://doi.org/10.1177/1529100615569721

Huang, R. H., Liu, D. J., Tlili, A., Yang, J. F., Wang, H. H., et al. *Handbook on Facilitating Flexible Learning During Educational Disruption: The Chinese Experience in Maintaining Undisrupted Learning in COVID-19 Outbreak.* Beijing: Smart Learning Institute of Beijing Normal University, 2020.

Koehler, M. J., & Mishra, P. What happens when teachers design educational technology? The development of Technological Pedagogical Content Knowledge. *Journal of Educational Computing Research,* 32(2), 2005: 131–152.

Kukulska-Hulme, A., Norris, L., & Donohue, J. *Mobile Pedagogy for English Language Teaching: A Guide for Teachers.* London: British Council, 2015.

Saad, M. M., Barbar, A. M., & Abourjeili, S. A. R. Introduction of TPACK-XL: A transformative view of ICT-TPCK for building pre-service teacher knowledge base. *Turkish Journal of Teacher Education,* 1(2), 2012: 41–60.

CHAPTER 2

Blended Learning in Teaching History: Contents and Paradigms

NGUYEN THI BICH[1*] and NINH THI HANH[2]

[1]*Hanoi National University of Education, Cau Giay District, Hanoi, Vietnam*

[2]*Hanoi Pedagogical University, Cau Giay District, Hanoi, Vietnam*

[*]*Corresponding author. E-mail: bichnt@hnue.edu.vn*

ABSTRACT

Blended learning is a modern learning model and an inevitable trend in the digital era all over the world. In Vietnam, blended learning has not been commonly used yet and has been encountering a variety of difficulties. With the increasing number of Internet users and educational reforms and investments in recent years, reality shows that blended teaching has many development opportunities in schools in Vietnam. Based on clarifying the concept of blended learning, this chapter focuses on analyzing the role and meaning of blended learning, and proposes the process of using this model in teaching History—one of the subjects facing many difficulties in the innovation process in high school so that students can be more creative, critical, and interactive and collaborate.

2.1 INTRODUCTION

The explosion of science and technology from the 1970s of the twentieth century significantly influenced the development of humankind. "We are standing in the technological revolution that fundamentally changes the way we live, how we work, and how we connect with each other. The speed of this revolution is an unprecedented breakthrough in human history" (Schwab,

2017). Education is no exception; the power of technology makes access to knowledge more comfortable and more equitable. Technology paves the way for many new teaching models, in which technology itself becomes a useful support tool to combine face-to-face (F2F) teaching with the increasingly popular online teaching method in the world. In Vietnam, technology applied to educational practices is still new and faces specific difficulties. The use of the blended learning (BL) model in the subjects at high schools has become one of the solutions to shape the form of the learning organization, increase the efficiency of using technology facilities, and develop learners' abilities. Applying the BL model in teaching History provides an opportunity to overcome one of the "severe drawbacks of subject education which is the majority of students do not like History, considering it a subject of events, subjects of memory, dry, boring" (Le, 2012). The chapter focuses on solving the following issues:

1. General analysis of the BL model in teaching History
2. The theoretical basis of the BL model in teaching History
3. Introduce some models of applying the BL model in the History subject
4. Discussion and conclusions.

2.2 LITERATURE REVIEW

BL is not a new concept. Lectures, libraries, books, or handouts in teaching BL are part of the traditional classroom. In the years 2002–2003, the concept of BL is understood according to four main ideas (Driscoll, 2002):

1. Combining or blending web-based technology models (e.g., live virtual classroom, self-study instruction, collaborative learning, video, audio, and text streaming) to fulfill educational goals.
2. Combine different pedagogical theories (e.g., constructivism, behavioral theory, and cognitive theory) to achieve optimal learning outcomes whether or not technology is used in teaching.
3. Combining any form of technology use in teaching (e.g., videotapes, CD-ROMs, web-based teaching, and movies.) with F2F teaching.
4. Combining or blending educational technology with practical tasks to create a harmonious effect between learning and work.

The concept of teaching BL changes over time with the development of technology, mobile devices, and the Internet. Today, BL, which is widely used in the world and Vietnam, is a combination of F2F teaching with online teaching (via the web) in different combinations through technology tools such as websites, lectures, audio, interactive forums, and online submissions and assessments to ensure the highest educational effectiveness (Bonk & Graham, 2006; Garriso & Vaughan, 2008; Giang & Nam, 2019; Hung, 2018; Hoang & Quang, 2011; Voos, 2003).

BL is becoming a popular in global education. Many studies have shown the effectiveness of using the BL teaching model to train and develop higher order thinking for learners, such as critical thinking (Akyuz & Samsa, 2009), problem-solving and creativity (Yen & Lee, 2011), collaborative learning (Schweizer et al., 2003). However, most research on the BL teaching model focuses on higher education. Lesmana and Basiran (2017) showed that the application of the BL model in teaching the History subject at the high school level in the world as well as in Vietnam seems very less. The US Department of Education studies conducted by the Office of Planning, Evaluation and Policy Development and Policy and Program Studies Service between 1996 and 2006 found no studies or trials and experiments comparing the effectiveness of online and F2F learning for high school students (K-12). From 1996 to July 2008, the expanded statistical scope found 1132 abstracts of online and combined learning, but only 9 studies for high school (OPEPD, 2010). In Indonesia and Malaysia, the application of teaching BL at the high school level has not gained much attention, like at the university level (Lesmana & Basiran, 2017; Nordina & Alias, 2013). The above fact is true for Vietnam. In December 2015, the Ministry of Education and Training, Vietnam, held a seminar for the first time: "The application of information technology (IT) in combined teaching to meet the requirements of the new general education program." A brief survey of some journals such as the Education Review; *Journal of Vocational Education; Scientific Journal of Pedagogical Universities* confirms that. In Vietnam, articles are available on IT applications in teaching. Besides, a few studies have already applied the BL model in the subjects of Natural Sciences, Foreign Languages, and Informatics, the majority of the remaining ones focus on the proposal and testing of the BL model in university training. Particularly, some inter-level schools such as The Olympia School and FPT High School have implemented the BL teaching model for a few years and had positive results (FPT Education, 2019; The Olympia Schools, n.d.)

Thus, the application of the BL model in teaching has not been emphasized at high schools, especially in the History subject. However, it does not mean that the application of the BL model is not effective. It is important to identify mechanisms and conditions for the implementation of the BL model clearly and have a reasonable plan to apply it in teaching subjects.

2.3 THEORETICAL BASIS

2.3.1 THE ROLE OF BLENDED LEARNING IN TEACHING HISTORY

BL teaching models have three outstanding advantages compared with traditional teaching models. First, they help to expand the classroom space, not limited to a four-wall and 8-hours class, and broaden the opportunities for communication and social sharing of learners. Second, by expanding the learning content, learners can experience and have access to diverse learning content, knowledge, and updates outside textbooks in a fair manner. Third, personalize the learning process. BL model teaching creates opportunities for learners to learn according to their needs, interests, and personal abilities.

In addition to the above role, for teaching History in high school, the model of teaching BL has many positive meanings (Hanh, 2016). First, in accordance with the characteristics of History knowledge, which is about past and is nonrepeatable, the unity between "history" and "discussion" helps learners visualize vividly about the past, concretizes events, overcomes the situation of "modernizing" history, and creates interest in learning. Second, it helps overcome limitations in the curriculum, textbooks, and the distribution of time for history in high school. The current History curriculum is thought to be cumbersome, and textbooks are dry, almost summarizing the history of adults and forcing students to study biology(Le, 2012; Vy, 2012). Compared with the goal, the content of the course and the distribution of the current History curriculum (Basic Board) at high school are built in 139 class hours (52 hours in grade 10, 35 hour in grade 11, and 52 hour in grade 12), averaging 1–2 hour/week (depending on each term) is not adequate. Third, developing common and subject-specific competencies: output-oriented characteristics built on technology, combined teaching, and learning have the advantage of developing self-study and using IT, cooperation, creativity, and specific capabilities of History subjects such as collecting and processing information about historical events and phenomena; re-constitute historical events and phenomena; explain and evaluate events and phenomena from a historical perspective; applying historical knowledge to create real products. Fourth, the development of information and communication technology

(I.C.T.) skills is still less compared to the subjects of Natural Sciences such as Chemistry, Physics, Biology, and Mathematics (Huong, 2012) for teachers of History in high schools. There are not only challenges but also an opportunity for teachers to develop professional skills under the technological pedagogical content knowledge model in the 4.0 technology era, starting from simple training on complicated technological tools to design learning lessons and managing classroom.

2.3.2 CONDITIONS AND PRINCIPLES OF APPLYING BLENDED LEARNING IN TEACHING HISTORY

Conditions for applying the BL model in teaching the History subject: implementing the model of BL teaching in Vietnamese education should ensure four conditions with core characteristics.

First, regarding the class facility: Ensure basic equipment for teaching with technology integrated. In addition to the general standards prescribed by the Ministry of Education and Training on subject classrooms and the list of minimum teaching equipment at the high school level for the History subject, it is necessary to equip other supporting devices: projectors, a large screen or T.V. with an HDMI port connected to a laptop or desktop computer, Internet or stable WiFi. Computers can run some basic applications such as web browser, MS PowerPoint, Kahoot, Canva, Easelly, Padlet, Edmodo, and Mentimeter.

Second, regarding the student's learning medium: Upon participating in learning using the BL model, besides textbooks, notebooks, and basic learning materials, each student or group (4–6 students) has a computer or smart device connected to the Internet or WiFi. Students create individual or group accounts to log in to the applications: Canva, Easelly, Kahoot, Padlet, Edmodo, and Mentimeter. In case students do not have computers, smartphones or tablets, teachers can instruct students to use the school's computer room and computer lab to prepare and participate in the lesson.

Third, it is important to have the agreement, support, and encouragement of high school managers. In order to deploy teaching BL, the principal must be the leader, facilitator, or driver of innovation. The Vietnam Principles of General Education Institution 2018 has two criteria to evaluate the level of management and facilitate the application of IT in general and the use of I.C.T. in schools in particular (MOET, 2018)

Fourth, knowledge and skills in using IT: This is one of the conditions determining the success of the lesson using the BL model. Teachers know how to solve basic computer problems; can use IT to support the presentation of lively and interesting lesson content; can understand and be able to use basic applications and software (Internet, MS PowerPoint; MS Word; Canva, Padlet.) in the lesson; can be able to learn and update IT, can enhance teaching methods of history, and can achieve lesson objectives. These are basic requirements for teachers to teach effectively with the support of IT in general and to use the BL model in particular. Application of IT, exploitation, and use of technological equipment in education have become a standard for assessing the professional competence of high school teachers (MOET, 2018).

THE PRINCIPLES OF APPLYING THE BL MODEL IN TEACHING HISTORY

There are three key principles of applying the BL model in teaching History (Garriso & Vaughan, 2008):

1. Flexible integration between F2F and online learning (O.L.)
2. Design the lesson/course that optimally combines the learner's participation
3. Restructuring and replacing traditional classroom communication is a systematic view of the lesson/course and online—direct combinations to redesign the structure and organization of teaching BL in History.

In teaching BL, complete integration is not a rigid, mechanical supplement to compensate for the disadvantages of traditional online or F2F. BL is a completely new model of teaching in terms of quality-fundamentally changing the perspectives on traditional teaching theory. The integration process ensures the flexibility to promote the strengths of each form of learning and reduce their limitations. For History, O.L. not only supports students to study simple knowledge of statistical content (such as dates, places, and historical figures) but also extends the approach to the problem through access to diverse historical documents, developing critical thinking and historical–critical thinking; in the context of the current History textbooks that lack updates (Le, 2012), historical documents are very Modest (Vy, 2012). Meanwhile, F2F is suitable for communication, direct interaction between teachers and students, and between students to exchange, explain, and discuss many interesting historical issues.

One of the important factors that determine the success of a BL model lesson/course lies in the appropriate planning and design process. The design of the BL lesson to ensure optimal participation of the learner is the change from "teacher-centered teaching" to "learner-centered teaching." In particular, learners actively participate in learning activities and enhance interaction between teachers and learners, learners with learners, learners with knowledge content, and learners with media, teaching materials.

The way teachers and learners communicate is an important part of the structure of BL lesson/course. The tools support communication in every class activity from F2F to asynchronous communication by using e-mail, forums, social networks (Facebook, Instagram, Twitter.), helping improve student learning performance.

2.3.3 DIFFERENT MODELS AND METHODS OF APPLYING BLENDED LEARNING IN TEACHING HISTORY

In 2011, BL teaching was proposed by Heathe Staker with six models. In 2014, there were four major models: Rotation, Flex, A La Carte, and Enriched Virtual.

The proposed model by Michael B. Horn and Heather Staker is commonly used to shape the way teachers organize teaching BL in a scientific and systematic manner. Model selection depends on teachers and their teaching context. Choosing the most appropriate design model for a blended course is a big challenge for many high school teachers because they are not familiar with the concept of teaching BL teachers to choose and effective combination based on the common combination method (Table 2.1).

TABLE 2.1 The Combined Method in Teaching (Alammary et al., 2014)

Level of Combination	Characteristics of Lesson/ Course	Characteristics of Teachers
Low	Add some online activities to existing traditional lessons/courses	Teachers do not have experience in designing and implementing BL; lack of knowledge and confidence in integrating technology into teaching; no instructions; support from the school yet
Medium	Replace some traditional lesson/course activities with online activities	Teachers have designed and implemented BL; have basic knowledge and confidence in integrating technology into teaching; get guidance and support from the school.
High	Redesign the entire lesson/ course in a combined format	Teachers with experience in designing and implementing BL; have good knowledge and confidence in integrating technology into teaching; get guidance and high support from the school.

In addition, to each subject, it is possible to base on the following criteria to choose an effective combination model and method:

1. By the student: regarding learning styles: learners in visual style, learners in the listening style, learners in the motor style, learners in the writing style; regarding learner's ability: good students, fairly good students, average students.
2. By the progress of organizing the lesson: start-up, learn new knowledge content; preliminary review and consolidation; practice and expand.
3. By lesson type and form of lesson organization: research paper on new knowledge content; review and preliminary review; project-based teaching; problem-solving; integrated teaching; WebQuest-based teaching.

With the current situation of applying BL teaching in Vietnam, the combination of low and moderate levels can be widely deployed with many public and private schools, some continuing education institutions where equipment, learning materials are not yet synchronized support. Teachers through professional activities or consultations from experts improve the experience of building and organizing BL teaching so that they can switch to integrated teaching at a higher level. The method of high combination can be suitable for international schools, inter-level schools, and specialized and gifted schools, where the system of equipment and technology is fully equipped and modern.

2.3.4 DESIGNING PROCESS AND DEPLOYMENT OF THE BLENDED LEARNING MODEL IN TEACHING HISTORY

To design and implement the BL model in teaching History, it is possible to follow the proposed process (Figure 2.1).

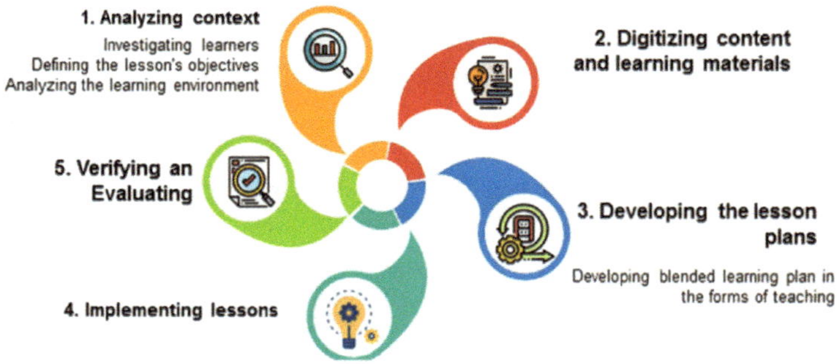

FIGURE 2.1 Designing process and deployment of the BL model in teaching History in Vietnam

Step 1: Context analysis is the first and important basis for determining the method of incorporation, the content, and the results of the lesson. Context analysis focuses on answering three main questions: Who is the learner? What are the goals of the lesson? What are the advantages and disadvantages of the environment/learning conditions for applying BL?

To answer the question, "What is the objective of the lesson?..." it is necessary to conduct a preliminary investigation of the learners' needs and competencies. The survey questions focus on five basic issues: student's level of interest and understanding of historical forms/knowledge; students' interest level when participating in IT-assisted lessons; proficiency and difficulty of students when using IT; and student's wishes/expectations for the subject. To facilitate the process of surveying learners and obtaining quick results, you can use any of the following applications: Google Forms, Survey Monkey, or Vizir to design online questionnaires. With these applications, students can answer online survey questions through various electronic devices such as computers and smartphones. Teachers will immediately receive answers in the form of lists or graphs of statistics so that it is easy to draw up plans for organizing learning activities suitable to learners.

To answer the question, "What is the objective of the lesson?," teachers base subjects and school programs on national programs to determine the required level of knowledge, skills, attitude, and competence in History lessons. Note that the most common and basic goal of each History lesson is that students are able to learn about historical events, characters, and phenomena through documentary exploration.

The learning conditions, in this case, are associated with the infrastructure, equipment, and learning facilities, including learning materials (digital learning materials) and technical facilities for teaching. Before designing and implementing a BL lesson, the teacher lists the necessary resources, determines which resources are available and what are additional resources, and plans to select technology applications to design the most appropriate integrated teaching website.

The popular teaching website design tool can be divided into the following two groups:

1. A group of highly secure tools for complete packaged products (e.g., Java or .Net is usually highly professional, requiring users to have specialized knowledge and skills in programming (professional programmers often choose this group).

2. A group of open-source tools and usually free. Easy to access, easy to use, and convenient in administration, very suitable for the majority of people who are not IT specialized, and do not require a high level of programming skills. The strong development of the Internet has created a variety of software and online applications: Joomla, Zoho, Moodle, Google Sites, Site123, Google Classroom, Class Dojo, and Edmodo.

Step 2: Content and learning material digitization. The selection of the combination method and the existing technology platform will determine the digital format of the learning material. Teachers can consult the table of support tools below:

FIGURE 2.2 Tools supporting content and learning material digitization by lesson progress.

The classification of the technological tool system by teaching progress is relative and based on the unique advantages of the tool because each tool can participate in many stages in the teaching process itself. In addition, the system of tools listed is suggested; choosing which tool is to be used not up to the teacher.

In teaching BL, teachers not only play the role of the executor but also the role of the director who develops a technology scenario and solve issues such as follows: Which content and materials should be digitized? Which technology is suitable for digitalization? Where to get the extended materials? How are they linked to lesson content? Is the learning activity with the content and materials in direct or online interaction?

The technology scenario is like an overview design, providing a complete and consistent view of the logic of the content and the structure of relevant information in the lesson. An effective technology scenario is a balance between technological factors and pedagogical factors. Technological factors should be applied in a calculated manner, avoiding the misuse of technological elements. The structure of the technology scenario is provided in Table 2.2.

TABLE 2.2 Technological Scenarios

Knowledge Content	**Format of Digital Learning Materials**				**Supporting Tools**	**Learning Activity (F2F or O.L.)**
	Document	**Picture, Chart, Diagram**	**Sound, Documentary**	**Study Form**		
Content 1						
Content 2						
Content 3						
Content n						

The steps for writing technological scripts are as follows:

1. Determining the content of key knowledge that requires supporting technology tools: based on knowledge standards, lesson skills/topics.
2. Ideas and selection of learning materials: text format, pictures, sounds, documentaries, and study cards. The types of digital learning materials selected must be consistent with the objectives, needs of the learners, and the content of the lesson.
3. With the text format, in addition to basic knowledge in textbooks, it is necessary to expand knowledge. The technology scenario must also contain additional knowledge. Teachers may choose text in the form of articles, specialized research, which are closely related to certain

content in the lesson and have clear instructions. With images, sounds, documentaries, etc., it is necessary to reflect the basic contents of events, historical phenomena, and ensure the quality requirements (copyright, sharpness, accuracy, and pedagogy). With study sheets, teachers can use simple software such as MS Word, MS PowerPoint, Canva, Easelly, and Design Bold to help increase interest in learning, thus encouraging student creativity.

4. Select applications to digitize content and learning materials. For each content, teachers select applications and materials to support digitization as suggested in Figure 2.2.
5. Identify learning activities and associated methods: individual tasks, groups requiring certain outputs, may also be activities that increase student–teacher interaction, student–student interaction: interactive quiz, online submission. To design an effective learning activity, teachers can answer the following five questions:

    1. Do the learning activities encourage learners to build knowledge?
    2. Do learning activities encourage learners' cooperation?
    3. Do learning activities encourage learners to use IT to design products?
    4. Do the learning activities require students to solve practical problems?
    5. Do students have the opportunity to self-assess and adjust their work? (Microsoft, 2012).

For each activity, the teachers determine between F2F and O.L.

Step 3: Planning of organizing BL in teaching History. Teachers understand the role of each phase of organizing learning activities related to determining pedagogical purpose and level of using online and direct phases in each learning activity.

- Determining the purpose of using each learning material and each phase of organizing the corresponding live or online activities is a mandatory requirement. In fact, the teacher is looking for answers to the following questions: How will using the online phase in the lesson increase the chances of IT application? Which requirements of the school program, curriculum, and formation and development of students' skills and competencies meet? The purpose of using the online phases—directly in the lesson/topic or each part of the lesson/topic is different.

- Determine the use of online and offline details to help teachers take an active role in organizing lessons in class:

Firstly, about the level of instructional materials used in the lesson.

For the F2F study, using too many instructional materials in one lesson hour is difficult for students to focus on the content of basic knowledge; even many instructional materials become a "burden." Teachers should not use more than five photo materials or do not use videos longer than 6 min for the content of knowledge.

In the combined study phase, the teacher proactively bases his/her study materials on the content and type of lesson, needs, the capacity of learners and facilities, and the existing technology background to determine the appropriate level of learning materials.

Secondly, about teaching methods.

During the F2F learning phase, the teacher pays attention to flexible teaching methods, group discussions, using visual aids, using words (characteristics of figures and events), role play, and positive teaching techniques (puzzle techniques, Know–Want to know–Learned) but not only toward the goal of knowledge, practice skills of exploiting historical paintings, photos, maps, schemas, diagrams, and tables statistics, but also to form the capacity of thinking, evaluating, analyzing, and applying History.

In the BL phase, learning activities need to be implemented through active teaching methods through small projects, exploratory teaching, problem-solving teaching, integrated teaching, WebQuest.

Thirdly, regarding the knowledge content.

Not all knowledge content is suitable for the BL model. Some characteristics of knowledge content can use the BL teaching model:

- The content is open challenging, and the ability to connect with reality is high. For example, ancient nations in Vietnam, feudal China, the Industrial Revolution in Europe, Tay Son movement, and the cause of national reunification.
- Content is suggestive about the ability to create real products, which is meaningful to learners, capable of attracting learners' interests. For example, Ancient Eastern and Western nations; Kingdom of Cambodia and Kingdom of Laos; Geographic discoveries; Renaissance Culture Movement; review, preliminary, and final reviews.

- Content is interdisciplinary, integrated, or can be formulated into academic topics. For example, World Wars in history; the role of the Nguyen in the national history; major reforms in Vietnamese history.

In teaching BL, the two phases of F2F learning and O.L. have their own characteristics, so the use of digital materials and the learning website will have certain differences.

With F2F learning, teachers use individual digital materials on the website as a visual medium and a source of knowledge to support the implementation of history lessons.

With the BL phase, the learning website will be used as both a visual medium for providing learning resources and an environment for learning activities. You can refer to the organizational structure plan of BL in teaching the history subjects shown in Table 2.3.

TABLE 2.3 Plan of Blended Learning in Teaching History

Topic/Content A				
Organizing Process	**F2F Learning Phase**		**O.L. Learning Phase**	
	Teacher	**Student**	**Teacher**	**Student**
Activity 1	Values/Data	Values/Data	Values/Data	Values/Data
Activity 2	Values/Data	Values/Data	Values/Data	Values/Data
Activity 3	Values/Data	Values/Data	Values/Data	Values/Data
Activity *n*	Values/Data	Values/Data	Values/Data	Values/Data

Step 4: Deploying the BL model in teaching History. In order for the BL model implementation to be effective, teachers have specific plans and notices for students, familiarize students with necessary technology tools, and log into the course. At the same time, prepare a sufficient learning space: equipment, teaching facilities, and technological elements to implement learning activities.

We conducted a pilot model of teaching BL in 10 classes of grade 10 from 5 schools in the cities of Vinh Phuc, Hanoi, Ho Chi Minh City, with 336 students participating. The selected students belong to diverse areas in different regions: big cities, towns, rural areas, plain, and midland, including specialized schools, public schools, private schools, international schools. Teachers participating in the experiment have at least two years of seniority, in a spirit of voluntariness, studiousness, and have introductions and preliminary training on BL. Details of schools, teachers, and students participating in the test are shown in Table 2.4.

TABLE 2.4 List of Schools and Classes Joining the Test

No.	School	Class		Note
		TN1	TN2	
1.	Le Hong Phong Special High School, Ho Chi Minh City	10 V1	10A3	Special and public school
2.	Trieu Thai High School, Lap Thach, Vinh Phuc	10A3	10A4	Rural public school
3.	Quoc Oai High School, Hanoi	10A15	10A14	Public school, outskirts of Hanoi
4.	Vinschool High School, Times City, Hanoi	10A1	10A2	Inter-level and private school
5.	Hanoi FPT High School	10A2	10A3	Private school of FPT corporation

The test has some initial results as follows:

- In the first test, students take a lot of time to familiarize themselves with the lesson designed on the website, and teachers take a long time to guide students to perform technical operations. Time for technical operations is reduced gradually in the next test. Students and teachers focus on solving and performing academic and creative tasks. Students shared a number of common problems such as lack of equipment and facilities. An unstable Internet connection causes inconvenience during group tasks, especially disruption faced when performing interactive quizzes. Besides, there are some comments. For example, the first time it took a lot of time to get acquainted with accessing the learning website and the form of combined teaching.
- About students' interest in learning, 100% of feedback and interviews of 336 students who participated in the test show the students' interest in the lessons.
- Get feedback from teachers about the satisfaction level on the website according to technical and educational criteria; effectiveness of school hours; the appropriate level of trained skills; 87% of the feedback from 83 teachers who participated in the experiment want

to use the integrated teaching website in the next lessons. Detailed results are shown in Figure 2.3 and Table 2.5. At the same time, teachers share their concerns when regularly applying the BL model: need time to prepare for lessons; lack of valuable learning resources; teachers need training on teaching; BL and supporting technology tools to be able to design and implement the BL teaching model.

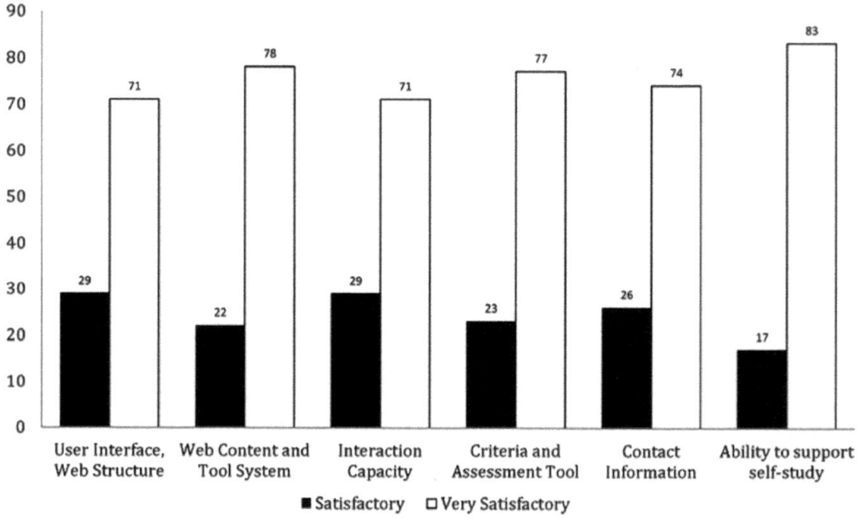

FIGURE 2.3 Chart of teacher's satisfaction on integrated teaching web (unit: %).

TABLE 2.5 Survey Results of Teachers' Satisfaction on the Lesson Efficiency and Specific Trained Skills in Integrated Class (Unit: %)

Efficiency of Lesson	**Level of Satisfaction (*n* = 83 Teachers)**				
	1	2	3	4	5
1. The participation of students in activities	0	0	0	57	43
2. Level of task completion	0	0	0	71	29
3. Level of student's interest	0	0	0	28	72
4. Level of memorization, understanding and knowledge use	0	0	12	24	64
Trained Skills	1	2	3	4	5
1. Team Work skill	0	0	0	14	86
2. Writing skill	0	0	29	43	28

TABLE 2.5 *(Continued)*

Efficiency of Lesson	Level of Satisfaction (*n* = 83 Teachers)				
	1	2	3	4	5
3. Self-study skill	0	0	0	13	87
4. IT skill	0	0	0	13	87
5. Skill of selecting and using information	0	0	14	14	72
6. Skill of analyzing, evaluating figures and events	0	0	0	40	60
7. Statistics and comparison skills	0	0	12	42	46
8. Skill of using history knowledge in real-life situation	0	0	0	43	57

Step 5: Validate and evaluate. Evaluation and improvement of lessons are the last stages that teachers often ignore after the organization and implementation of lessons are completed. Based on the student's feedback, the teacher proposes and improves the lesson in order to promote the areas that have been achieved and overcome limitations and shortcomings, providing learners with more appropriate and effective online and F2F learning phases

Teachers deploy assessments through two methods:

1. Through distributing questionnaires and in-depth interviews after class: Students finish F2F phase after class ends; and
2. Teachers create online questionnaires on the website so that the students can answer the questionnaire or provide feedback.

The content of the students' feedback focuses on the excitement with a class hour in general and with organized phases in particular. Students may come home to complete the questionnaires at their convenience, before the next History lesson.

2.4 BLENDED LEARNING MODELS IN TEACHING HISTORY

The BL model can be applied to a specific lesson: a new study and group of lesson/topic organized in the form of a project or WebQuest. For each type of article, select a different web-building tool as a suggestion for using technology tools to support BL

2.4.1 USING BLENDED LEARNING MODEL IN TEACHING LESSONS OF NEW KNOWLEDGE

There are three types of History lessons in high school: a new knowledge study; preliminary review and review, in which the new knowledge study accounts for 90%. Deploying effectively the BL teaching model on new knowledge study contributes to developing the necessary skills of the subject, equipping new knowledge and helping students love history.

The learning phases in the new knowledge study take place according to the basic five-step lesson progress (Table 2.6): prepare the lesson; launch; discover new knowledge; review and consolidate knowledge; check the review and expand.

TABLE 2.6 Plan of Organizing Teaching in Combination with Deploying New Knowledge Study

Topic/Content A				
Organizing Process	**F2F Learning Phase**		**O.L. Phase**	
	Teacher	**Student**	**Teacher**	**Student**
Prepare the lesson	Provide coursework's code and learning plan (in class)	State the problems (if any)	Support students if problems arise (off-class hour)	Log in, read materials following guidance, fill the survey form (off-class hour)
Launch (review previous lesson, introduce new knowledge) (in class)	Introduce content, test method; introduce new lesson	Answer the question: relate to the new lesson (if any)	Review, grade (if any)	Answer the questions or fulfill the test on website
Discovery new knowledge (in class)	Introduce the tasks and requirements for each task; support students directly	Discuss and question (if any)	Support through Internet; monitor the progress of individual and group	Do the following: complete the product, answer the questions…
Review, consolidate knowledge (in class)	State the requirements of the exercise; review the student's task	Discuss and question (if any)	Support through the Internet	Complete the exercise on Canva
Check, evaluate and expand (if any) (in class)	Guide the students in doing the multiple-choice exercise	Students state their opinions, questions (if any)	Customize the questions, time table (if necessary)	Join the multiple-choice test

Edmodo application is used to deploy integrated lessons in the new knowledge study model (Figure 2.4). Using Edmodo is relatively simple; Edmodo can link with Google Drive or One Drive for storage and make one able to connect with teachers from the same school and around the world. In particular, Edmodo has a friendly interface, close to learners, allowing connection with parents of students (if necessary). It also allows the management of regular participation of learners and the completion of learning tasks.

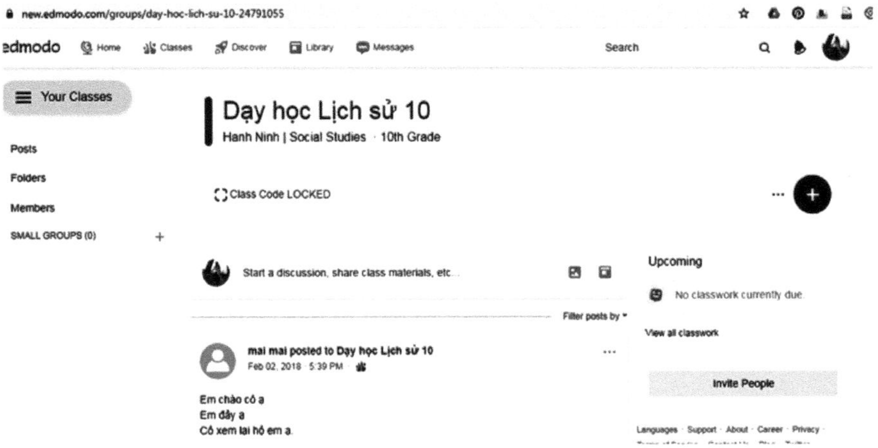

FIGURE 2.4 Graphic user interface of integrated class on Edmodo.

2.4.2 USING A BLENDED LEARNING MODEL IN PROJECT-BASED TOPIC TEACHING

In project teaching, learners carry out a complex learning task, which combines theory and practice and can create real products: articles, models, publications, and websites. Students also have the opportunity to play a role with different roles in society: scientists, editors, explorers, historians, or journalists. Project teaching in History subject is often suitable for topics/ special subject with a large amount of knowledge with proposed learning phases including (Table 2.7): select topics; build implementation plan; implement project; evaluate feedback.

TABLE 2.7 Plan of Organizing Step-wise Project-based Teaching Blended Learning

	Topic Content A			
Organizing Process	**F2F Phase**		**O.L. Phase**	
	Teacher	**Student**	**Teacher**	**Student**
Select topic	Discuss and agree on the name of topic/ project	All groups agree on topic, project	Survey the demands of students	Implement the survey
Build an implementation plan	Orient the major content, task, and requirements of each task	Discuss to plan the project implementation	Answer, support though Internet (e-mail, learning website)	Discuss with each other and with teachers to plan the project implementation through online tools (blog, e-mail, learning website.)
Implement the project	Support students to complete the product	Discuss and question (if any)	Monitor, help, evaluate student	Read materials, do exercise, share information, and complete product to show the learning results
Evaluate, feedback	Evaluate the student's real product based on the built criteria (real evaluation)	Present the product	Feedback in details about the activity of individual, groups	Self-evaluate own product and friend's product (peer-to-peer evaluation)

2.4.3 USING GOOGLE SITES TOOLS IN PROJECT-BASED TOPIC TEACHING

Google Sites is a Google service that allows users to create free online websites with a simple design interface (Figure 2.5). Google Sites allows inserting gadgets into websites: Google Calendar, Google NewsShow, Clock & Date, Weather, Google Clock, thereby especially helping teachers to follow and support learners when needed. Activity is notified and synchronized with a personal e-mail.

FIGURE 2.5 Graphic user interface of integrated class on Google sites.

2.4.4 *USING THE BLENDED LEARNING MODEL IN TEACHING TOPIC IN THE WEBQUEST FORM*

WebQuest is a "Discovery on the Web" teaching method—a special form of teaching that uses Internet access. The general structure of WebQuest consists of the following six parts:

1. Introduce: State the problems; state the meaning of the problem;
2. Task: Describe clearly the tasks students must perform; state the end result;
3. Process: Clearly state the steps learners must take to complete the task; provide the necessary instructions;
4. Resources: Textbooks, websites that have been listed to help learners complete their tasks; encourage the additional sources of information outside the website;
5. Evaluation: Design tools to guide students for self-assessment, usually rubric;
6. Conclusion: Review the basic content of the lesson, encourage learners to broaden their own problems.

However, depending on the content or specific learning object, teachers can be flexible in building the WebQuest structure on the basis of a common structure. In the process of deploying BL in historical teaching topics using WebQuest, we propose a structure of the following five parts: introduction, general procedure, task, test, and conclusion/discussion (Table 2.8).

TABLE 2.8 Plan of Organizing an Integrated Class Using WebQuest

	Topic/Content A			
Organizing Process	**F2F Phase**		**O.L. Phase**	
	Teacher	**Student**	**Teacher**	**Student**
Introduction	Introduce the content, test method; introduce the content of the new lesson	Answer the question; relate to the new lesson	Comment, mark (if any)	Answer the question or complete the test on Mentimeter
General procedure	Introduce the task and measure requirements of each task; support student directly	Discuss and state the question (if any)	Support via the Internet; monitor the progress of an individual or a group	Implement the task; complete the product; answer the questions on Canva
Task	State the requirement of the exercise; evaluate the student's work	Discuss and question (if any)	Support via the Internet (e-mail, learning website)	Complete the exercise on Canva
Test and evaluate	Guide the student to do the test	Discuss and question (if any)	Customize the questions, time table (if necessary)	Join the Kahoot test
Conclusion/ discussion	Review, conclude, and broaden the knowledge	Discuss and question (if any)	Support through the Internet (if any)	Self-study, self-research (widen the knowledge)

2.4.5 USING THE SITE123 TOOL IN TOPIC TEACHING USING THE WEBQUEST FORM TO DESIGN THE INTEGRATED LEARNING WEB

Compared with the similar tools, Site123 has four outstanding advantages: easy to use; provides modern templates for all types of websites; provides multimedia options for displaying galleries of images, videos, and embedded links; can integrate social networking sites (Facebook, Twitter, LinkedIn, and Pinterest); and has an option to automatically switch web pages to the mobile version (Figure 2.6). The disadvantage of Site123 is that it lacks a feature that supports synchronized classroom management. Teachers need integrated tools to support classroom management, to control students' O.L. processes, such as Padlet, Edpuzzle, and Kahoot.

FIGURE 2.6 Graphic user interface of integrated class on Site123.

2.5 DISCUSSION

The application of the BL model in teaching is becoming popular in the world in the context of integrating technology into education is an irreversible trend. However, for various subjective and objective reasons, the new BL model is mainly applied at the university level. The reason for this can be the less educational practice at the high school level, especially in the case of social sciences, including History. Some issues raised by the teachers in the context of using the BL model are as follows:

1. In order to implement the BL model that is feasible and more popular in high schools, it is necessary to synchronize the solutions: from investing in infrastructure, equipment, to fostering and improving competence and awareness of management staff, subject teachers; encourage and motivate technology experiences for all teachers; strengthen the professional role of professional organizations; improve the relationship of pedagogical schools with high schools; and even propose a model in which key lecturers on BL will be on supporting the development of design skills and implementation

of teaching BL for a group of high school teachers with the same expertise in a local high school cluster.

2. For the application of the BL model in teaching subjects in general and teaching History subjects in particular, it is necessary to show the flexibility and elasticity from the selection of technology tools, BL model, communication method in the BL classroom, to assess the learning outcomes of learners to suit different contexts and teaching subjects. Because the most important thing when applying teaching models, including BL in education, is not to measure the "golden ratio," but the important thing is learners' experience and progress; and the interaction and happiness of learners with teachers and friends.

2.6 CONCLUSION

The real experiences of studying and applying BL in teaching show the appropriateness of it at high schools level in general and History subject in particular. Applying the BL model in teaching History subject provides the learners with the opportunities to experience technology and an approach to historical truths by exploring a rich and diverse system of learning materials and it helps the learners mature in many competencies when participating in interactive and positive learning activities. To achieve these, teachers can use the procedure and models proposed above as a practical suggestion. However, each procedure or model itself does not change the quality of education; the important thing lies in the activeness and studiousness, the spirit of lifelong research, learning, and innovation of each teacher.

KEYWORDS

- **blended learning**
- **teaching History**
- **contents**
- **paradigms**

REFERENCES

Akyuz, H. I., & Samsa, S. (2009, January 30). The effects of blended learning environment on the critical. *Procedia-Social and Behavioral Sciences*, *1*(1), 1744–1748. https://doi.org/10.1016/j.sbspro.2009.01.308

Alammary, A., Sheard, J., & Carbone, A. (2014). Blended learning in higher education: Three different design approaches. *Australiasian Journal of Educational Technology*, *30*(4), 440–454.

Bonk, C. J., & Graham, C. R. (2006). *Handbook of blended learning: Global perspectives, local designs*. Pfeiffer.

Driscoll, M. (2002). Blended learning: Let's get beyond the hype. *e-Learning*, *3*(3), 1–4.

FPT Education. (2019, March 20). *Learning history 4.0, teachers set up a website to connect with students' graphic exercises*. Retrieved from https://fpt.edu.vn/tin-tuc/26759/hoc-lich-su-thoi-40-thay-giao-lap-website-ket-noi-bai-tap-do-hoa-cua-hoc-sinh

Garriso, D. R., & Vaughan, N. D. (2008). *Blended learning in higher education: Framework, principles, and guidelines*. John Wiley & Sons.

Giang, V. T., & Nam, N. H. (2019). B-learning: A suitable learning modality for higher education in Vietnam at digital age. *Educational Sciences*, *64*(1), 165–177.

Hanh, N. T. (2016). Applying blended learning model to teaching Vietnamese history (1858–1884), grade 11 high school. *Research and teaching history in the current context* (pp. 304–314). The Publishing House of Political Theory.

Hoang, L. H., & Quang, L. X. (2011, March 10). *E-learning its application in teaching*. VVOB.

Horn, M. B., & Staker, H. (2014). *Blended: Using disruptive innovation to improve schools*. Jossey-Bass.

Hung, T. V. (2018). Blended learning model in training university-level teachers. *Journal of Vocational Education Science*, *52*, 66–74.

Huong, N. M. (2012). *On training skill of applying information technology for pedagogical students at the Faculty of History, Hanoi University of Education* [National Science Conference on Teaching History at High Schools, pp. 796–806], Ha Noi.

Le, P. H. (2012). *How to improve the quality of History education in high schools* [National Science Conference on Teaching History in High Schools, pp. 6–12], Ha Noi.

Lesmana, A., & Basiran. (2017). *Use of blended learning in teaching Islamic history: The case of the 8th grade of Sukma Bangsa School in Aceh* [Master's thesis in education]. University of Tampere.

Microsoft. (2012). *21st century learning design rubrics*. Microsoft Partner in Learning.

MOET. (2018a). *Promulgating the standards for principal of general education institutions* [Circular No. 14/2018/TT-BGDĐT].

MOET. (2018b). *Promulgating professional standards for teachers of general education institutions* [Circular No. 20/2018/TT-BGDDT].

Nordina, A. B., & Alias, N. (2013). Learning outcomes and student perceptions in using of blended learning in history. *Procedia-Social and Behavioral Sciences*, *103*, 577–585.

OPEPD. (2010). *Evaluation, and policy development evaluation of evidence-based practices in online learning: A meta-analysis and review of online learning studies*. U.S. Department of Education.

Schwab, K. (2017). *The fourth industrial revolution*. Retrieved from https://www.weforum.org/about/the-fourth-industrial-revolution-by-klaus-schwab

Schweizer, K., Paechter, M., & Weidenmann, B. (2003). Blended learning as a strategy to improve collaborative task performance. *Journal of Educational Media*, *28*(2–3), 211–224.

Staker, H. (2011). *The rise of K-12 blended learning: Profiles of emerging models.* Innosight Institute.

The Olympia Schools. (n.d.). *Application of ICT in blended learning: Breaking up classroom space*. Retrieved from https://theolympiaschools.edu.vn/

Voos, R. (2003). Blended learning: What is it and where might it take us? *Sloan-C View*, *2*(1), 2–5.

Vy, N. D. (2012). *Some initial thoughts on the orientation of building programs and textbooks for history subject at high school level after 2005* [National Science Conference on Teaching History at High Schools, pp. 48−57], Ha Noi.

Yen, J. C., & Lee, C. Y. (2011). Exploring problem solving patterns and their impact on learning achievement in a blended learning environment. *Computers and Education*, *56*(1), 138–145. https://doi.org/10.1016/j. compedu.2010.08.012

GLOSSARY

THPT: Level 3, one of the three levels in the Vietnam educational system, higher than the elementary level (Level 1), secondary level (Level 2), lower than the college and university level. THPT last for three years (from Grade 10 to Grade 12). To graduate this level, the student must pass the National High School Graduation Examination.

SGK (Sách giáo khoa): textbook.

CHAPTER 3

Digital Journalism for Communicative Competency and Literary Skills in English: Scope, Challenges, and Implications

S. RUKMINI

School of Social Sciences and Languages, VIT University, Vellore 632014, Tamil Nadu, India

**Corresponding author. E-mail: rukminikrishna123@gmail.com*

ABSTRACT

In recent years, digitalization has revolutionized all the sectors, particularly journalism. Higher education also witnesses the digital shift, resulting in teachers using digital tools as an aid for teaching. Due to this shift, English teaching has benefited tremendously by using the digital medium. Especially, digital journalism has contributed to teaching English for specific purposes (ESPs) further, instilling interest in the students for their communication competency and literary skills. Given the importance of communication competency and literary skills in ESPs, the chapter will propose the integration of digital journalism as a new teaching methodology by designing a model curriculum for undergraduate courses. The chapter will also analyze the scope and challenges of using the digitalized print media for teaching ESPs and its implications for communication competency and literary skills by using digital journalistic content for better results.

3.1 INTRODUCTION

Journalism has always been a boost to all those who wish to learn a new language, whether it is the lingua franca or any other language in the world.

After the Macaulay's minutes in 1857, English education has become mandatory in India. Since then, English education has taken center stage to date. More so, in the globalized educational scenario, English education has become part and parcel in everyone's formal life. During the pre-independent era, it is the journalistic writings of the national leaders, especially Sri Aurobindo, the then lecturer of English and principal of Baroda College and later a seer-poet and yogi, who awakened the Indian masses from their slumber. Through his journalistic writings, he reminded the Indians of their ancient prowess, empowered them, and brought the whole nation to one platform. He was a polyglot—well versed in all the Western classical languages and Sanskrit. His linguistic expertise and literary skills are seen in his journalistic writings. Unfortunately, none of Sri Aurobindo's writings are included in the curriculum at the school and college levels, and thus he is less known to the student folk. He was the man behind the screen in India's independence.

Sri Aurobindo was one of the forefathers and pioneers of the nationalist movement whose diligence was a giant leap in gaining India's independence from the British rule. Few people know that it was Sri Aurobindo who termed the freedom movement as "Independence." Sri Aurobindo's penmanship in the journals was well appreciated by the British themselves. The then government felt shaken as Sri Aurobindo enticed the masses through his penmanship through his journalistic writings such as *Bande Mataram*, empowering the masses of the ancient prowess that India held—his intention to unify the country. Thus, the government repudiated Sri Aurobindo as they considered his writings to be an obstacle for their continuation here. As to what place *Bande Mataram* used to hold in the Indian hearts and the English men can be seen in the letter written by Retcliff who was the editor of *The Statesman* of Calcutta to the Manchester Guardian. To quote Retcliff, "We know Aurobindo Ghose only as a revolutionary nationalist and editor of a flaming newspaper, which struck a ringing new note in Indian daily journalism." The editor of the Statesman seems to have bitterly complained that, although the editorial articles in the *Bande Mataram* were diabolically clever and crammed full of sedition between the lines, the paper was still legally unassailable because of the superlative skill of the writing. The government too must have said this view, for they did not venture to prosecute the paper for its editorial or other articles, which was from Sri Aurobindo's pen. Even today if one reads his journalistic writings, one gets not only empowered with the ancient nationalistic pride that India holds but also is carried away by his communicative competence and literary skills. Journalistic writings have always been a torchbearer in one's path toward excellence in linguistic competence and literary skills. Especially, Sri Aurobindo's journalistic writings have given a wider scope to do research for the covers ancient Indian literature, English literature, philosophy, history, culture, and the situation of British India; and his versatility in all the Western and Indian classics enriches us with good literary vocabulary and narrative techniques. Further, his journalistic writings have become an area of research under Indian writing in English today.

From the above discussions, we understand that since Macaulay's days, English journalistic writing skills of Indians have been applauded by the British themselves who introduced these in India. Print media has always been a source of inspiration and a teaching resource in English language classrooms. In the past, when English was taught for general purposes (EGPs), all information in the print media was used by the language teachers to impart general knowledge and current affairs to the students.

3.1.1 ENGLISH FOR GENERAL PURPOSES TO ENGLISH FOR SPECIFIC PURPOSES

Today, the shift from EGP to ESP has made teachers go ahead in using the print media to impart specific skills such as language skills, scientific and technical writing skills, literary skills, and soft skills such as group discussion skills, negotiation skills, and case studies required to sustain in the global competitive environment. It updates the students and teachers with the researches done in the specific fields, be it science or arts. Provided if the teachers focus on the imparting of content-based instruction (CBI) along with the imparting of the aforementioned specific skills, it will make the students more confident in speaking and writing, thus facilitating them to excel in competitive exams and build a successful career and carve a niche of their own in this global and cut-throat competitive markets.

To impart CBI through experiential learning theory, journalistic writings (articles) are an effective resource material to be taken to the classroom as it fetches varied current topics related to varied disciplines. The advancement of science and technology has brought in vogue many technology-integrated teaching methodologies, and technology-integrated CBI definitely serves the purpose of dissemination of language skills, linguistic competency skills, communication skills, and literary skills. More so digitalization of journalism has facilitated the English language teachers immensely.

For the advantages of online journalism, web journalism as a teaching resource facilitates the teachers to disseminate updated knowledge, in general, and knowledge in instilling specific interest in the students to have an in-depth knowledge of any specific area of their interest. Web journalism also helps teachers initiate some interdisciplinary, multidisciplinary, and transdisciplinary projects that is based on an empirical study and investigational approach, through conducting a survey, as digital journalism has broken the contours of journalism in the print media for it is multilinear and hypermedia in nature, while the print media is unilinear. Further, web journalism helps introducing citizen journalism where the consumers themselves can become the authors of particular news.

The author draws a difference between the print media and online journalism by giving a detailed description of journalistic writings of both the media in the impartation of communication competency and literary skills. An article from the newspaper, such as *The Hindu* (print media), is chosen as a teaching tool to enhance the students' literary skills.

Tables 3.1 and 3.2 give the detailed descriptions of how each column of *The Hindu* (print media) is used to teach literacy skills and language skills, respectively, in the classroom.

TABLE 3.1 Newspapers in Print Media as a Teaching Methodology to Enhance the Learner's Literary Skills

Newspaper Column	Literary Skills
Editor's column	Letter writing
Interviews	Dialogue writing and building a conversation and updating oneself in that specific field
Speaking Tree	Instilling moral and spiritual values
Updated articles in research	Research writing skills
Reading for Pleasure	Narrative techniques in developing stories
Literary Review	To update the information in the literature field • Enhance one's literary vocabulary • Will know who's who in the literary field • Learn good conversation techniques from interviews with eminent personalities • Gives scope for further research in literature
Education Plus	Articles in education and updated information of the universities
Cartoons and graphs depicting poling	Information transfer and visual cultures

TABLE 3.1 *(Continued)*

Newspaper Column	Literary Skills
Pictures depicting natures, historical monuments, etc.	• Geography • History • Arts • Philosophy • Culture
Business column	• Business communication • Management skills
Weather column	Locating places in the map
Book review column	Review writing skills

TABLE 3.2 Newspapers in Print Media as a Teaching Methodology to Enhance the Learner's Language Skills

Newspaper Column	Language Skills
Science and Technology	• To update the knowledge in science and technology • To enhance one scientific and technical writing skills • Update oneself in scientific research
Sudoku	Vocabulary exercises and mathematical applications
Know your English	Vocabulary, tenses, idioms, and phrases
Opportunities column	Soft skills and the recent trends and situations at the workforce
General articles	Reading for comprehension
Word puzzles	Vocabulary building

This type of teaching through journalistic writings enhances the students with sufficient content knowledge and skills. All skills can be taught using print media. Print media having digitalized in contemporary times due to the technological advancement has bestowed many an advantage to hone the English language skills.

3.2 DIGITALIZATION OF THE PRINT MEDIA AND ITS IMPACT IN HIGHER EDUCATION

Due to the advancement in the field of science and technology, digitalization has taken over all the sectors, be it agriculture, banking, aviation, automobile, etc. More so, in the educational sector, digitalization has been doing wonders. It has brought drastic changes in distance education by facilitating and transforming the older version of distance education where resource material in print was used to be sent to the learners who had no contact with the teachers in person into massive open online courses where the students can even reach any topmost universities in the world and talk to subject experts across the world in their respective field. To exemplify, the online courses such as Udemy, Coursera, and Edx are working well, connecting the teachers and learners across the world. Webinar sessions have done wonders. Digital humanities is a new branch that emerged in the School of Languages and Social Sciences. Apart from these, the digitalization of journalism has also facilitated the field of education. It has facilitated the teacher in all respects—in handling the classes very effortlessly and bringing out better results. Information is lively and current enough to command the attention of restless youth. The use of current publications goes beyond keeping students up to date on politics and exposed to the newest literature. Due to the advancement of science and technology, technology-integrated CBI would definitely do wonders in the classroom.

We understand that digitalization has affected all the sectors and the same effected journalism too for today we find web journalism, online journalism, etc. Introducing online journalism in ELT classrooms really does wonders. Unlike in the print media, the students can respond as a citizen journalist to various news on the web where the consumers themselves can become the authors of a few news writings.

Language teachers have benefited immensely from it. With digital teaching software such as Hi Class, Thaliyola, and Learning Management systems, teaching platforms such as Schoology, Edmodo, and Easy class teaching large classes have become very easy. Digital journalism is facilitating the teachers, especially the English teachers, in imparting language and literary skills. The digital shift in journalism has made the teachers overwhelm and not to leave any leaf unturned in the digital print media as a teaching tool to enrich the literary and language skills of their students. For it saves money, time, and makes easy to use the digitalized articles in a smart classroom using Smart Boards and PowerPoint presentations.

3.3 DIGITAL JOURNALISM AS A TEACHING TOOL IN ENGLISH LANGUAGE TEACHING CLASSROOM

In print media, teachers used to distribute handouts to students to handle the reading for comprehension and writing skills. Listening as well as speaking skills could not be handled through journalistic writings in the print media. But today, the emergence of online journalism has made English teachers benefit immensely from it. Unlike the print media, where only reading and writing skills could be imparted, digital media has facilitated the teachers to impart listening and speaking skills as digital journalism is embedded with multimedia, that is, text embedded with image, audio-video, and animation.

3.3.1 WHAT IS DIGITAL JOURNALISM OR ONLINE JOURNALISM?

Digital journalism is also called online journalism. Is digital journalism just an extension of print journalism? To quote Massey and Levy,

> Content that has been "re-purposed from parent news organizations," "legacy" or "traditional media," operation. Admittedly, publishing re-purposed news stories on the world wide web of computer networks stitched together through the Internet is an example of journalism being done online, but arguably in the most basic sense. (Massey & Levy, 1999)

It is misconstrued that online journalism is a mere extension of print journalism. At the same time, it possibly is more than transferring (disbanding) legacy news stories into Netspace. The most distinctive contribution of online journalism is the technical capability for interactivity that is inherent in the architecture of its delivery system.

Online journalism possibly emboldens its audience. To quote Khoo and Gopal, "They call it 'prosumers' or the act of content producers ceding to content consumers the power to control, to varying degrees, their interaction with the news." Online journalism's contribution thus is initiated in its capacity to free readers from the monotonous linearity and limitations of news space that traditional media holds.

The chief advantage of online journalism lies in its latent for interpersonal interactivity. The advantages of digital journalism are the availability of complexity of content choice and the diversity of content such as entertainment content, commercial content, hyperlinking out of everyday news story to such a background content as the same day or stored (archived) article, and multimedia content. All web newspapers publish articles on local, regional, national, and international events, business and sports weather news, cartoons as entertainment content, and commercial content as product/service advertisements.

Newspapers approved readers to leap by hypertext link from an everyday story to some system of on-site background content or to a relevant external website. In a way, they provided user-searchable archives. The complexity of content choice, subdimension of news, was the strongest. In a hypermedial world to keep oneself update with technology and the world, digital journalism has become an inevitable tool. It makes wonders in the English language classroom as a pedagogical and andragogical tool. As described in Tables 3.3 and 3.4, digital natives and techno-savvy students feel at home with this tool. They are very enthusiastic about participating in every task and assignment given in the classroom.

TABLE 3.3 Digital Newspapers as a Teaching Methodology to Enhance the Learner's Literary Skills Embedded with Technology

Newspaper Column	Literary Skills
Editor's column	Letter writing with no editorial deadlines
Interviews	Dialogue writing and building a conversation and updating oneself in that specific field
Speaking Tree	Instilling moral and spiritual values
Updated articles in Research	Research writing skills
Reading for Pleasure	• Narrative techniques in developing stories • No word limit for stories
Literary Review	• To update the information in the literature field • Enhance one's literary vocabulary • Will know who's who in the literary field • Learn good conversation techniques from interviews with eminent personalities • Gives scope for further research in literature
Education Plus	• Articles in education and updated information of the universities • Anyone can publish/user generated content
Cartoons and graphs depicting poling-(infographic journalism)	• Information transfer and visual cultures • Animated, live example of machines, etc.
Pictures depicting Natures, historical monuments, etc.	• Geography • History • Arts • Philosophy • Culture All covering global, regional, national, and local
Business column	• Business communication • Management skills encompassing cross-cultural barriers of communication

TABLE 3.3 *(Continued)*

Newspaper Column	Literary Skills
Weather column	• Locating places in the map • Animated updated statistics and weather predictions each minute
Book review column	• Review writing skills • Anyone can write a review for a book/article, etc

TABLE 3.4 Digital Newspaper as a Teaching Methodology to Enhance the Learner's Language Skills Embedded with Technology

Newspaper Column	Language Skills
Science and Technology	• To update the knowledge in science and technology with animations or pragmatic communications in the labs • To enhance one's scientific and technical writing skills • To update oneself in scientific research with all the back-up using hyperlinks
Sudoku	Vocabulary exercises
Know your English	Vocabulary, tenses, idioms and phrases, linguistic analysis from the syntactic web to the semantic web
Opportunities column	Soft skills and the recent trends and situations at the workforce worldwide (global opportunities)
General articles	Reading for comprehension with hyperlinks and archives
Word puzzles	Vocabulary building
Listening skills	Pronunciation, stress, accent, and intonation
Citizen journalism (prosumers)	Recording of one's own speech-upload in YouTube

A digital environment for news production qualifies multiple media formats to be pooled and the content to be delivered through several corresponding distribution channels. Pretty often, the Internet is studied and understood as one identical medium instead of a multiform or meta medium with diverse types of online news.

Internet is a multifarious blend of varied sites, platforms, genres, and applications with a particular content, form, and structure. In the case of online news, there lies nothing such as one standardized online news. In online journalism, the news is offered through numerous news media such as general news sites, news blogs, web TV, web radio, RSS feeds, and news alerts. The main features of online media, that is, multimedia, interactivity, and hyper textuality differ between these different online news media. Each news medium has its exclusive characteristics regarding form (multimedia), structure (hypersexuality), and possible manipulations and processes (interactivity).

The characteristics of online journalism are as follows:

1. Global–geographical borders do not limit one online media. They can be accessed anywhere and anytime.
2. Fast information can be provided without delay, no editorial deadline, no fixed printing or broadcast times (live streams, breaking news, etc.).
3. No limits: there is no maximum length for stories.
4. No barriers: anyone can publish/user-generated content.
5. Nonlinear: the user decides the order of stories and how deep he/she wants to get into the topic.
6. Hyperlinks: HTML makes linking possible; stories and documents can be connected (net).
7. Multimedia: text, pictures, videos, audios, animation, etc.
8. Interactive: direct dialogue between journalists and users or among users is possible.

3.3.2 EXPLORATION OF THE VARIATIONS IN NEWSMAKING IN THE DIGITAL WAY (MEDIA)

The criteria for analyzing newsmaking in the digital media are as follows: time (instantaneous, uninterrupted renovating); the number and category of interlocutors (interactivity); the textual format and its extension (hypersexuality); the level of automation of the operations; the method of archiving, examining, and handling texts (memory); and the wealth and assortment of signals that are of text and image (multimodality) barring style, subject, and function, while all the other criteria being common to the two fields.

The time-based dimension includes phenomena such as synchrony and asynchrony, the constant makeover of information through substitution or accumulation, duration of time for which it is apt for the news to persist on the web, or when it is linked to the other subsequent news, memory, and personalization. The criteria of the preponderance of definite themes in each text are completed with the analysis of the hypertextual structures of each item analyzed and the employment of the multimedia, interactive, and time-based prospects.

In this way, the inherent physiognomies of digital texts, which are unique of the new model, characterize the analysis of their basic variables and the genres linked to them. The rhetorical criteria take up pitches of journalism, editing in which the science of the text has notable resonances, basic advances based on the operations propositioned by

1. Topic and quantities of discourse
2. Narrative genres: multilinearity and polyacroasis
3. Interpretative genres: hypertextual structures
4. Dialogic genres: interactivity.

Hypertext is the capacity to connect to a series of documents, associations, and explanations-both from one's own archive and from external providers in a gateway that offers an essential resource. The hypertext is used ingeniously as it offers a wide variety of advantages over print journalism, namely it has no space and time boundaries, offers a variety of outlooks, has direct access to sources, adapts paths of news perception and reading, contextualizes breaking news, and simultaneously targets a different group of readers, for example, those only inclined in the headlines and those inclined in the deeper layers of information and sources.

Interactivity meant two people or things working together, affecting each other, that is, the capacity of a computer to respond to a user's input. In online journalism, interactivity is segmented into the following three types:

1. Navigational Interactivity
2. Functional Interactivity
3. Adaptive Interactivity.

It lets the website to familiarize to itself and the behavior of the visiting browser. It is shown that the more interactive opportunities websites offer to users, the more involved are the users of that particular website.

Interactivity requires deal and preparation but involves and informs the user in a way other media cannot. Further, providing a "long tail" resource that engenders recurrence visits over a long-time scale. A flash interactive may consume some time to produce but can offer a captivating amalgamation of hypertext, video, audio, animation, and databases—a form that provides a place for people to gather and post experiences and information:

- Argumentative genres: multimedia resources
- Memory: memory

Genres of online journalism are as follows:

1. News Genres: news item
2. Interpretative genres: reportage (reportage of current events, special reports, documentary dossiers)
3. Dialogue genres
   a. interview
   b. forum, debate
   c. chat (online interview, interaction with characters, interaction amongst users); live chats can permit users, direct access to newsmakers, journalists, and experts
   d. survey
   e. opinion genres, namely traditional, editorial, comment, criticism, letters to the editor, article, column, cartoons, networked debates, forums, chat
4. Infographics-individual and collective infographics.

Graphic journalism: It is a combination of words and images to express ideas and information in ways, unlike in the text or illustrations. The best graphics journalism aids reviewers, understand and picture the information providing insights into the topic and opening up possibility where the audiences are able to find meanings and interpretations beyond those envisioned by the journalists. Journalistic infographics, unlike encyclopedic infographics, are positioned on the explanations of a universal character and are further concerned with features that are closer to singularities, and fairly common in cases such as accidents, elections, etc.

Tables 3.5–3.14 depict how digital journalism can be used as a teaching tool in an ELT and literature classroom.

This pattern can be tailored accordingly in a management classroom and other classrooms designing it as per the specific subject areas. Management-related articles and stories of various management gurus can be taken as teaching material under CBI.

Digital journalism can be used to disseminate not only language skills, competency skills, and literary skills but also academic research writing skills and techniques, conversational skills (dialogue) that has emerged in digital literature (hypertextuality) and how to pen down findings, observations, and suggestions apart from the pragmatics of written communication, discourse analysis, critical analysis of discourse, cultural studies, the theory of interaction, and the classification of theories.

TABLE 3.5 Use of Digital Journalism as a Teaching Tool in a BCom Classroom

Social Sciences Courses	Listening	Speaking	Reading	Writing
B.A. (Economics/ Commerce)	They were listening to biography, lectures by Nobel laureates in the field of economics, commerce and their electives. YouTube lectures on budgets, economic summit, GST, income tax, EPF schemes can be given to listen. Live interviews can be given, and assessment can be done through mock interactive live chatrooms	Group discussions, oral presentations can be conducted Skits on the biographies of economics or eminent personalities in commerce such as chartered accountants, ICWAs, etc. can be performed. Having listened to the lectures in the listening sessions group discussion, extempore, the monolog can be conducted on budgets of various countries, GST, income tax, EPF schemes, bank interest rate schemes, etc. Participating in live chat rooms, forums, etc., using hyperlinks, one can read the entire literature of the aforementioned topics.	Several passages from the biography of eminent personalities in economics can be given for imparting the skimming and scanning method. Articles from newspapers and journals can be given for reading comprehension using the syntactic web to the semantic web.	Based on the listening, speaking, and reading classes, students can be asked to write an essay on the life of their favorite economists, bankers, entrepreneurs, and Nobel laureates and their relevance in the present global scenario. Students can be asked to develop a case study based on their LSR classes. They can be asked to prepare a questionnaire and interview the eminent personalities. What impact they had on their lives, etc. To write a blog/digital story, write a review, write an article, or to give an opinion in the editorial.

TABLE 3.6 Use of Digital Journalism as a Teaching Tool in a BBM Classroom

Social Sciences Courses	Listening	Speaking	Reading	Writing
BBM (HR, Finance, and Marketing)	Listening to biography, lectures by famous management gurus in the field of HR, marketing, and finance. Audio/video lectures/ documentaries on business ethics and the hardships of persons who became famous entrepreneurs can be made to listen. Live interviews by financial experts, marketing experts, and HR experts can be given. Polyacroasis can be given. Polyacroasis means the multiplicity and diversity of the receipt of the speeches and is a vital concept to understand the persuasive nature of advertising discourse.	Group discussions, oral presentations can be conducted. Skits on the lives of great management gurus and eminent entrepreneurs can be performed. Certain management-related words can be given, and students can be asked to use it as a noun and a verb in a sentence. They are participating in live chats and forums in financial discussions.	Several articles from newspapers written by eminent entrepreneurs on management sutras can be given for imparting the skimming and scanning method using hyperlinks. The lives of entrepreneurs can be read and analyzed. Articles from business magazines and journals can be given for reading and comprehension	Based on the listening, speaking, and reading classes, students can be asked to write an essay on the life of their favorite economists, bankers, entrepreneurs, and Nobel laureates and their relevance in the present global scenario. Students can be asked to develop a case study based on their LSR classes. They can be asked to prepare a questionnaire and interview the eminent personalities. What impact they had on their lives, etc. To write a blog/digital story, write a review, write an article, or to give an opinion in the editorial.

TABLE 3.7 Use of Digital Journalism as a Teaching Tool in B.A. History and Political Science

Social Sciences Courses	Listening	Speaking	Reading	Writing
B.A. (History, Political Sciences)	1. Listening to biography, lectures by great historians and politicians 2. Documentaries depicting various cultural and political background of the world in the past can be given to listen. 3. Listening skills can be imparted by making the students listen to historical and cultural stories that depict the country's past glory. 4. Listen to various documentaries from geographical channels about people living in various remote islands, areas, their lifestyles, etc., and learn the techniques of creating a documentary and its presentation	Group discussions, oral presentations can be conducted. skits on the lives of various eminent personalities in the respective field can be performed. Words describing certain life situations can be given, and students can be asked to use it as a noun and verb in a sentence using interactivity platforms like live chats and forums.	Several passages from the biography of eminent personalities in the respective field can be given for imparting the skimming and scanning method. Articles from newspapers and journals can be given for reading and comprehension using archives and hyperlinks .finding out the difference in language, style, diction in the older usage of the language, and the contemporary journeying from the syntactic web to the semantic web.	Based on the listening, speaking, and reading classes, students can be asked to write an essay on the life of the eminent historians and politicians, and their relevance in the present global scenario Students can be asked to develop a case study based on their LSR classes. They can be asked to prepare a questionnaire and interview historians and cultural, religious, and political leaders. What impact they had on their lives. To write a review for books/articles written by historians and politicians.

TABLE 3.8 Use of Digital Journalism as a Teaching Tool in a B.A. Literature Classroom

Social Sciences Courses	Listening	Speaking	Reading	Writing
B.A. (Literature)	Listening to the biography of literary giants, lectures by Nobel laureates in the field of Literature. Socio-economic and political background of the world during their lives and propounding of various theories can also be listened to as documentaries. Digital storytelling to be given. Audiobooks in fiction/ autobiography watching genres of literature depicted through visual cultures.	Group discussions, oral presentations can be conducted. Skits on various fictions, drama can be performed. Literary words can be given, and students can be asked to give an extempore. To attend to interviews/ participate in live chats with eminent literary giants.	Several passages from fiction can be given to analyze the situation of that particular period. Literary analysis can be done by imparting the skimming and scanning method. Articles from newspapers such as literary review and journals can be given for reading comprehension.	Based on the listening, speaking, and reading Classes, students can be asked to write a short story/play depicting the present global scenario. Students can be asked to develop a blog of their own. They can be asked to prepare a questionnaire and interview the literary giants. What impact they had on their lives.

TABLE 3.9 Use of Digital Journalism as a Teaching Tool in B.A. Sociology

B.A. (Sociology)	Listening	Speaking	Reading	Writing
	Listening to a biography of secular, national, religious leaders, lectures by eminent sociologists like Karl Marx, etc. Socio-economic and political background of the world during their various sociology theories propounded can also be listened to as documentaries. Listening to various interviews using hyperlinks and archives.	Group discussions, oral presentations can be conducted. Skits on the contemporary times religious, secular democratic, and political situations can be performed. Technical words from sociology can be given, and students can be asked for an extempore. Participating in live chat rooms with sociologists dealing with global sociological issues.	Several passages from the biography of eminent personalities in sociology can be given and for imparting the skimming and scanning method using hyperlinks. Articles from newspapers and journals can be given for reading and comprehension using archives and hypertexts, comprehending the articles specifically on sociological issues of the world.	Based on the listening, speaking, and reading classes, students can be asked to write an essay on the life of the sociologists and his/her period and their relevance in the present global scenario. Students can be asked to develop a case study based on their LSR classes. They can be asked to prepare a questionnaire and interview the current sociologists and what impact they had on their lives. To write a review opinion on articles that deal with sociological issues.

TABLE 3.10 Use of Digital Journalism as a Teaching Tool in BTech (IT and Computer Science)

Science and Technology Courses	Listening	Speaking	Reading	Writing
BTech (IT and Computer Science)	Listening to biography, lectures by Nobel laureates in the field of science and technology, especially in the IT world. Listening to the updates in IT and the status of IT in the present scenario.	Group discussions, oral presentations can be conducted. Skits on the biographies of IT experts or and entrepreneurs or eminent personalities in the IT industry. To participate in live chats and interview panels and try to give a live mock presentation in the classroom.	Several passages from the biography of eminent personalities in IT and computer science can be given for imparting the skimming and scanning method. Articles from newspapers and journals that help one to update their knowledge in the specific area.	Based on the listening, speaking and reading classes, students can be asked to write an essay on the life of their favorite economist and Nobel laureates and their relevance in the present global scenario. Students can be asked to develop a case study based on their LSR classes. They can be asked to prepare a questionnaire and interview the eminent personalities. What impact they had on their lives, etc.

TABLE 3.11 Use of Digital Journalism as a Teaching Tool in BTech (Electrical and Electronics, Electrical and Communication Engineering)

Science and Technology Courses	Listening	Speaking	Reading	Writing
BTech (Electrical and Electronics, Electrical and Communication Engineering)	I was listening to biography, lectures by Nobel laureates in the field of electrical and electronics, and computer science. Audio/video and panel discussions can be given for listening.	Group discussions, oral presentations can be conducted. Skits on the biographies of eminent personalities in electrical and electronics/ECE. Having listened to the lectures in the listening sessions group discussion, extempore, monologue can be conducted on the same and attempt to conduct and participate in live chat rooms/interview panels.	Several passages from the biography of eminent personalities in economics can be given for imparting the skimming and scanning method. Articles from newspapers and journals can be given for reading comprehension. Read the archive material on ECE, EEE.	Based on the listening, speaking, and reading classes, students can be asked to write an essay on the life of their favorite economist and Nobel laureates and their relevance in the present global scenario. Students can be asked to develop a case study based on their LSR classes. They can be asked to prepare a questionnaire and interview the eminent personalities. What impact they had on their lives, etc. To write a review for an article or book in the specific area.

TABLE 3.12 Use of Digital Journalism as a Teaching Tool in BTech (Mechanical Engineering)

Science and Technology Courses	Listening	Speaking	Reading	Writing
BTech (Mechanical Engineering)	Listening to biography, lectures by Nobel laureates in the field of mechanical engineering using audio/visual and hyperlinks YouTube lectures on mechanical engineering. Listening to various documentaries on the history of inventions and discoveries of various mechanical engineers.	Group discussions, oral presentations can be conducted. Skits on the biographies of mechanical engineering or eminent personalities in mechanical engineering. Having listened to the lectures in the listening sessions group discussion, extempore, monolog can be conducted. participation in live chat rooms and panel discussions can be done. Mock presentations can be done in the class.	Several passages from the biography of eminent personalities in mechanical engineering can be given for imparting the skimming and scanning method. Articles from newspapers and journals can be given for reading comprehension using hyperlinks and hypertexts.	Based on the listening, speaking and reading classes, students can be asked to write an essay on the life of their favorite economist and Nobel laureates and their relevance in the present global scenario. Students can be asked to develop a case study based on their LSR classes. They can be asked to prepare a questionnaire and interview the eminent personalities. What impact they had on their lives, etc. To write a review for a book/article in the specific area.

TABLE 3.13 Use of Digital Journalism as a Teaching Tool in BTech (Biotechnology)

Science and Technology Courses	Listening	Speaking	Reading	Writing
BTech (Biotechnology)	Listening to biography, lectures by Nobel laureates in the field of biotechnology. Updates with the latest technologies in the area of biology and medicine.	Group discussions, oral presentations can be conducted. Skits on the biographies of biotechnologists or eminent personalities in biotechnology having listened to the lectures in the listening sessions group discussion, extempore, monolog can be conducted. Participation in live chat rooms, chatbots, blogs, interview panels. Mock presentations in classrooms can be done.	Several passages from the biography of eminent personalities in economics can be given for imparting the skimming and scanning method. Articles from newspapers and journals can be given for reading comprehension using hyperlinks.	Based on the listening, speaking, and reading classes, students can be asked to write an essay on the life of their favorite economist and Nobel laureates and their relevance in the present global scenario. Students can be asked to develop a case study based on their LSR classes. They can be asked to prepare a questionnaire and interview the eminent personalities. What impact they had on their lives, etc. To write a review on a book/article in the specific area.

TABLE 3.14 Use of Digital Journalism as a Teaching Tool in BTech (Agricultural Sciences)

Sciences Courses	Listening	Speaking	Reading	Writing
B.Sc. (Agricultural Sciences)	Listening to biography, lectures by Nobel laureates in the field of Agricultural Sciences. YouTube lectures on the latest technology used in agriculture.	Group discussions, oral presentations can be conducted. Skits on the biographies of agricultural scientists having listened to the lectures in the listening sessions group discussion, extempore, monolog can be conducted. Participation in the live chats and interview panels.	Several passages from the biography of eminent personalities in economics can be given for imparting the skimming and scanning method. Articles from newspapers and journals can be given for reading comprehension using hyperlinks.	Based on the listening, speaking, and reading classes, students can be asked to write an essay on the life of their favorite economist and Nobel laureates and their relevance in the present global scenario. Students can be asked to develop a case study based on their LSR classes. They can be asked to prepare a questionnaire and interview the eminent personalities. What impact they had on their lives, etc. To write a review for a book/article in the specific area.

Aforementioned is a model for curriculum to impart the language skills such as listening, peaking, reading, and writing skills, literary skills focusing on the targeted audiences' needs (students of specific courses) for the incorporation of CBI in ELT and literature classrooms. This type of teaching through digital journalism is thought-provoking and stirring for the digital natives. It charms to the distinctive learning styles of the students honing communication skills and multimedia skills. It is emboldening, self-motivated, and self-directed. It involves students in their individual learning process, offers accurate material, and helps students to cultivate planning skills, their creativity, and innovation skills.

Students want innovations and new ways of learning. Journalistic writings have always remained an abundant source for teaching in all innovative ways the teacher can, as they change thought, topic, and ingredients daily. Newspapers as teaching tools have facilitated teachers to a maximum extent. It helps the students to hone their communicative competence and literary skills. The shift from EGPs to ESPs has given many a scope to use digital journalism for the study of English language for specific purposes, for it contains articles covering from all disciplines. The use of journalistic writings to impart reading and writing skills in the classroom makes the class livelier. It provides the students with a specific subject of their interest and read, analyze, and also come out with their own ideas and go ahead doing further research in it.

3.4 CONCLUSION

To infer, we understand that journalistic writings have always been a teaching material to students irrespective of their age or qualifications to learn English. No matter in whichever form (print media or digital) it is, it has always been helping the teaching fraternity to impart literary and language skills to their students both in EGP and ESP platforms. When compared to a prescribed textbook for literature or language, it is only journalism that is catering to the educational needs of the students from all the four dimensions of the individual, that is, the physical, mental, emotional, and spiritual. Digital journalism is a real boost for the language teachers who use the latest digital learning platforms in the classroom, for it facilitates one to reach a wider audience very easily, at the same time giving personal attention to each student in a mixed ability classroom. Further, digital journalistic writings give many a scope for pursuing research through archival research method, interview method, discourse analysis, textual analysis, in general, and intertextuality and hypertextuality in specific. Digitalization is a boon for researchers in digital humanities. Digital journalism is a boon to the literary community as it introduces the recent trends in technology that brought a sea change in the literary genres in the twenty-first century. The characteristics of DJ the hypertext, interactivity, and intertextuality act as a boost to the researcher's community. For it facilitates the researchers in the investigation of the previous research is done in their specific field from the columns such as literary review, scientific, and technology articles, socio-economic and political articles, and business: columns and financial articles and HR reviews. Further, the digital archives help in facilitating the resource materials pertaining to finance, business, and marketing. All the language styles of genres of literature have exposed in digital journalistic writings. Students of literature and language can benefit from it immensely. To infer, the journey of print journalism has come a long way where digital journalistic writings and platforms help researchers from any discipline to pursue his empirical study using the archives and the latest updates. Digital journalism has given a boost to digital education pedagogy and andragogy. It hones critical skills that facilitate to pursue research identifying the research gaps in particular areas of their interest. It boosts up the teachers of English to use the latest visual cultures pertaining to various subject areas that help the students to better their visualization and imagination.

KEYWORDS

- **digitalization**
- **English teaching**
- **communication competency**
- **literary skills**
- **teaching methodology**
- **journalistic content**

REFERENCES

Fenholt, J. (1985). *The newspaper: Turkey to better living, using the newspaper in adult education classes*. American Newspaper Publishers Association (ANPA) Foundation. [ED 299 546]

Massey, B. L., & Levy, M. R. (1999). Interactivity, online journalism, and English-language web newspapers in Asia. *Journalism & Mass Communication Quarterly*, *76*(1), 138–151.
Sanderson, P. (2002). *Using Newspapers in the classroom*. Cambridge University Press.
www.onestopenglish.com
www.edwrite.com

CHAPTER 4

Using GeoGebra Software in Teaching and Learning Space Geometry

NGUYEN THI HONG HANH[1], TA DUY PHUONG[2], NGUYEN THI BICH THUY[3], and TRAN LE THUY[4]

[1]*Quang Trung High School, Hanoi, Vietnam*

[2]*Hanoi Institute of Mathematics, Vietnam*

[3]*University of Science, Vietnam National University, Hanoi, Vietnam*

[4]*University of Education, Vietnam National University, Hanoi, Vietnam*

[*]*Corresponding author. E-mail: tdphuong@math.ac.vn*

ABSTRACT

GeoGebra is free software, that is available in and in Vietnamese. It can be used in teaching and learning in all areas of high school and it is very effective. In this chapter, we show how to use *GeoGebra* in teaching and learning space geometry, especially how to draw a section of a polyhedron and a conical circle, as well as, how to cut them by a plane, and related problems (calculating volume and ratio of volumes, area of cross-section, etc.). Calculating a cross-section is a difficult part of the mathematical program of geometry (Grades 11–12 of the current Vietnam High School Program), and *GeoGebra* proved effective in teaching and studying cross-sections.

4.1 INTRODUCTION

Throughout the world, as well as in Vietnam, the use of information technology in teaching and learning has become popular and increasingly widely applied. Many mathematical software programs/packages have appeared and helped effectively in teaching and learning. *Scientific Calculator* becomes an indispensable tool for every Vietnamese student in learning. In recent years, popular math software such as *Cabri* (*Cabri 3D*, *Cabri II Plus*), *The Geometer's Sketchpad*, and *GeoGebra* got popularized and widely used in Vietnamese schools and universities (see, e.g., Dien et al., 1998; Ha, n.d.; Phuong, 2006; Du Sinh, 2004; Vui and Hung, 2006, 2007a, 2007b, 2009; Khalil et al., 2019). In the program and content of the new textbooks that are currently being developed, the use of information technology in learning, including mathematical software, is not only considered but also a requirement that each student needs to achieve (see Ministry of Education and Training, 2018).

Currently, most of the Vietnamese high schools, especially city schools, are equipped with computers, projectors, computer labs with enough computers for a class, and thus have facilities to implement the application of information technology in teaching, including instruction on calculating and drawing figures on the computer, greatly improving the quality of math teaching and learning. *GeoGebra* software can be used in all areas of the math program (see, e.g., Ha, n.d.; Khalil et al., 2019; Kovacs and Solyom-Gecse, 2016; Phelps, 2011; Project Maths Development Team, 2013; Saha et al., 2010; Gerry Stahl and The VMT Project Team, 2013; Venema, 2013; Yi and Leong, 2016). However, there is not much literature on the use of *GeoGebra* in teaching and learning the space geometry. The purpose of this chapter is to show how to use of *GeoGebra* mathematical software in teaching and learning the space geometry, especially in the problem of cross-sections of a solid figure, when cutting by a plane and related problems (calculating the volumes or ratio of volumes, calculating the area of the cross-section). Space geometry, especially the cross-section, is a difficult part of math in Grades 11–12 of high school as well as in the analytical geometry of the technical university. In this chapter, we try to show that the help of mathematical software *GeoGebra* is necessary and really effective.

4.2 GEOGEBRA SOFTWARE

4.2.1 SOME ADVANTAGES OF GEOGEBRA

GeoGebra software has many advantages in teaching and learning. First, it is free software, which can be easily downloaded and used. Moreover, it can be installed with the Vietnamese language, so it is easy to use for all teachers and students, including students in remote areas of Vietnam, where English proficiency is limited. Then second, indifferent from some other software, either just for use in algebra and calculus (*Maple*) or more convenient to use in geometry (*Cabri*, *Geometer's Sketchpad*), in *GeoGebra*, algebraic calculations and calculus be combined harmoniously with geometry in a unified system. For example, after drawing a picture, *GeoGebra* also immediately gives results of calculations (of volume, area, etc.). Third, unlike traditional geometry teaching (by the board or on paper), when drawing one picture is difficult and time-consuming, *GeoGebra*, a dynamic geometry, is used where drawing can be moved and changed (position and size) by mouse or slide. Animation is an advantage of *GeoGebra*, thus allowing students to see an image in many different angles, following the movement of drawings. In *GeoGebra*, lessons become more visually vivid, students understand more, and are interested in learning math. *GeoGebra* also helps student self-study and discover math very well.

4.2.2 GEOGEBRA'S INTERFACE

Once installed, the *GeoGebra* icon will appear on the computer screen. Double click on the icon to start. The screen shows the main interface of the *GeoGebra Classic* version 6.0. The main interface usually consists of a toolbar, some working areas (Figure 4.1).

You can switch to the Vietnamese language by going to the Language box and dragging the mouse to find Vietnamese. To hide or change some workspaces, click the delete button ×.

To hide the coordinates (for airy), click the button . Figure 4.2 is a Vietnamese screen with hidden coordinates.

4.2.3 ANIMATIONS ON GEOGEBRA

An important advantage of *GeoGebra* is animation. To move the object, we usually click on the point and move the mouse. Other objects associated with that point will change accordingly.

To resize the image, we use the slider. For example, the size of the image can vary from 0 to 9 on the slider (in Figure 4.3, $r = 2.7$, and in Figure 4.4, $r = 5.6$).

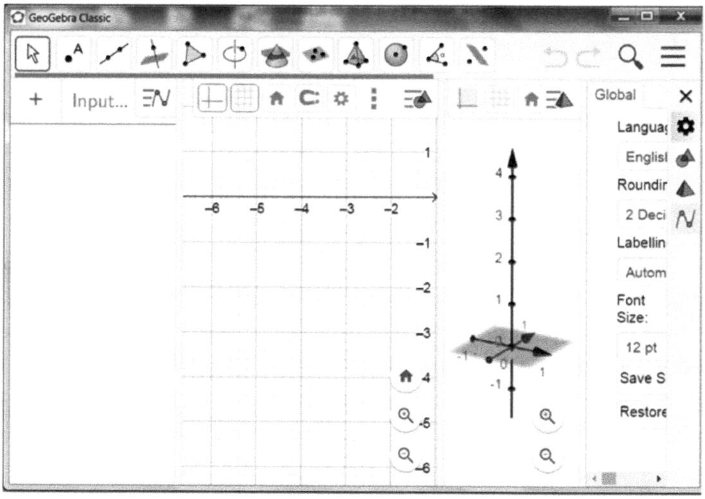

FIGURE 4.1 The basic interface of *GeoGebra.*

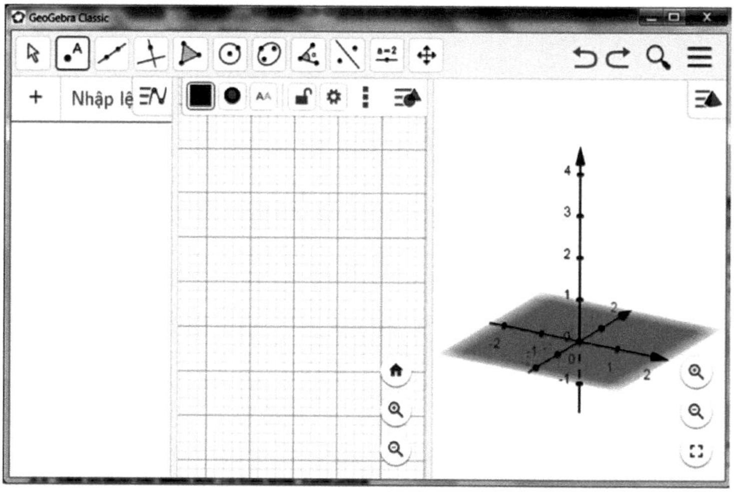

FIGURE 4.2 Vietnamese interface of *GeoGebra.* (instead of *input* is Vietnamese: *Nhập lệnh*)

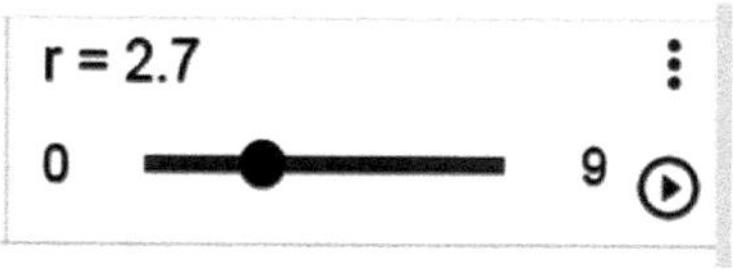

FIGURE 4.3

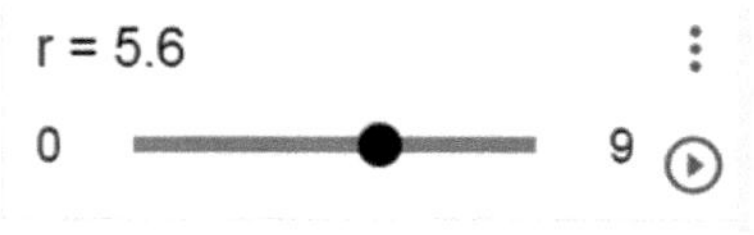

FIGURE 4.4

4.3 USING *GEOGEBRA* IN TEACHING AND LEARNING SPACE GEOMETRY

4.3.1 SOME BASIC GEOMETRIC MANIPULATIONS ON GEOGEBRA

There are several tools available on the toolbar to draw basic geometric objects. To draw a plane geometry, go to "View" to find "Graphic," the screen shows a toolbar, as shown in Figure 4.5.

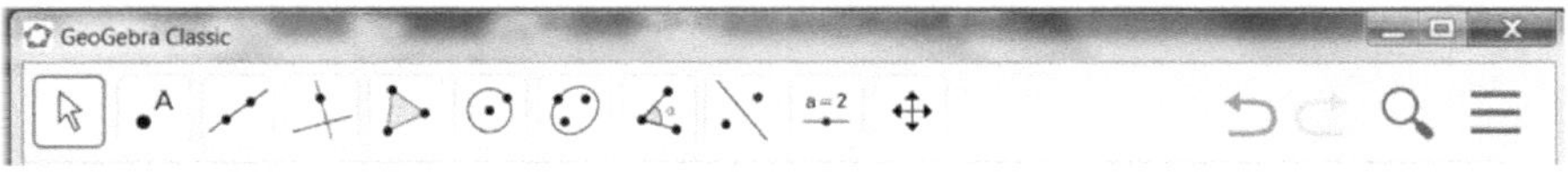

FIGURE 4.5 Toolbar for drawing a plane geometry.

For drawing a space object, go to **View** to find **3D Graphic**, the screen shows a toolbar as shown in Figure 4.6.

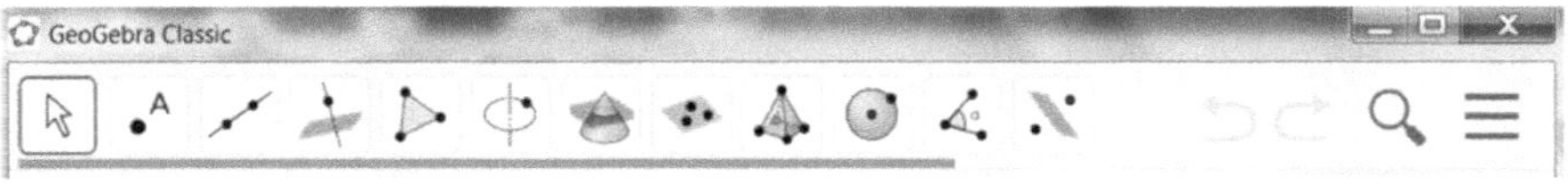

FIGURE 4.6 Toolbar for drawing a space geometry.

To have an object, we just click on the corresponding icon. For example, to draw a point, you need to click on the icon •A (point) and move the mouse over the Workspace to get point A. With other objects, we can do the same. The figure shows the results of drawing point A, segment BC, triangle DEF, and singular pentagon HIJKL.

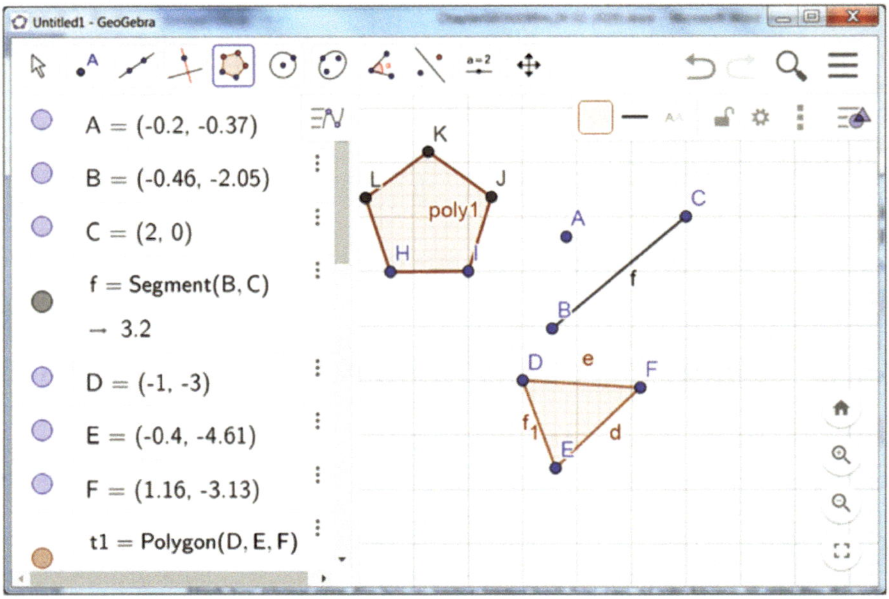

FIGURE 4.7 Point A, segment BC, triangle DEF, and HIJKL pentagon with the corresponding coordinates.

Although the coordinate system is hidden, when hovering the mouse to draw the figure, *GeoGebra* immediately declares the coordinates of point A, coordinates of points B, C, and denotes the line segment BC is f, ...(Figure 4.7).

4.3.2 SOME BASIC SPACE GEOMETRY PROBLEMS ON GEOGEBRA

Space geometry requires a high geometric imagination, so it is not easy for most students. *GeoGebra* software can assist effectively in teaching concepts and solving basic problems of space geometry. In this section, we present some illustrative examples in teaching and learning space geometry, thanks to *GeoGebra*.

Problem 4.3.2.1. *Calculating the angle between two planes: The concept of the angle between two planes (the angle between two lines in the two planes and perpendicular to the intersection line) is one of the basic concepts of space geometry.* GeoGebra *helps students understand well and reinforce this concept.*

Example 4.3.2.1. *Let the pyramid S.ABC,* $SA \perp (ABC)$, $SA = AB = 2a$, $AC = a$ *is given. Calculate the angle between SBC and ABC planes. Calculate the areas of the triangles ABC and SBC.*

Solution 4.3.2.1. *After drawing the pyramid S.ABC, with the given data, GeoGebra draws the angle between the planes SBC and ABC as the angle SIA automatically calculates and gives the areas of the triangles ABC and SBC (Figure 4.8). Mobile point S or point B by mouse, drawings, and calculations immediately made by GeoGebra and produce new results (Figures 4.9 and 4.10).*

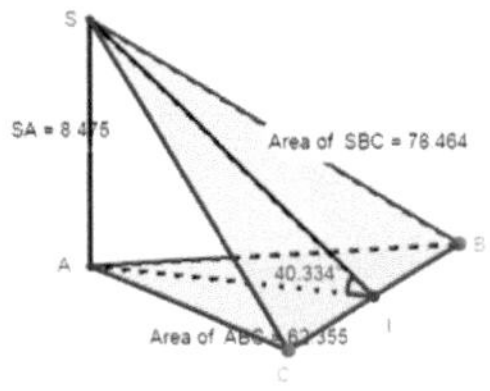

FIGURE 4.8

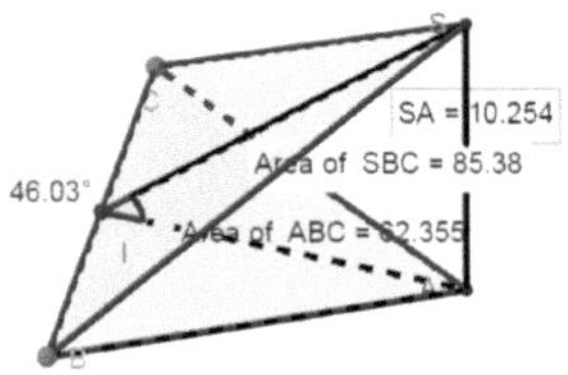

FIGURE 4.9

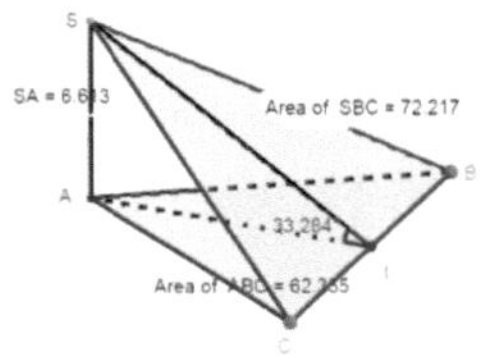

FIGURE 4.10

Problem 4.3.2.2. *Calculating the projection area of a polygon: When you project a polygon down a plane in the perpendicular direction, you get the projection as also a polygon. Let S be the area of the polygon, S′ its projection area, and* α *the angle between the polygon plane and the projection plane. Then we have the following formula:*

$$S' = S.\cos\alpha. \qquad (4.1)$$

For example, the pentagon BCFGH, with a perpendicular projection down to the plane P, is a pentagon BCDEA. *GeoGebra* helps to reinforce knowledge about the projection of polygons. The figures allows understanding the concept of projection, calculating the value of the angle between two planes (e.g., $\alpha = 51°5'$, the area of the polygon BCFGH and the area of BCDEA (Figure 4.11) and verify the formula (4.1). We can change the position of the BCFGH polygon, but the angle and polygon area remain unchanged (Figure 4.12). Convert *r* on the slider; we can now change the position of the planes, resulting in a change of the projection surface and the angle between two planes (Figure 4.13).

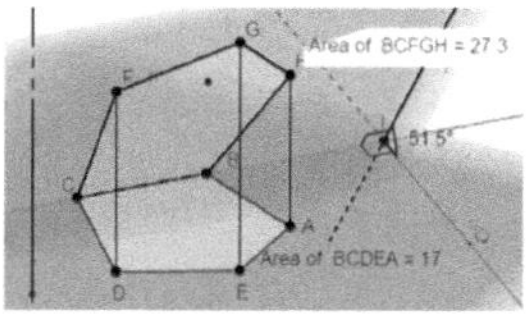

FIGURE 4.11

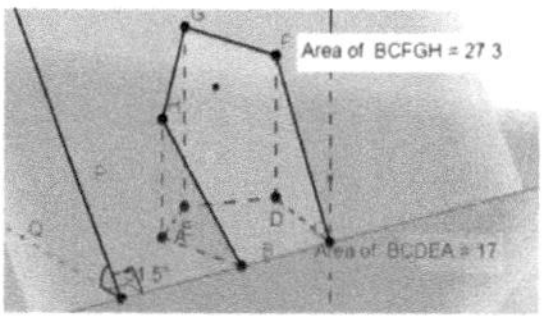

FIGURE 4.12

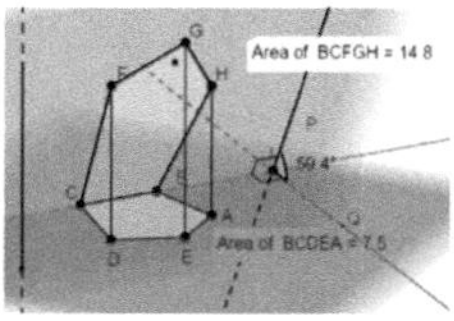

FIGURE 4.13

Problem 4.3.2.3. *Creating a rotating surface: To draw a rotating surface, we must create the axis and the generating line (Figure 4.14). Rotating the generating line around the axis, we have a rotating surface (Figure 4.15).*

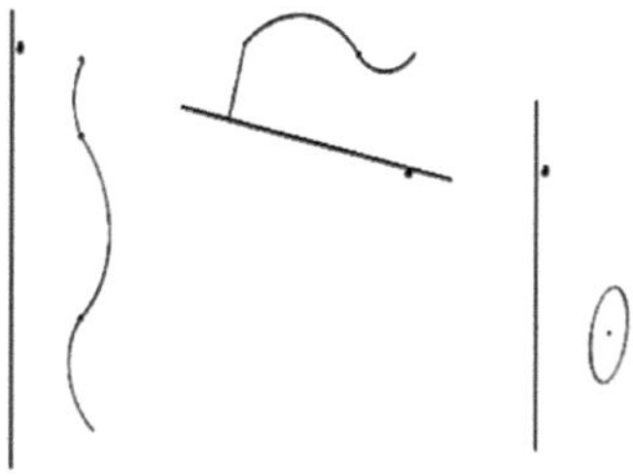

FIGURE 4.14

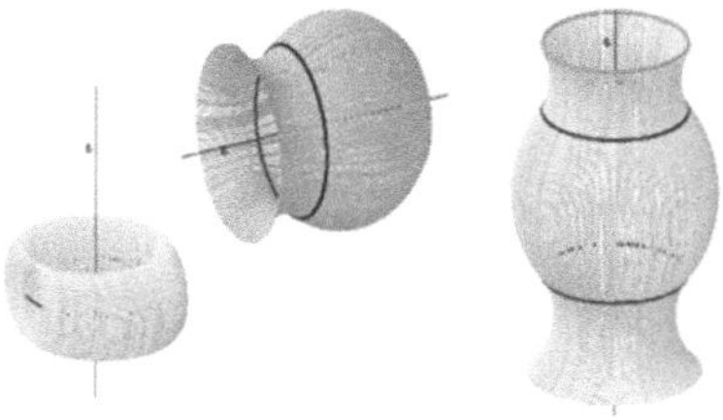

FIGURE 4.15

Problem 4.3.2.4. *Creating a circular conical surface: To create a circular conical surface, we draw two intersecting lines a (the axis) and d (the generated line), the angle between a and d is α, 0° < α < 90° (Figure 4.16).*

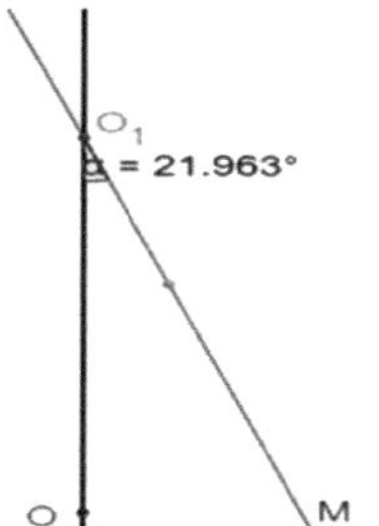

FIGURE 4.16

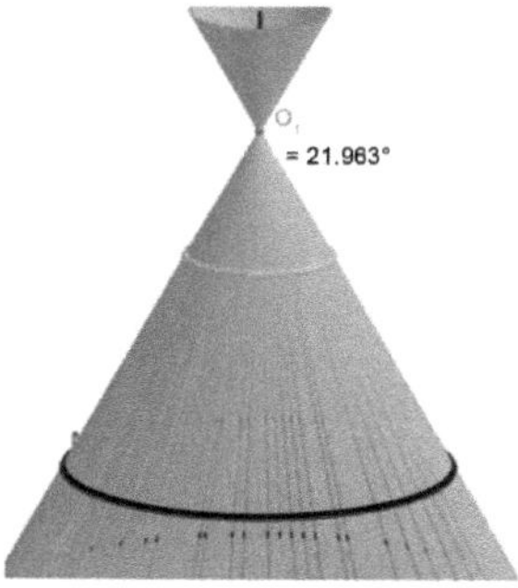

FIGURE 4.17

Rotate line d around line a and then line d produces a lateral face of cones. Point M creates a circle (Figure 4.17).

Problem 4.3.2.5. *Creating a circular cone with a right triangle: Draw a right triangle ABC at B (Figure 4.18). Turning the hypotenuse AC (the generated line) around the perpendicular edge AB (axis), we obtain the cone with vertex A and B as the center of the base (Figure 4.19). Draw a right triangle ABC at B with the height BI (Figure 4.18). Turning the edge AB (the generating line) around the AC hypotenuse (axis), we obtain two cones with the same base, the center of the base is I (Figure 4.20).*

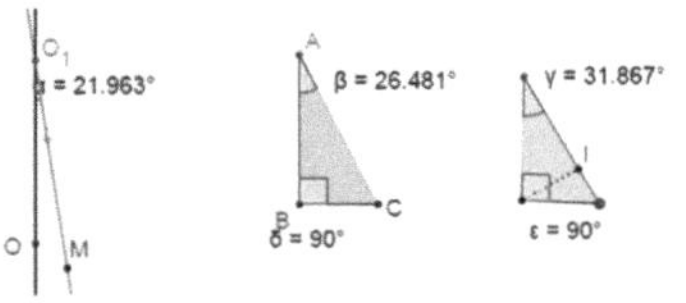

FIGURE 4.18

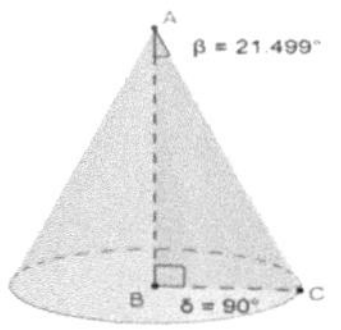

FIGURE 4.19

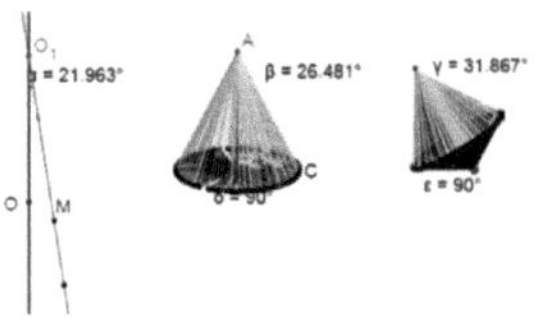

FIGURE 4.20

Problem 4.3.2.6. *Calculate the volume of a circular cone: For each cone with a radius of base r and height h,* GeoGebra *calculates the volume of the cone according to the formula* $V = \frac{4}{3}\pi r^3 h$ *(Figure 4.21).*

Using the slider to change the radius or height (move the top O), we can change the volume of the cone (Figure 4.22).

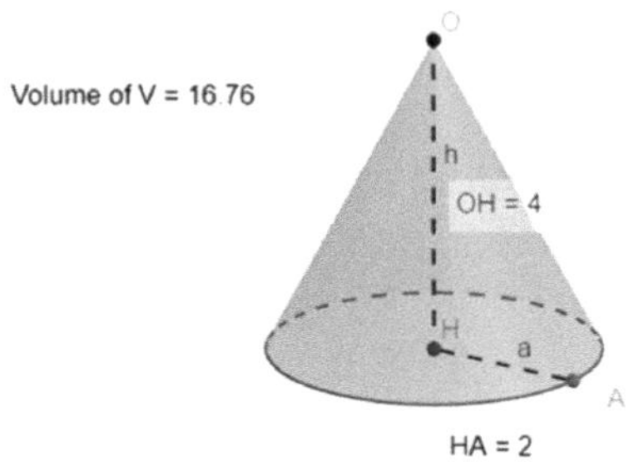

FIGURE 4.21

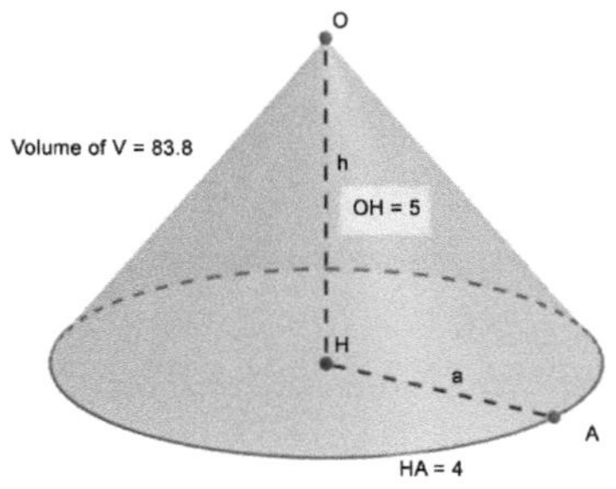

FIGURE 4.22

Problem 4.3.2.7. *Creating a cylindrical surface: Draw two parallel lines a (axes) and b (the generated line).*

Draw a rectangle ABCD (Figure 4.23). Give the DC generated line to rotate around the AB axis, and we have part of the cylinder or the cylinder (Figure 4.24).

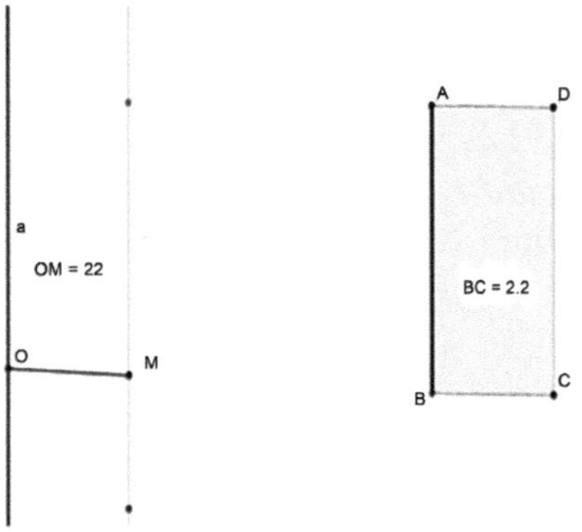

FIGURE 4.23

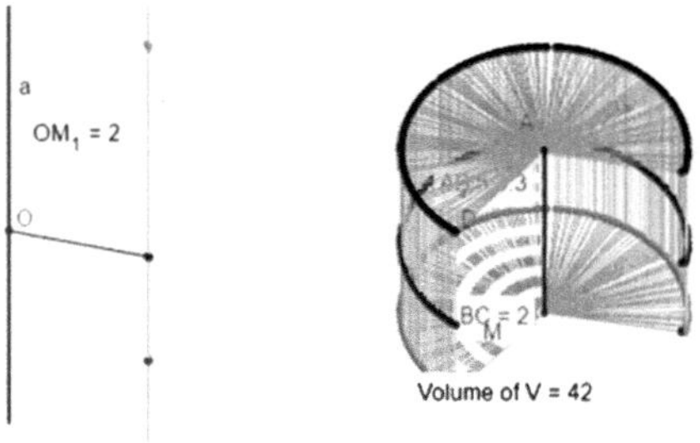

FIGURE 4.24

Each point on the line draws a circle (Figure 4.25). You can change the lengths of the edges AB and BC of the rectangle to get different volumes (see Figures 4.26 and 4.27).

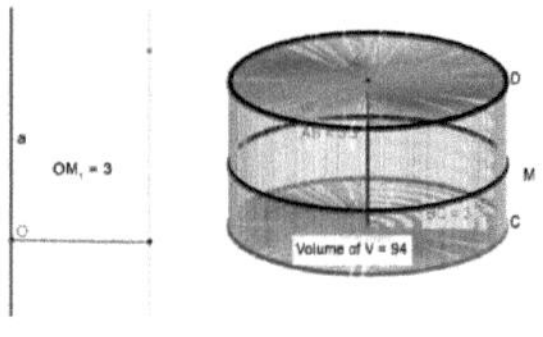

FIGURE 4.25

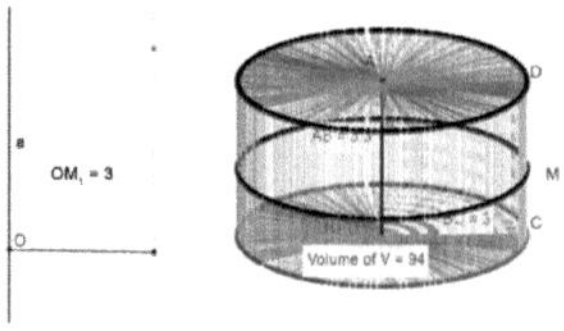

FIGURE 4.26

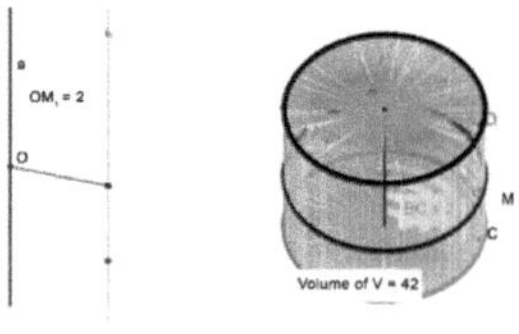

FIGURE 4.27

Problem 4.3.2.8. *Creating a sphere: Draw a half-circle (generated line) of a given diameter (axis, Figure 4.28). When the half circle rotates around the diameter (axis), we receive a sphere (Figure 4.29). Each point on the generated line draws a circle (Figure 4.30).*

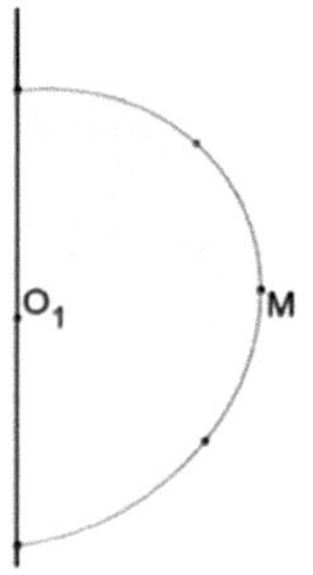

FIGURE 4.28

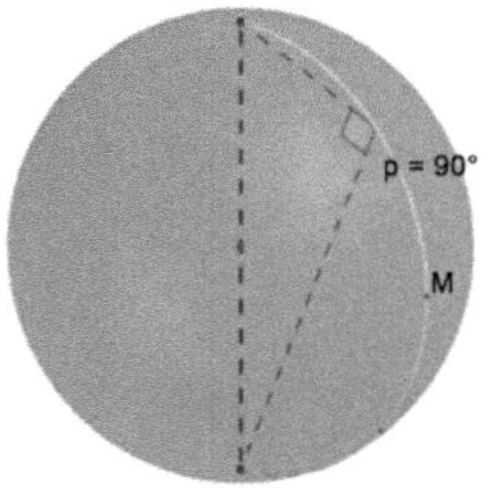

FIGURE 4.29

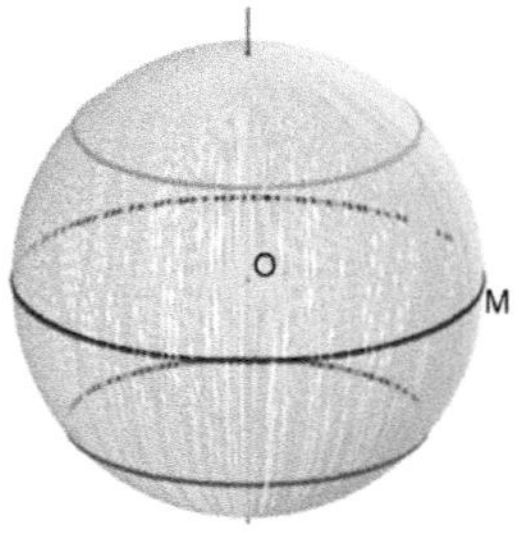

FIGURE 4.30

GeoGebra calculates the volume of a sphere according to the formula $V = \frac{4}{3}\pi r^3$ (Figure 4.31). Change the radius of the sphere by slide; the volume of the sphere will change (Figure 32).

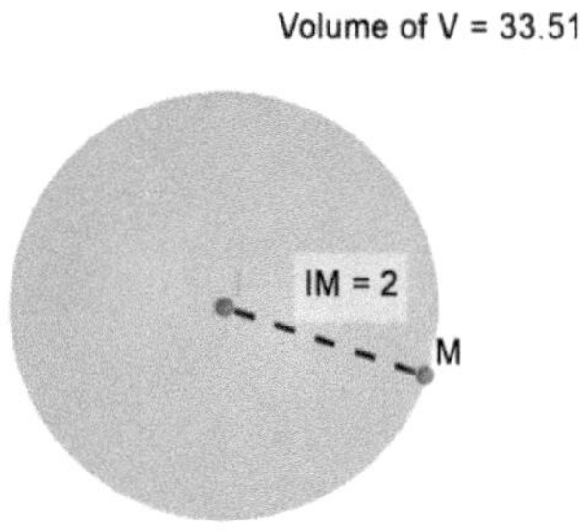

FIGURE 4.31

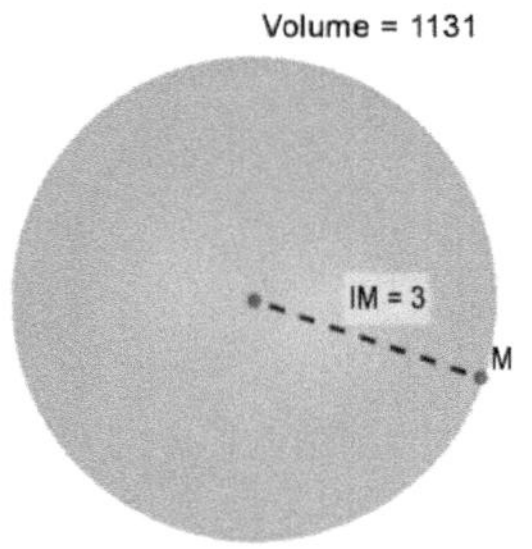

FIGURE 4.32

Problem 4.3.2.9. *Nested figures:* GeoGebra *is a great help for drawing and solving problems with nested figures. Changing the position of our points, we receive different figures (Figures 4.33 and 4.34).*

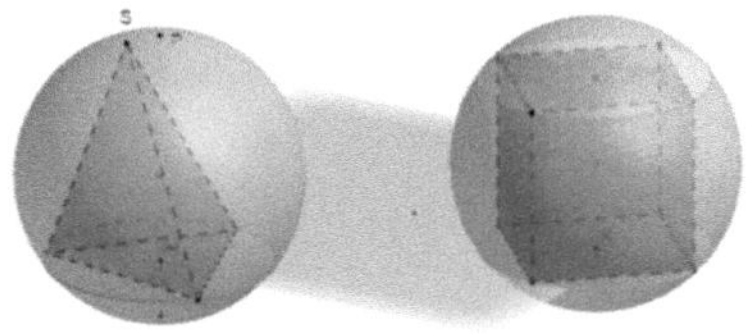

FIGURE 4.33

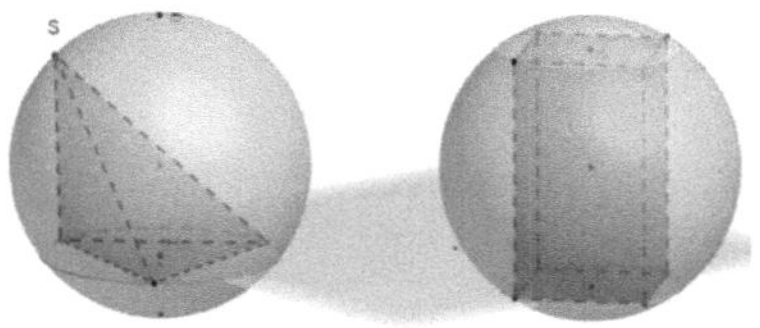

FIGURE 4.34

Changing the positions of the points M, P, and J, we get different figures (Figures 4.35 and 4.36).

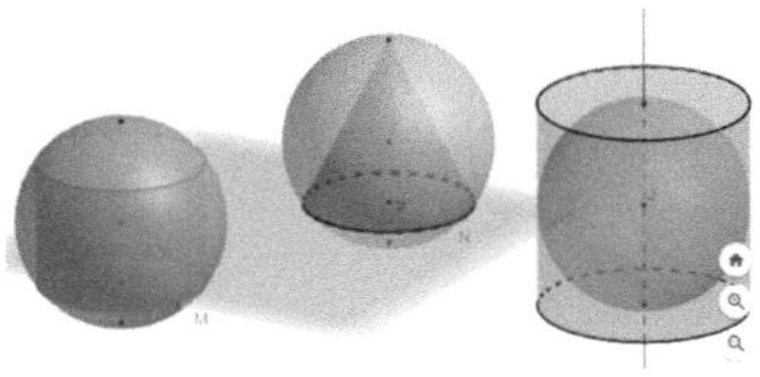

FIGURE 4.35

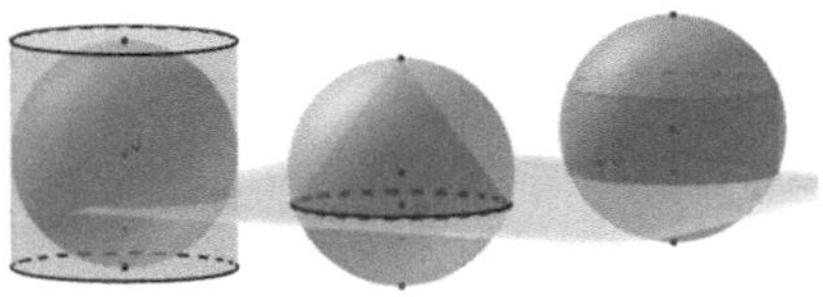

FIGURE 4.36

Problem 4.3.2.10. *Draw and calculate the volumes of the cylinder and the inscribed cone:* GeoGebra *can draw and calculate the volumes of the* *cylinder and the inscribed cone by a given height and radius (Figure 4.37). Change height (moving point A by mouse) or change the radius of the base (according to the sliding scale),* GeoGebra *recalculates the volume of the cone and the cylinder (Figure 4.38).*

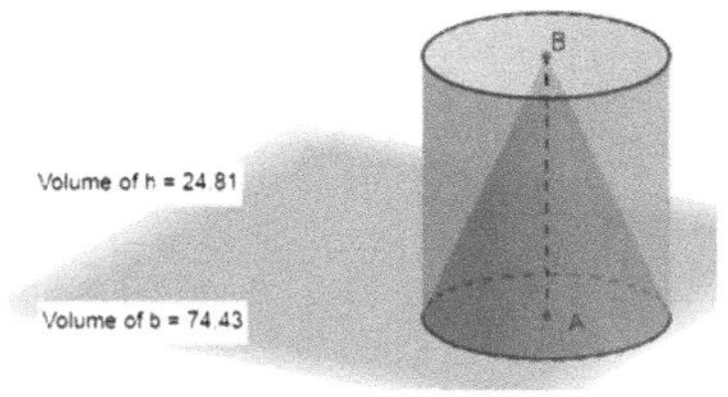

FIGURE 4.37

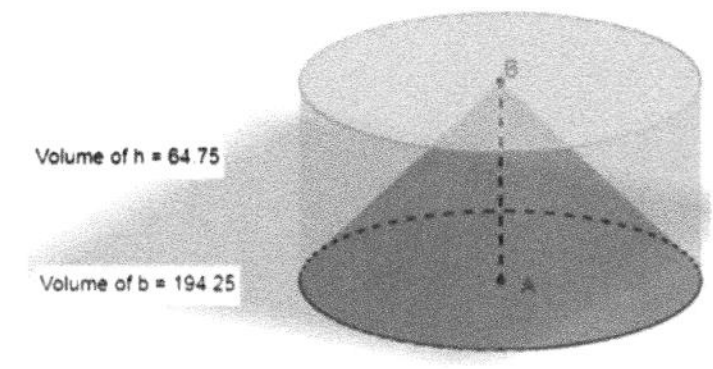

FIGURE 4.38

Problem 4.3.2.11. *Calculate the volume of the regular pyramid and the volume of the two inscribed and circumscribed cones: Draw the regular pyramid S.ABCD with two inscribed and circumscribed cones (Figure 4.39). Changing the height and the bottom radius, the volumes also change (Figure 4.40).*

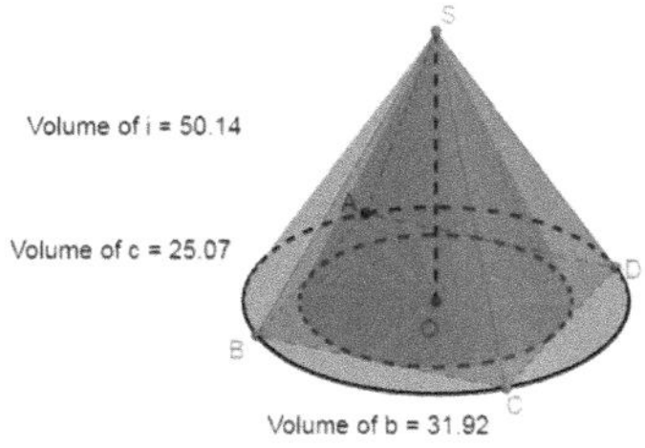

FIGURE 4.39

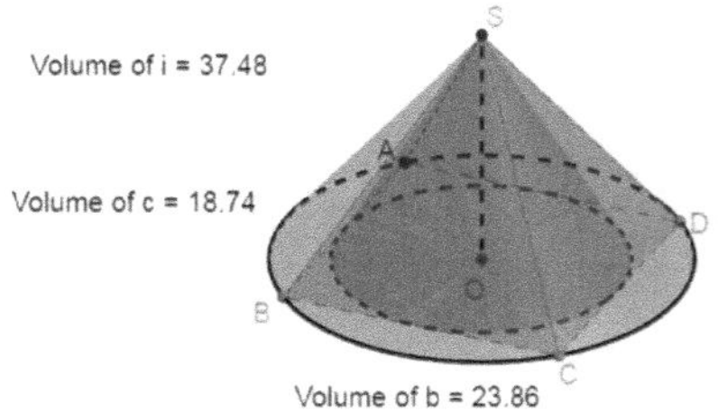

FIGURE 4.40

Problem 4.3.2.12. (Exercise 16, page 54, *Geometry Vietnamese Textbook*, Grade 12): *Three tennis balls of the same size are placed in a cylindrical box with a bottom of a large circle of tennis balls and a height equal to 3 times the diameter of a tennis ball.*

1. *Let S_1 be the total area of three tennis balls and S_2 the lateral area of the cylinder. Calculate the ratio $\frac{S_1}{S_2}$*
2. *Let V_1 be the total volume of three table tennis balls, V_2 is the cylinder volume. Calculate the ratio $\frac{V_1}{V_2}$.*

Using GeoGebra *software, draw a cylinder of radius r height* 6*r. Draw three spheres of radius r, inscribed in a cylinder. Use CAS to calculate the area and volume (Figure 4.41).*

Use the slider to change the radius of balls. Then the area, the volume of three tennis balls, and of the cylinder will change, but the ratio will be the same: $\frac{S_1}{S_2} = 1$ *and* $\frac{V_1}{V_2} = \frac{2}{3}$ *(Figure 4.42).*

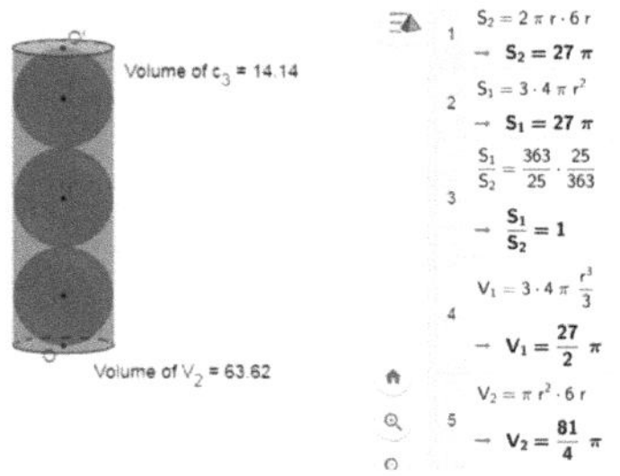

FIGURE 4.41

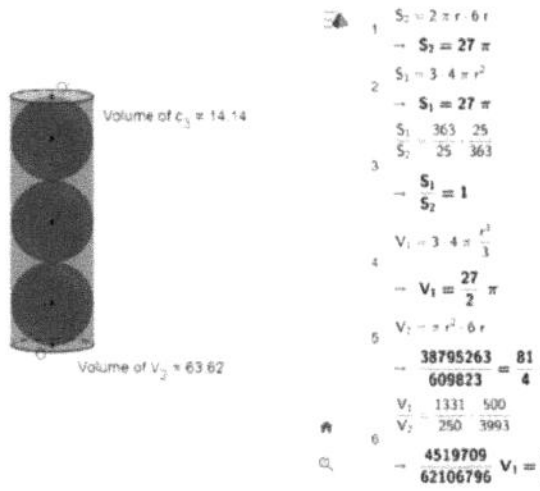

FIGURE 4.42

Thereby, the formulas of area and the volume of cylinder and sphere are repeated and reinforced. Students can easily solve problems with general data (for all r) as follows. From the formulas for cylinder and sphere,

$$V_c = \pi r^2 h; S_c = 2\pi rh \; V_s = \frac{4}{3}\pi r^3; S_c = 4\pi r^2, \tag{4.2}$$

we have

$$\frac{S_1}{S_2} = \frac{3 \times 4\pi r^2}{2\pi r \times 6r} = 1; \; \frac{V_1}{V_2} = \frac{3 \times \frac{4}{3}\pi r^3}{\pi r^2 \times 6r} = \frac{2}{3}.$$

This example shows that GeoGebra *can be considered as a very good experimental tool to find and verify mathematical formulas (4.2).*

Problem 4.3.2.13. *(Exercise 17, page 54,* Geometry Vietnam Textbook, *Grade 12): Put seven balls with the same radius r in a jar so that the seven balls are in contact with the base, the middle ball is in contact with all six balls. Every six balls come in contact with the lateral surface of the cylinder. Calculate the total area and the volume of the cylinder.*

Solution. GeoGebra *allows us to draw seven spheres within a cylinder. For change (thanks to the slider), we get the total area and the cylinder volume (Figures 4.43 and 4.44).*

After that, students can calculate the lateral area, the volume of the cylinder, and the volume of the sphere by the formula (4.2) to verify the results of GeoGebra.

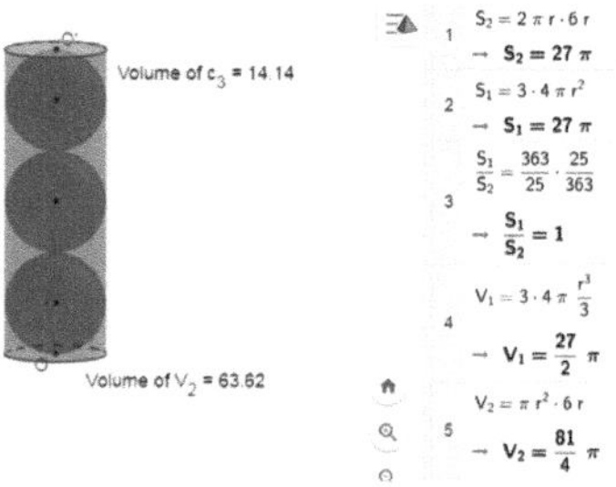

FIGURE 4.43

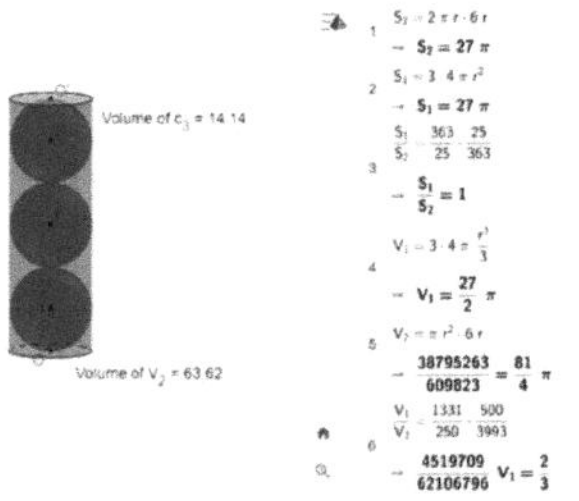

FIGURE 4.44

Problem 4.3.2.14. *(From the entrance exam for Grade 10 of High School, Thua Thien Hue Department of Education and Training, 2019–2020): A cylindrical glass cup is filled with water, with a height of 6 cm, the radius of the base is 1 cm. One slowly drops water into a cup, a spherical ball, and a cone. The water in the cup overflows. Calculate the volume of the remaining water in the cup (know that the diameters of the ball, conical bottom, and of the beaker are about the same; ignore the thickness of the glass cup).*

Solution. *Using* GeoGebra *to draw a cylinder of a radius of base r, and height 6r. Using GeoGebra, draw a sphere and an inscribed cone in a cylinder (Figure 4.45). Then, the volume of the cylinder is equal to* $V = \pi r^2 6\pi = 6\pi r^3$. *The volume of ball is* $V_1 = \frac{4}{3}\pi r^3$ *and the volume of cone* $V_2 = \frac{1}{3}\pi r^2 4r = \frac{4}{3}\pi r^3$. *The volume of water overflow*

$$V_3 = V - V_1 - V_2 = 6\pi r^3 - \frac{4}{3}\pi r^3 - \frac{4}{3}\pi r^3 = \frac{10}{3}\pi r^3.$$

GeoGebra *gives more than that: Changing by the slider the radius r, we can calculate the volume of the cylinder, cone, ball, and the water flowing out according to the above formulas (Figure 4.46).*

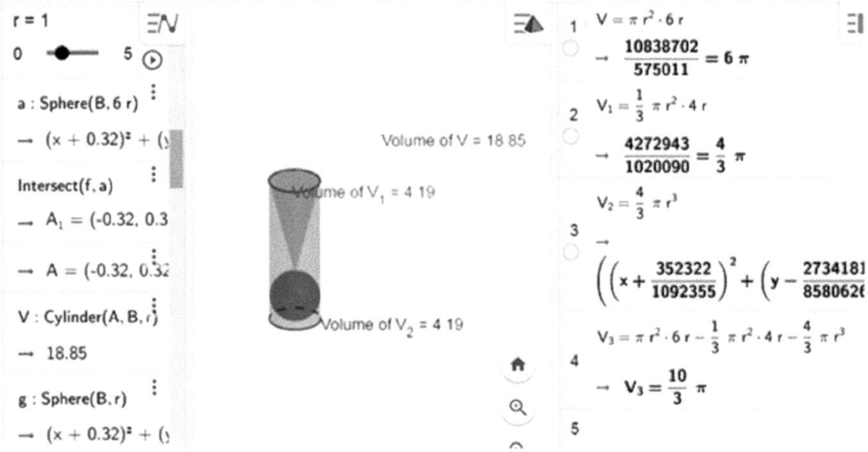

FIGURE 4.45

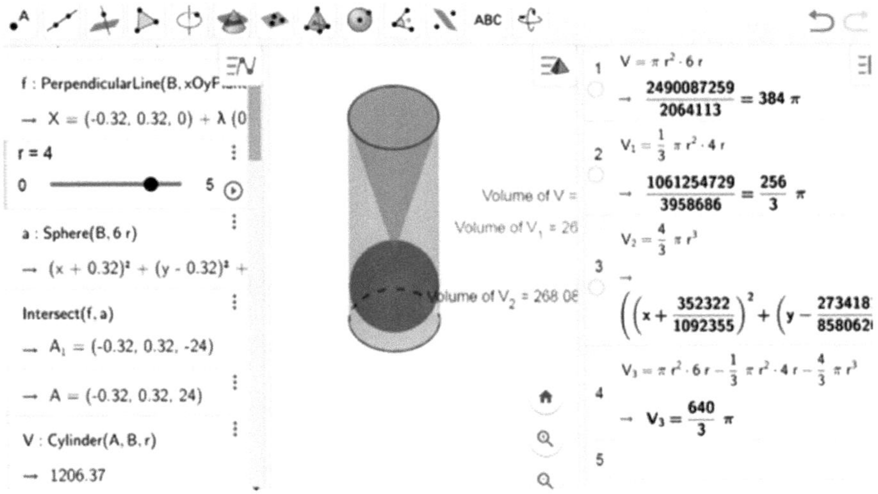

FIGURE 4.46

Problem 4.3.2.15. *A cone water container (without base) is filled with water; its height is 3 times the radius of the base. One cylinder dropped into the cone, and the volume of water overflowed measured is equa tol* $\frac{16\pi}{9}$ dm^3. *Considering that one base of the cylinder is on the base of the cone and the height of the cylinder is the same as the diameter of the base, calculate the lateral area of the cone.*

Solution. *Using* GeoGebra *for drawing a cone with the radius of base r, r, height 3r, and the volume of a cylinder equal to the volume of water overflowed,* $\frac{16\pi}{9}$ dm^3, *the radius of the base is* $\frac{r}{3}$, *and the height of the cylinder is 2r Draw the cylinder inscribed in the cone. Change the radius r of the cone (thanks to the slider, see in the left column of Figure 4.47), for example, when* r = 1.5, GeoGebra *will calculate and show the volume of the cylinder, the volume of the cone, and the generated line and literal area of the cone in the CAS workspace (right of Figure 4.47).*

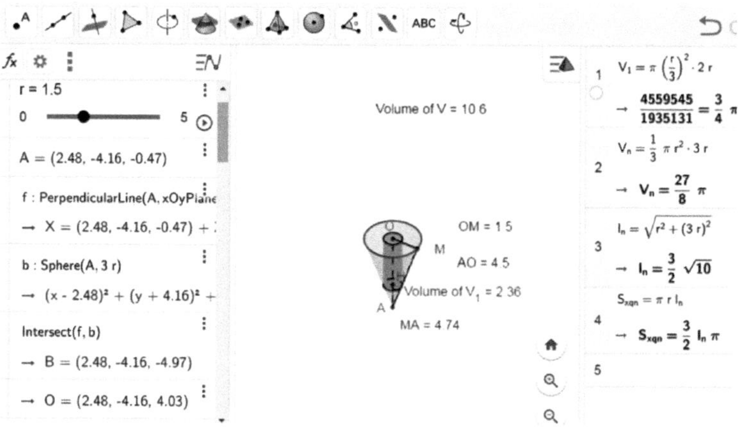

FIGURE 4.47

Every time you go to the slider and select r, GeoGebra *draws and recalculates the whole, for example, with r = 2 and r = 4, we get the results in Figures 4.48 and 4.49.*

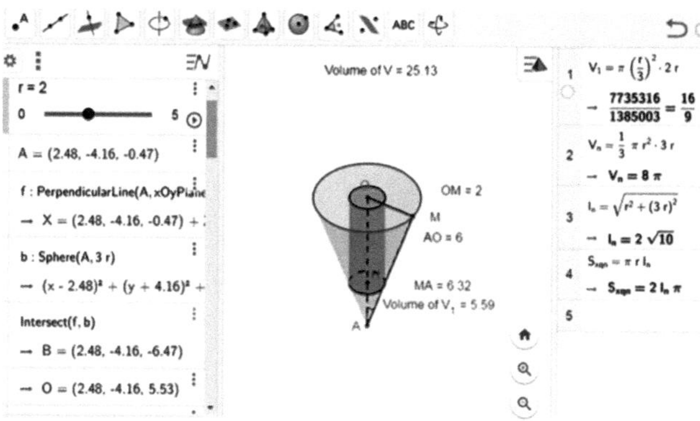

FIGURE 4.48

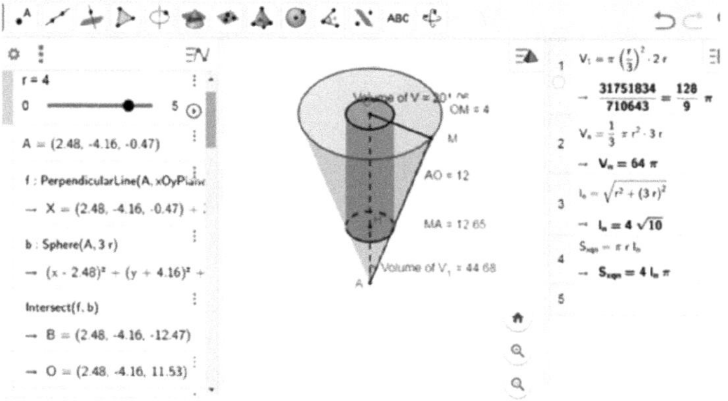

FIGURE 4.49

4.3.3 *USING GEOGEBRA SOFTWARE IN TEACHING AND LEARNING CROSS-SECTION*

A plane is completely defined if one of the following is known:

1. Three points are not collinear.
2. A point and a line do not contain that point.
3. Two parallel lines.
4. The plane goes through a point and parallel to a given plane.

A cross-section is the part of a plane received when cutting a solid figure by a plane. In order to construct a cross-section, it is required not only for the knowledge of space geometry but also for a good imagination of space. In addition to constructing a section, it is also required to calculate the cross-sectional area, volumes, and ratio of the two volumes of parts divided by the plane. *GeoGebra* has great support for teaching and learning cross-section. The following are illustrative examples.

Problem 4.3.3.1. *Construct and calculate the area of the cross-section obtained by cutting an S.ABC pyramid by a plane passing through point M on the side AB and parallel to the plane SIC, where I is the midpoint AB.*

Get the point M on AB. The section passing through M parallel to SIC is the triangle MPN.

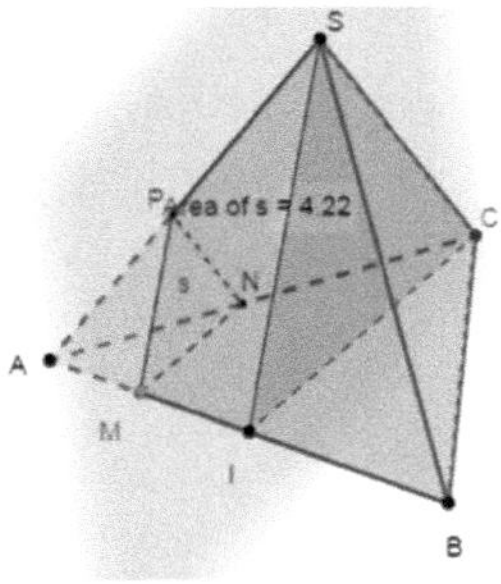

FIGURE 4.50

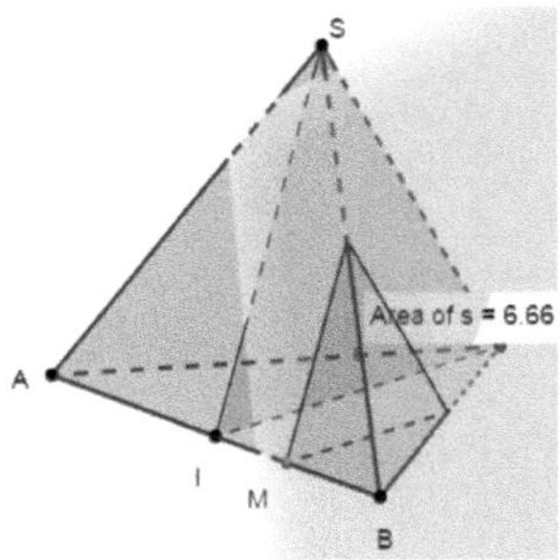

FIGURE 4.51

Let M move on AB, and we have cross-sections with different areas (Figures 4.50 and 4.51).

Problem 4.3.3.2. *Cut a pyramid S.ABC by a plane passing through three points M ∈ SA, N ∈ AB, P ∈ SC, we have a quadrilateral MNQP (Figure 4.52).*

When M and N are fixed, P moves to C along with SC; Q also moves to C along with BC. When P coincides with C, and Q also coincides with B, the cross-section is triangle BMN (Figure 4.53).

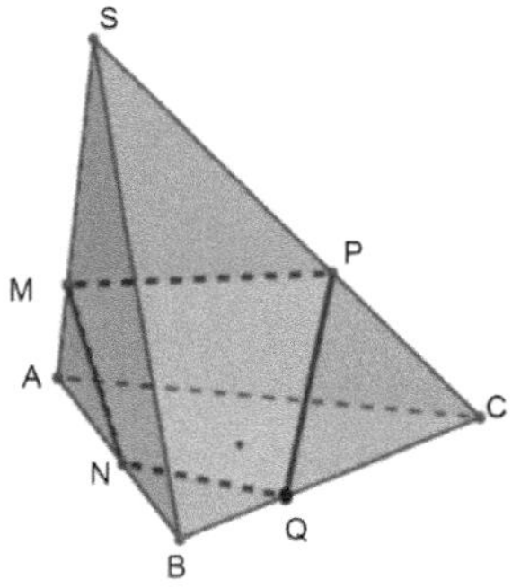

FIGURE 4.52

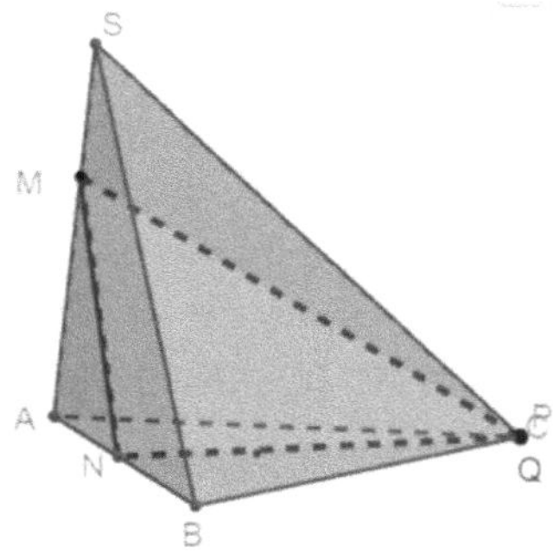

FIGURE 4.53

Problem 4.3.3.3. *Cut a pyramid S.ABC by a plane through three points M ∈ SA, N ∈ SB, and P ∈ SC, we have the cross-section is the triangle MNP (Figure 4.54). When the points M, N, and P change, the area of the MNP triangle and the volumes of the pyramid S.MNP and S.ABC also change (Figure 4.55). Change the slide angle; we can now cut the tip of the pyramid separately (Figure 4.56).*

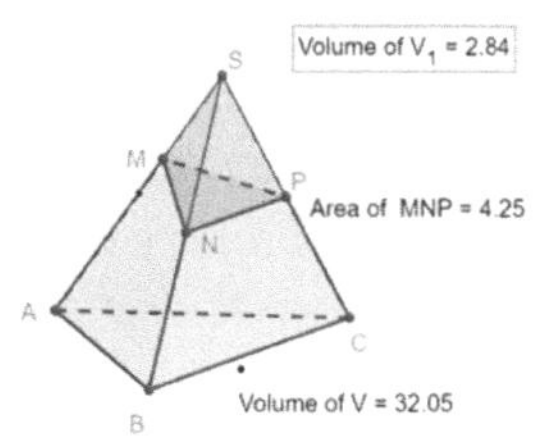

FIGURE 4.54

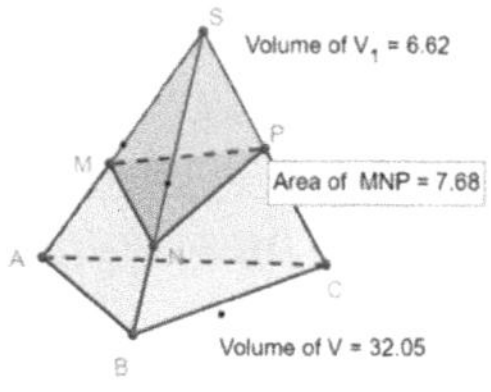

FIGURE 4.55

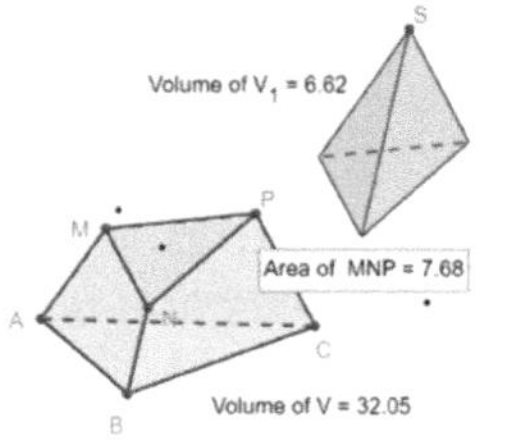

FIGURE 4.56

Problem 4.3.3.4. *Cut a pyramid S.ABC by a plane passing through A and parallel to BC. Then, the cross-section is the triangle AEH. Calculate the volumes of S.ABC and S.AEH.*

Solution. *We have EH parallel to BC. Using GeoGebra, we can draw and calculate the volumes of S.ABC and S.AEH (Figure 4.57). Changing the position of S by translating or changing B, C by rotation, we obtain different figures with different volumes (Figures 4.58 and 4.59).*

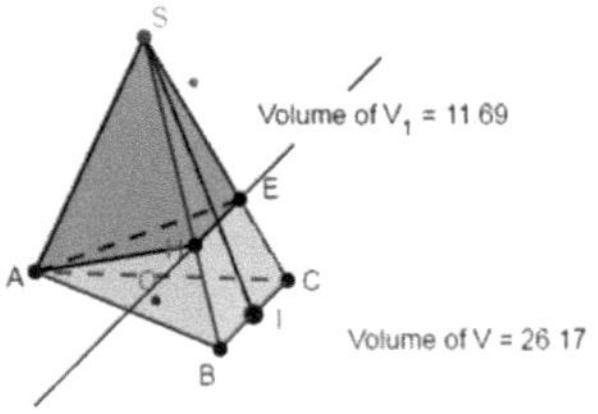

FIGURE 4.57

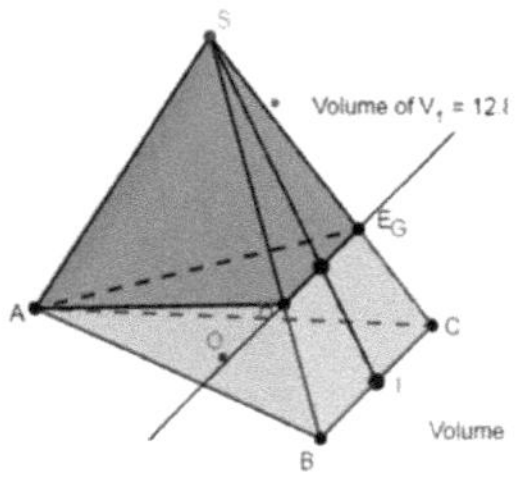

FIGURE 4.58

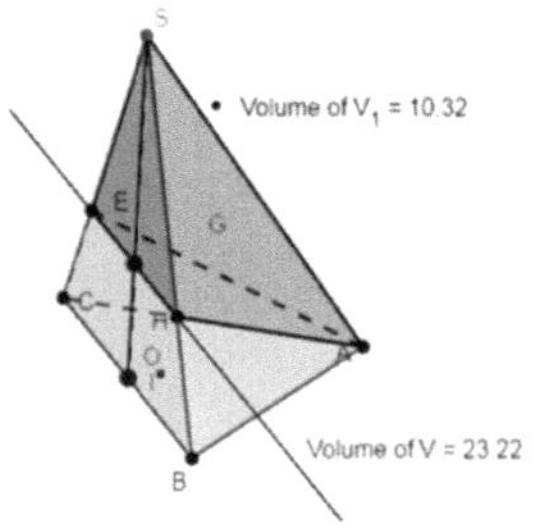

FIGURE 4.59

Problem 4.3.3.5. *Cut a quadrilateral pyramid S.ABCD by a plane parallel to the base. The section is a quadrilateral EFGH similar to the base ABCD (Figure 4.60). We are using the slider icon to build rotations. Change the slide angle $\alpha = 30^o$, we can cut the tip of the pyramid separately (Figure 4.61).*

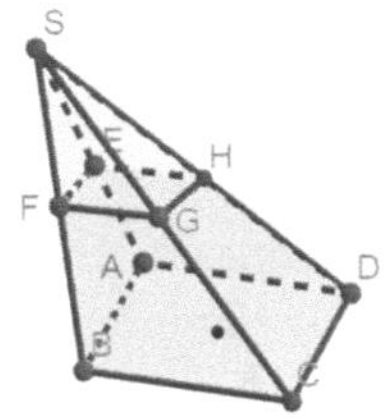

FIGURE 4.60

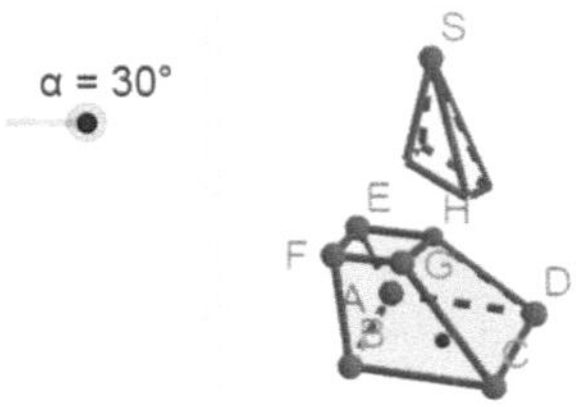

FIGURE 4.61

Problem 4.3.3.6. *Cut a quadrilateral pyramid S.ABCD by the plane passing through three points M ∈ BC, N ∈ (ABCD), P ∈ SA. The cross-section is a pentagon PQMEF (Figure 4.62). When M is moving on BC, P and N are fixed, and Q, E, and F also change accordingly. When M coincides with B, the cross-section is quadrilateral PBEF (Figure 4.63). When M and N are fixed, P is mobile and coincides with S, and Q and F also coincide with S. The cross-section is the triangle SME (Figure 4.64).*

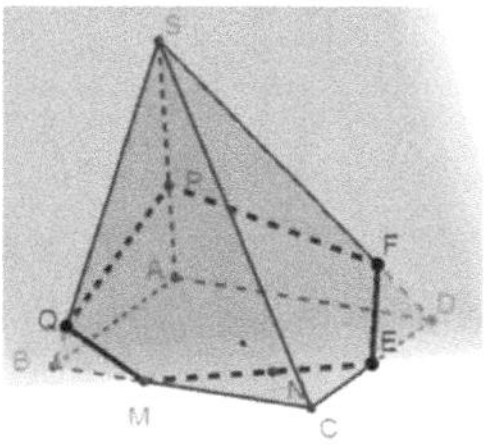

FIGURE 4.62

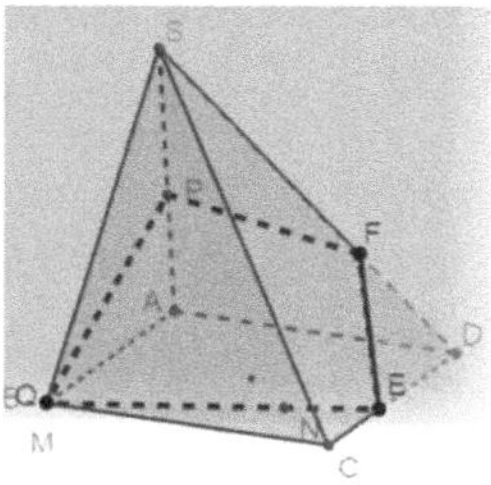

FIGURE 4.63

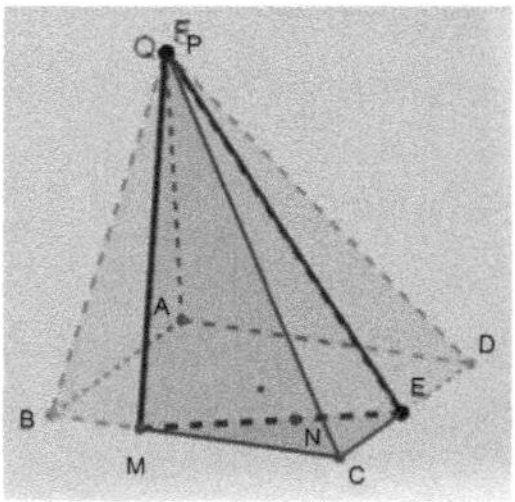

FIGURE 4.64

Problem 4.3.3.7. *Cutting the quadrilateral pyramid S.ABCD by the plane passing through the fixed point P ∈ SA and a straight line d orbiting the M ∈ (ABCD), we receive a pentagon PFEGH (Figure 4.65). When d rotates around M, the cross-section also moves and can become quadrilateral PFEG (Figure 4.66).*

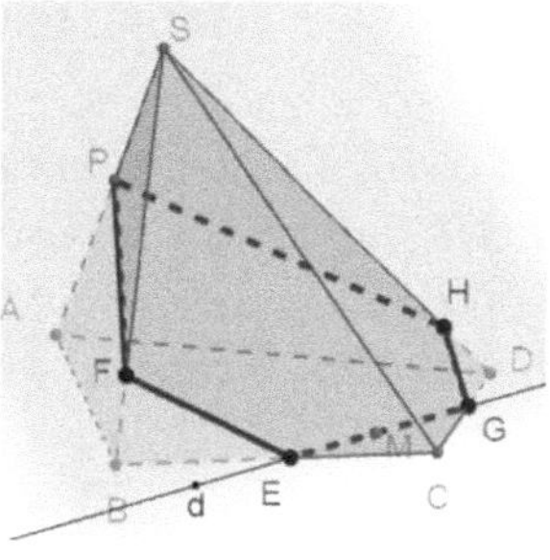

FIGURE 4.65

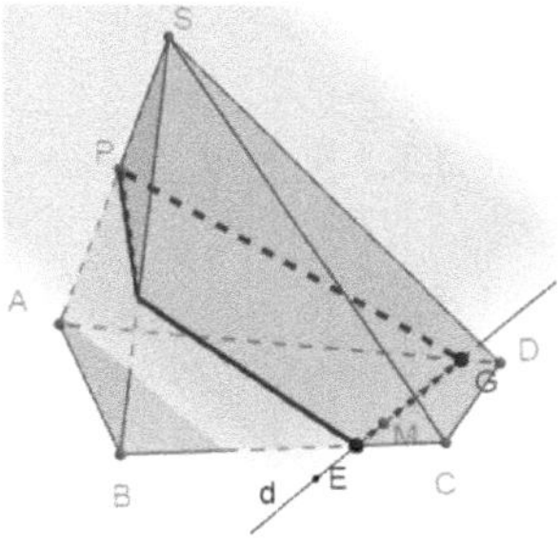

FIGURE 4.66

Problem 4.3.3.8. *Let the regular quadrilateral pyramid S.ABCD, O, be the center of the base, AB = a, SO = 2a. Cut S.ABCD by the plane P passing AB and perpendicular to the plane SCD, and the cross-section is an isosceles trapezoid ABEF. Use the slider. With each value of a,* GeoGebra *calculates* *the area of the section. For example, for a = 5 cm, the area of the cross-section is s = 22.83 cm² (Figure 4.67). For a = 3 cm, s = 8.22 cm² (Figure 4.68).*

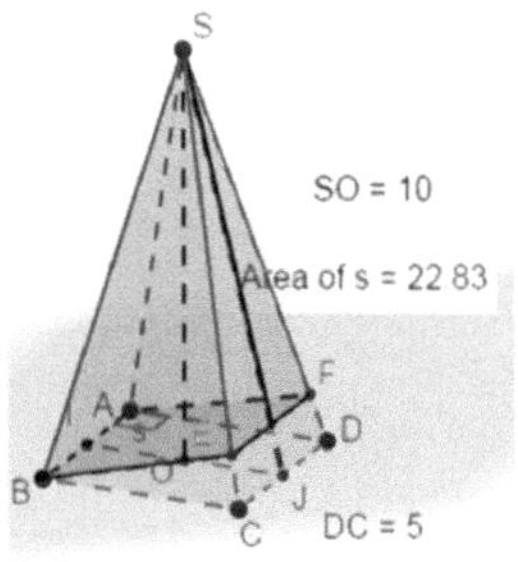

FIGURE 4.67

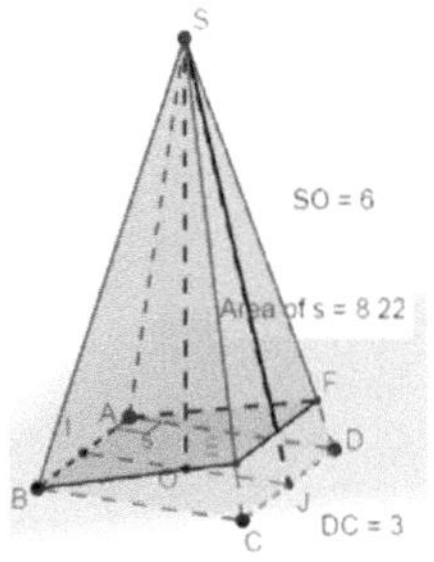

FIGURE 4.68

Problem 4.3.3.9. *Let the vertical prism ABC.A'B'C'. When M moves on A'C', the plane MAB cuts the prism along with the trapezoid ABGM.* GeoGebra *easily calculates the area of the cross-section AGMB and the volume of a prism (Figure 4.69). When M coincides with C', the cross-section is the triangle C'AB (Figure 4.70).*

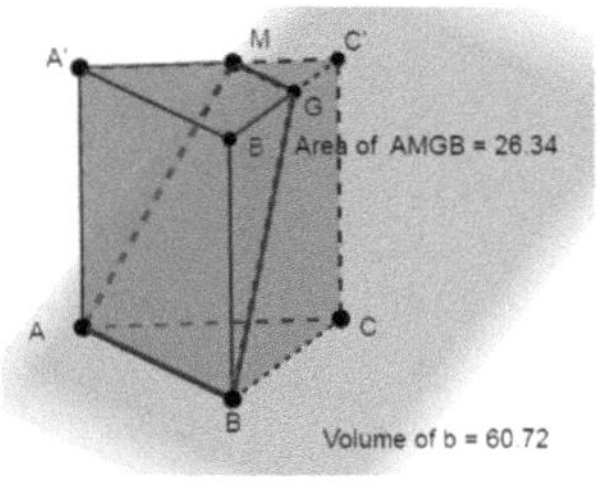

FIGURE 4.69

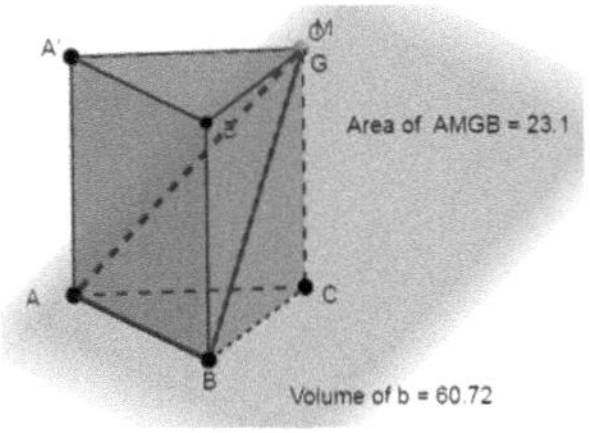

FIGURE 4.70

Problem 4.3.3.10. *Given an oblique prism ABC.A'B'C', whose base ABC is a right triangle at A, AB = c, AC = b, AA' = a. Construct and calculate the area of the cross-section created by the plane P passing through A' and perpendicular to BC.*

Solution. *Construct the prism ABC.A'B'C' and the two straight lines A'F and FE perpendicular to B'C'. Because the angle of B'A'C' is right, F is inside the B'C' segment.*

The cross-section is trapezoid A′FED (Figure 4.71). Let a, b, and c be given; then, GeoGebra *calculates the area of the cross-section. Changing the position of the figures and a, b, c, the cross-section is still trapezoidal but has an area with different measurements (Figure 4.72).*

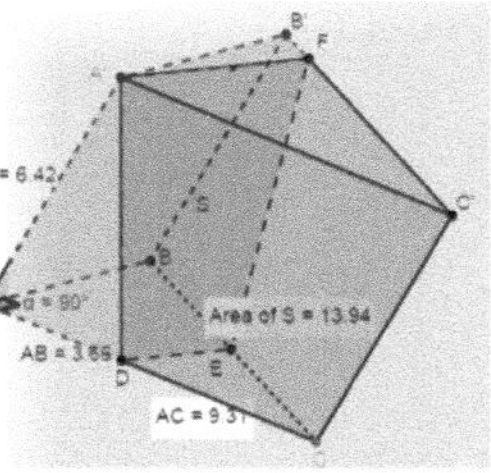

FIGURE 4.71

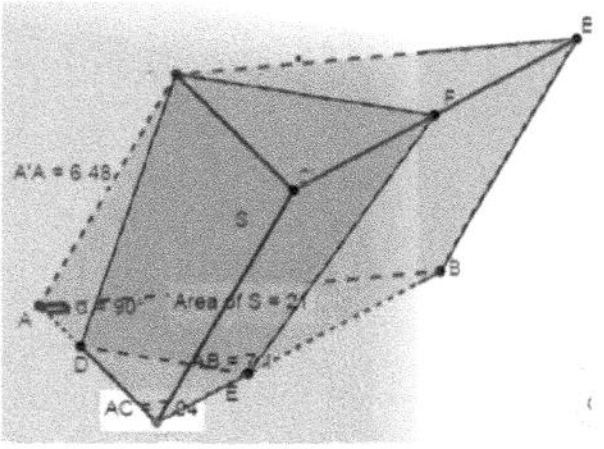

FIGURE 4.72

Problem 4.3.3.11. *Let the vertical prism ABC.A′B′C′ have the base ABC which is a right triangle at A, with AB = c, AC = b, AA′ = h. What shape does the plane P go through A and perpendicular to B′C cut the prism?*

Solution. *Draw a vertical prism ABC.A′B′C′. Change the slider or move vertex A by mouse. Then we will get different cross-sections: trapezoid (Figure 4.73) or triangle (Figures 4.74 and 4.75).*

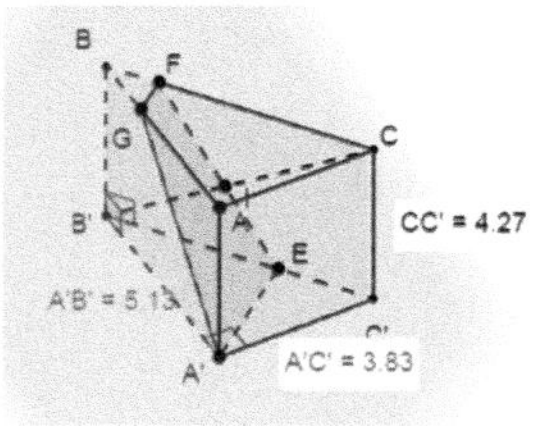

FIGURE 4.73

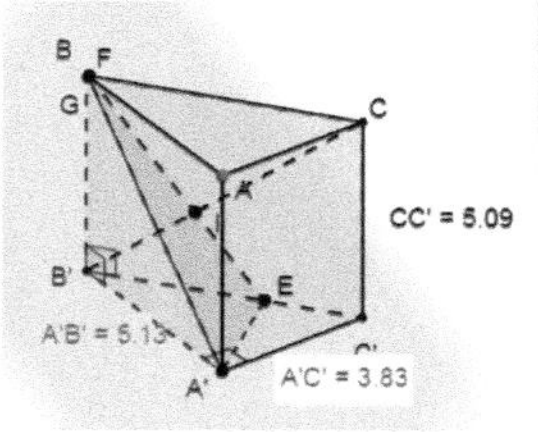

FIGURE 4.74

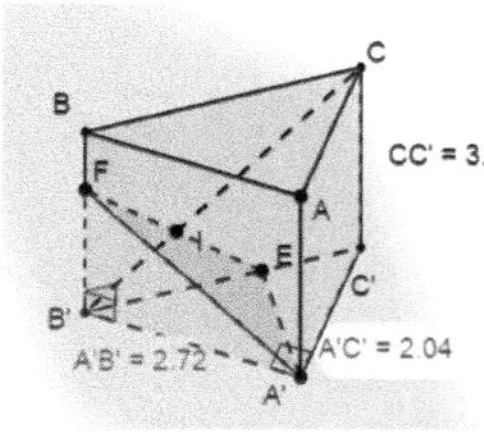

FIGURE 4.75

Problem 4.3.3.12. *Cutting the cone face by a plane passing through the top, we get the triangle. Cutting the cone by a plane parallel to the base we get a circular cross-section (Figure 4.76). Moving points L and M, we get other sections (Figure 4.77).*

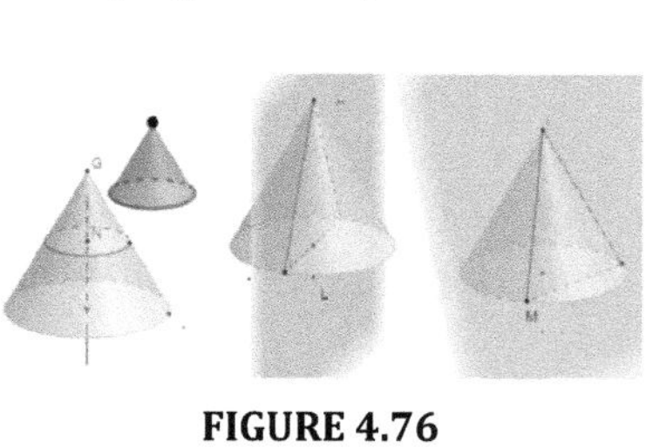

FIGURE 4.76

FIGURE 4.77

Problem 4.3.3.13. *Cutting the cone by a plane that does not pass through the vertex gives us quadratic curves: ellipse, parabola, and hyperbole. Changing the plane position, we get the shape changes (Figures 4.78 and 4.79).*

FIGURE 4.78

FIGURE 4.79

Problem 4.3.3.14. *The intersection of a sphere I, radius r with a plane is a circle with center H and radius HK. GeoGebra calculates the radius and area of the cross-section according to IK = r and IH (Figure 4.80). Change the position of the plane (move point H), the area of the cross-section changed (Figure 4.81). Changing the slider, we change the sphere radius (IK = 3, Figures 4.82 and 4.83).*

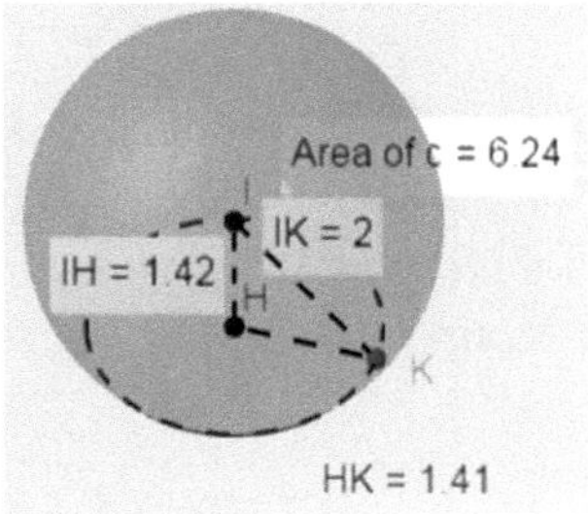

FIGURE 4.80

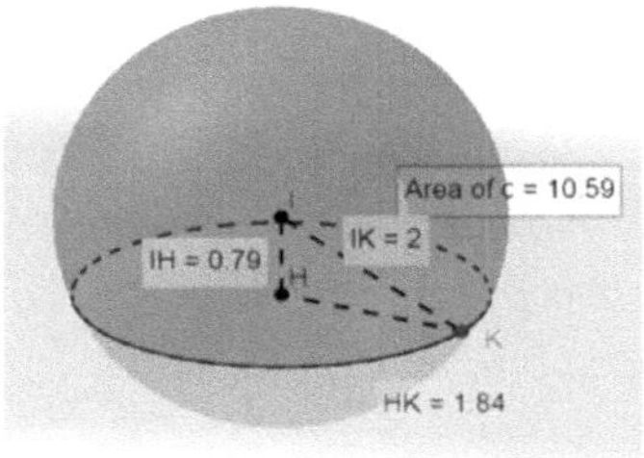

FIGURE 4.81

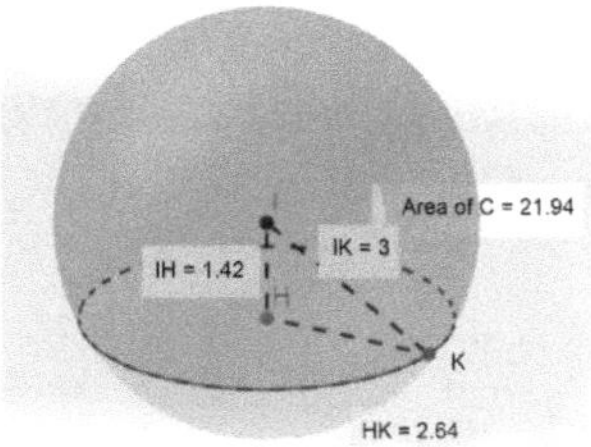

FIGURE 4.82

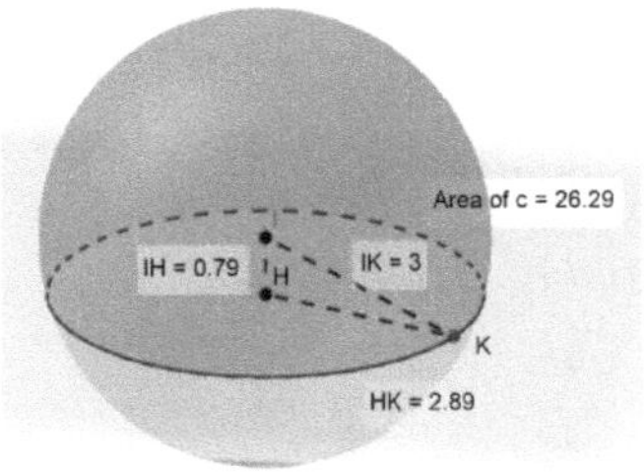

FIGURE 4.83

The area of the cross-section varies, although IH is constant (compare calculation results in Figures 4.80 and 4.82 when IH = 1.42 and in Figures 4.81 and 4.83, when IH = 0.79).

4.3.4 SOME PROBLEMS ON THE CROSS-SECTION

Problem 4.3.3.15. *Given a pyramid S.ABCD whose base is a parallelogram, the point P is the midpoint of SC. A plane α through AP. Cut SD and SB, respectively, at M and N (Figures 4.84 and 4.85). Suppose that the volume of the S.AMPN pyramid is V, and the volume of the S.AMPN pyramid is* V_1. *Then, find the smallest value of the ratio* $\frac{V_1}{V}$..

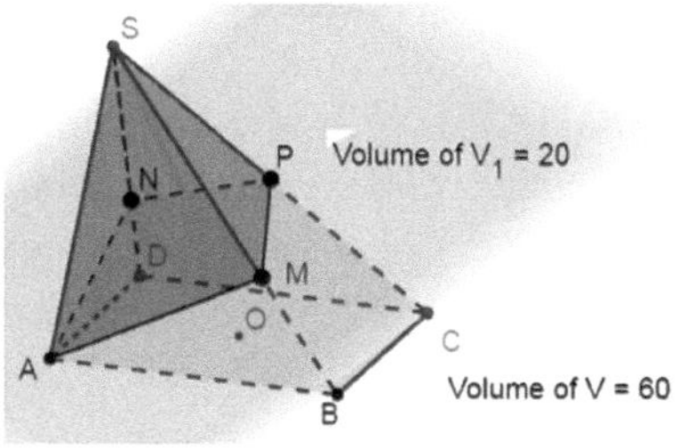

FIGURE 4.84

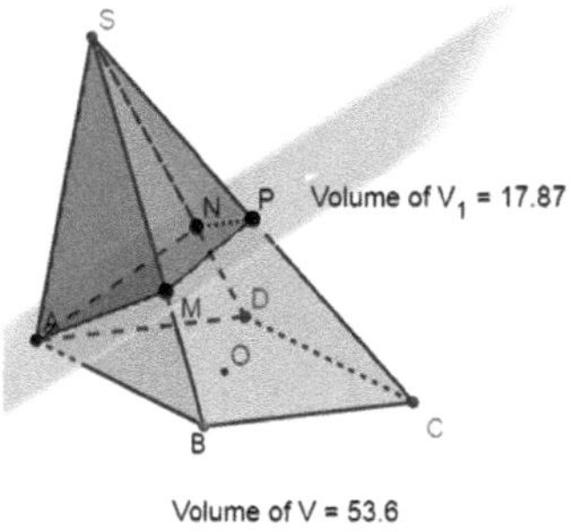

FIGURE 4.85

Problem 4.3.3.16. *Given a rectangular prism ABCD.A'B'C'D', AB = a, BC = 2a, AC' = 3a. Point N is on the edge of BB' so that BN = 2NB', point M is on the edge DD' so that D'M = 2MD. The plane A'MN divides the prism into two parts. Calculate the volume of the section containing the point C'. Choose one from A, B, C, D:*

A. $4a^3$ **B.** a^3 **C.** $2a^3$ **D.** $3a^3$

Problem 4.3.3.17. *Given a pyramid S.ABC with a volume of V. Let G be the center of the triangle SBC. The plane P passes through two points A and G, and is parallel to BC. The plane P intersects the edges SB and SC at points M and N, respectively. Calculate the volume of the pyramid S.AMN. Choose one from A, B, C, D:*

A. $\frac{V}{9}$ **B.** $\frac{V}{2}$ **C.** $\frac{4V}{9}$ **D.** $\frac{V}{4}$

Problem 4.3.3.18. *Given the tetrahedron ABCD, on the sides BC, BD, and AC, take the points M, N, and P, respectively, so that BC = 3BM,* $BD = \frac{3}{2}BN$ *and AC = 2AP. The plane MNP divides the tetrahedron ABCD into two parts with volumes* V_1, V_2. *Calculate the ratio* $\frac{V_1}{V_2}$. *Choose one from A, B, C, D:*

A. $\frac{V_1}{V_2} = \frac{26}{19}$ **B.** $\frac{V_1}{V_2} = \frac{3}{19}$ **C.** $\frac{V_1}{V_2} = \frac{15}{19}$ **D.** $\frac{V_1}{V_2} = \frac{26}{13}$

Problem 4.3.3.19. *Given a pyramid S.ABCD, whose base ABCD is a diamond with sides a, angles BAD = 60°, and SA is perpendicular to the plane ABCD. The angle between the planes SBD and ABCD is equal to 45°. Let M be the symmetrical point of C through B and N be the midpoint of SC. The plane MND divides the S.ABCD pyramid into two polyhedrons where the polyhedron contains the vertex S with volume* V_1 *and the other has volume* V_2*. Calculate the ratio* $\frac{V_1}{V_2}$*. Choose one from A, B, C, D:*

A. $\frac{V_1}{V_2} = \frac{1}{5}$ **B.** $\frac{V_1}{V_2} = \frac{5}{3}$ **C.** $\frac{V_1}{V_2} = \frac{12}{7}$ **D.** $\frac{V_1}{V_2} = \frac{7}{5}$

Problem 4.3.3.20. *For pyramid S.ABCD, whose base ABCD is the square with the edge a, SA = 2a. The planes SAB and SAD are perpendicular to the plane ABCD. The plane P through A and perpendicular to SC, cut the edges SB, SC, and SD, respectively, at B', C', and D'. The volumes of the pyramid S.AB'C'D' and the polyhedron ABCD.D'C'B' are* V_1 *and* V_2*. Calculate the ratio* $\frac{V_1}{V_2}$*. Choose one from A, B, C, D:*

A. $\frac{V_1}{V_2} = \frac{8}{15}$ **B.** $\frac{V_1}{V_2} = \frac{8}{7}$ **C.** $\frac{V_1}{V_2} = \frac{32}{13}$ **D.** $\frac{V_1}{V_2} = \frac{1}{2}$

Problem 4.3.3.21. *Given a pyramid S.ABCD, whose base ABCD is a parallelogram. Let M, N, P, Q are the focus of the triangles SAB, SBC, SCD, SDA. Let O be any point on the base plane ABCD. Know the volume of pyramid ONMPQ by V. Calculate the volume of pyramid S.ABCD. Choose one from A, B, C, D:*

A. $\frac{27}{8}V$ **B.** $\frac{27}{2}V$ **C.** $\frac{9}{4}V$ **D.** $\frac{27}{4}V$

Problem 4.3.3.22. *The base ABCD of the pyramid S.ABCD is a parallelogram, let M is the midpoint of SC. The plane P containing AM and parallel BD cuts the edges SB, SD at P, Q. Knows the volume of the pyramid S.ABCD is V. Calculate the volume of the pthe edges SB, Choose one from A, B, C, D:*

A. $\frac{V}{4}$ **B.** $\frac{V}{8}$ **C.** $\frac{V}{3}$ **D.** $\frac{V}{6}$

Problem 4.3.3.23. *The base ABCD of the pyramid S.ABCD is a parallelogram. The pyramid S.ABCD has the volume is equal to 1. Let M be the symmetrical point of C through B, and N is the midpoint of SC. The plane MDN divides the S.ABCD into two polyhedrons. Calculate the volume of the polyhedron contain vertex S. Choose one from A, B, C, D:*

A. $\frac{5}{6}$ **B.** $\frac{5}{8}$ **C.** $\frac{12}{19}$ **D.** $\frac{7}{12}$

Problem 4.3.3.24. *Given the pyramid S.ABCD, whose base ABCD is square, the side SAB is an equilateral triangle in the plane perpendicular to the plane ABCD and has an area equal* $\frac{27\sqrt{3}}{4}$. *The plane P passes through the center of the triangle SAB and parallel to the base plane ABCD divide the S.ABCD pyramid into two parts. Calculate the volume of the polyhedron contain vertex S. Choose one from A, B, C, D:*

A. $V = 24$ **B.** $V = 8$ **C.** $V = 12$ **D.** $V = 36$

Problem 4.3.3.25. *Given a pyramid S.ABCD, whose base ABCD is a parallelogram. Let N be the midpoint of SB, P be on the side SC so that SP = 2PC, M is on the side of SA such that* $SM = \frac{4}{5}MA$. *The plane MNP cuts SD at Q, NP cuts BC at E, CQ cuts DP at R. Know the volume of pyramid EPQR equals 18 cm*^{3}. *Calculate the volume of the pyramid S.MNPQ.*

Choose one from A, B, C, D:

A. 65 cm^3 **B.** $\frac{260}{9}$ cm^3 **C.** 75 cm^3 **D.** 70 cm^3

4.4 EXPERIMENTAL RESULTS

The basic content of this chapter and some of the extensions have been implemented by the teacher Nguyen Thi Hong Hanh (one of the four authors of the chapter) in some teaching hours for some classes 12, Quang Trung High School, Dong Da District, Hanoi, Vietnam. The results showed that the math lesson becomes lively, fun; students learn to be interested and understand the lesson much faster than traditional methods. *GeoGebra* has become a game, a passion for some students. Information technology lessons (using *GeoGebra*) have been highly appreciated by teachers as a new teaching method in Vietnam, not only in mathematics but also in other subjects.

An image in math class. The lessons "Cone" using *GeoGebra* in class 12D5 of Quang Trung High School, Dong Da, Ha noi (Figure 4.86). Students practice drawing and calculate lateral areas and volumes, draw a cross-section, and calculate the area of a cross-section and the ratio of volumes on *GeoGebra*. After that, they check the calculation results on paper according to the formula. Another group first calculates on paper and then practices on *GeoGebra*.

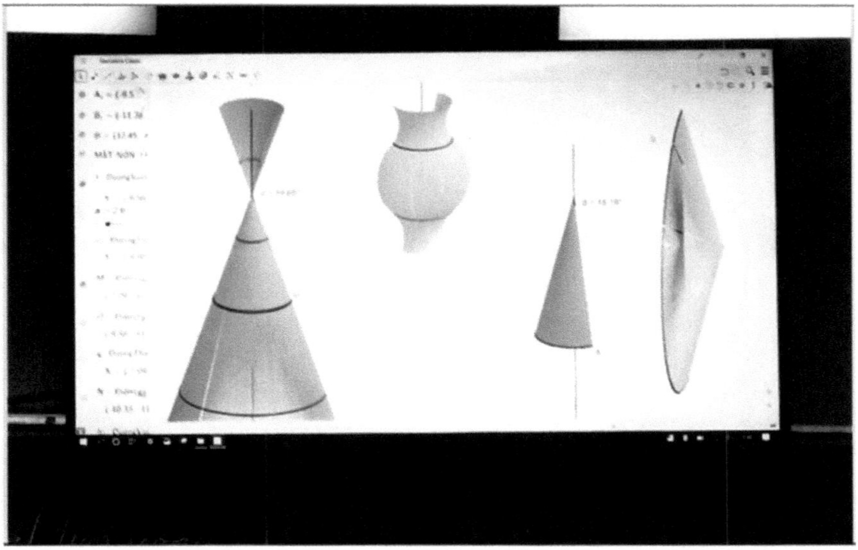

FIGURE 4.86 The lesson "Cone" using *GeoGebra*.

The practice lesson "Cone" in class 12D5 of Quang Trung High School, Dong Da, Hanoi (Figure 4.87). Students use real visual aids "Vietnamese hat" (Vietnamese cone). Students are divided into groups and practice: Measure the length of the generated line, the length of the circumference of the base of the "Vietnamese hat." From the measurement data (the length of the generated line, the length of the circumference), they calculate the radius, height, base area, literal area, and volume of a cone, area of cross-section, etc.; and compare the results with results on *GeoGebra*.

FIGURE 4.87 The practice lesson "Cone."

4.5 CONCLUSIONS

The chapter presents some initial attempts of the authors in the use of information technology, namely the use of *GeoGebra* mathematical software in teaching space geometry at a high school. The content of the chapter has been taught experimentally in several classes of high schools with positive results, attracted the interest of students, and evaluated by math teachers as a new teaching method in high school in Vietnam. In order to bring information technology into teaching, it requires teachers to try very hard and elaborate on lesson planning, so it can be said that many profound aspects of using *GeoGebra* in teaching the space geometry are not mentioned in this chapter. Hopefully, some problems, for example, promoting mathematical creativity, discovering new mathematical problems, thanks to *GeoGebra* as an experimental tool, will be mentioned in other articles in the future.

KEYWORDS

- **GeoGebra**
- **cross-section**
- **teaching, learning**
- **mathematics**

REFERENCES

Boo Jia Yi and Kwan Eu Leong. Teaching and learning of geometry in primary school using GeoGebra. *Procceding of the 21st Asian Conference on Technology in Mathematics*, 2016, pp. 289–320.

Bui Viet Ha. *Geogebra Software Training Lecture* (in Vietnames), https://www.academia.edu/

Gerard A. Venema. *Exploring Advanced Euclidean Geometry with GeoGebra*, the Mathematical Association of America (MAA), USA, 2013.

Gerry Stahl and The VMT Project Team, The Math Forum at Drexel University. *Dynamic-Geometry Activities with GeoGebra for Virtual Math Teams*, 2013. gerrystahl.net/vmt/activities.pdf

Ministry of Education and Training. *Mathematics General Education Program*, Hanoi (in Vietnamese), June 2018.

Muhammad Khalil, Umair Khalil, and Zahoor ul Haq. GeoGebra as a scaffolding tool for exploring analytic geometry structure and developing mathematical thinking of diverse achievers. *International Electronic Journal of Mathematics Education*, Vol. 14, No. 2, pp. 427–434, 2019.

Pham Huy Dien, Dinh The Luc, and Ta Duy Phuong. *Guiding practice of calculation on Maple V*, Education Publishing House, Hanoi (in Vietnamese), 1998.

Project Maths Development Team. *GeoGebra Geometry*, 2013. https://www.projectmaths.ie/documents/PDF/GeoGebraForGeometry.pdf

Royati Abdul Saha, Ahmad Fauzi Mohd Ayub, and Rohani Ahmad Tarmizi. The effects of GeoGebra on mathematics achievement: Enlightening coordinate geometry learning, *Procedia Social and Behavioral Sciences*, Vol. 8, pp. 686–693, 2010.

Steve Phelps. *An Introduction To GeoGebra*, 2011. https://www.math.utah.edu/~emina/teaching/5270s13/Intro_to_Geogebra.pdf

Ta Duy Phuong. *Math problem solving on electronic computers*, Education Publishing House, second edition (in Vietnames), 2006.

Tran Du Sinh. *Geometer's Sketchpad User's Guide*, Vietnam National University Press, Hanoi (in Vietnames), 2004.

Tran Vui and Le Quang Hung. *Discovery Geometry 10 with The Geometer's Sketchpad*, Education Publishing House, Hanoi (in Vietnames), 2006.

Tran Vui and Le Quang Hung. *Discovering Algebra 10 with The Geometer's Sketchpad*, Education Publishing House, Hanoi (in Vietnames), 2007a.

Tran Vui and Le Quang Hung. *Exploring Algebra and Calculus 11 with The Geometer's Sketchpad*, Education Publishing House, Hanoi (in Vietnames), 2007b.

Tran Vui and Le Quang Hung. *Explore Geometry 12 with The Geometer's Sketchpad*, Education Publishing House, Hanoi (in Vietnames), 2009.

Zoltan Kovacs and Csilla Solyom-Gecse. *GeoGebra Tools with Proof Capabilities*, arXiv:1603.01228v1 [cs.AI], pp. 1–23, March 2016.

CHAPTER 5

The Effect of Using E-Books in Teaching at High Schools in Hanoi, Vietnam

NGUYEN THI THUY QUYNH[1*], HA THU LY[1], LAI PHƯƠNG LIEN[1], and NGUYEN MAU DUC[2]

[1]*VNU University of Education, Hanoi, VietNam*

[2]*Thai Nguyen University of Education, Vietnam*

[*]*Corresponding author. E-mail: quynhntt-bio@vnu.edu.vn*

ABSTRACT

An electronic book (e-book) is another version of the traditional print book combined with technology that has benefited to educators worldwide. This chapter is a part of a research effort of using e-book in science teaching at high school with the aim of investigating e-book effectiveness. One investigation is administered to biological teachers at more than 20 high schools in Hanoi, Vietnam. Results indicated that the IT literacy rate among investigators is high; the e-book is mostly used in high school but not fully utilized and adopted. The main reason for using an e-book is academic related. The use of e-book makes teaching methods more interesting and makes students take interest in the lesson and stimulate their thinking. The use of an e-book has also created a connection between the student–teacher and student–student interaction. Besides, feedback of respondents indicated positive awareness on using an e-book and ready to change of traditional teaching method. This study has given new insights into the effect of an e-book to science teachers, thus building a solid foundation for future research in e-book projects in Vietnam. At the same time, this research proposed a process and illustrated an example of designing e-book in teaching biology, physics, and chemistry subjects.

5.1 INTRODUCTION

Along with the development of science and technology, learning by an electronic book (e-book) has become an effective support tool for teaching and learning activities gradually. In recent years, many countries, such as Korea, Japan, and Malaysia, have explored and deployed information technology (IT)-based education activities and materials in schools to enhance the efficiency of the learning process (Luaran et al., 2016; Maciej, 2013; Yu et al., 2012). The purpose of using IT in high school is to change the traditional way of learning to more interactive and engaging, diverse, and creative learning, and create a qualified workforce of the 21st century (William et al., 2014). The teaching method with IT applications that teachers applied is a powerful factor that can affect students' motivation to learn and improve their achievement. Nowadays, the direction of innovating teaching and learning methods has been emphasized as an important task, in which promoting IT application in education has been directed and implemented by the Ministry of Education of Vietnam (Thai and Oanh, 2016). In addition, studies show that e-book can be a traditional replacement textbook owing to their compact features. One e-book can contain a lot of information on subjects with using handhold equipment such as computers, smartphones, and tablets (Mutalib et al., 2012). This chapter contributes new insights into using e-books for science teachers and introducing the applications of Kotobee software to construct e-book in science teaching at the high school.

5.2 WHAT IS AN ELECTRONIC BOOK?

An e-book is a type of technical book that contains multimedia programs. All documents in an e-book are presented and operated in a web browser; learners can read it online or even in an offline network environment. Therefore, e-book-based learning will be more flexible. The main purpose of the IT application in teaching is to facilitate high-quality learning and teaching (Kisnieriene et al., 2008; Nazia and Islam, 2015; Qaiser et al., 2011). Along with the convenience and modernity of available technologies, it supports the learning of difficult topics through 2D/3D images, animations, and video.

Jaime et al. (2011) claimed that an e-book was really easy to read and get information with technology. Additionally, the e-book can also be a replacement medium for the original book. This is the same as the e-book supplied by the Terengganu government to the students in the state (Luaran et al., 2016). The students were only required to bring their laptops or e-books to school as a replacement to the heavy textbooks. At present, an e-book is a handheld device that can be read on digital devices such as tablets and smartphones. Nowadays, the term "e-book" is no stranger to many people in Vietnam and is understood in many ways. Since the e-book was first introduced, most thought that e-books were an electronic format of printed books. From then to now, the e-book has developed to be more diverse and can be understood in more ways. Based on the interactive feature, Rhodes and Rozell (2015) defined e-books as interactive or technical books that have content documents in the digital format and can be viewed on electronic devices. Text and illustrations may be similar to printed books, but they incorporate multimedia technology such as audio, animation, 3D images, and music.

Although there are many ways through which an e-book can be understood, we may conclude the following common points regarding it:

1. An e-book is a format of a printed book not having a corresponding printed book;
2. It has diverse formats, not only expressed in the text but also in audio, images, and even video;
3. E-book is used by specialized and handheld devices.

5.3 E-BOOK AND EDUCATION

5.3.1 THE ACTUAL SCENARIO OF USING E-BOOK IN HANOI, VIETNAM

In this study, before proposing the concept of e-book design, we conducted a survey of the understanding of e-book usage from 125 biology teachers in 10 schools in Hanoi, Vietnam. The results show that most of the teachers who have the seniority of more than 15 years do not use e-book (35%). However, they whose seniority is between 1 and 5 years are the highest among the users of e-books (42.3%) (Table 5.1).

TABLE 5.1 Teachers' Opinion on Using E-Book in Biological Teaching

	Seniority			
	1–5	**6–10**	**11–15**	**>15**
Yes	42.3%	35.2%	18.3%	4.2%
No	20%	30%	15%	35%
No answer	30%	50%	20%	0%

The purpose of using e-books by the users who utilized them in teaching was investigated. The results indicated that most teachers with the seniority of 1–5 years used e-books in teaching and assigning, followed by them who used e-books as reference materials (66.7% and 36.1%, respectively), while teachers with a seniority of more than 15 years never used e-book (Table 5.2). Besides, teachers use e-book for entertainment such as reading stories and watching illustrated videos.

TABLE 5.2 Purpose of Using E-Book

	Seniority			
	1–5	**6–10**	**11–15**	**>15**
Used in teaching and assigning	66.7%	20.0%	13.3%	0%
Used as a reference	36.1%	39.3%	19.7%	4.9%
Used for entertainment (reading stories and watching illustrated videos in books)	16.7%	66.7%	16.7%	0%
Other	2.9%	0%	0%	0%

From the investigation results of the e-book understanding and the purpose of teachers, it was found that the percentage of people who knew e-books was quite high but only at the an application level. Most of the teachers were aware of using e-books only as reference resources, but not fully exploited their effectiveness as well as designed e-book. Therefore, we propose a model for help designing e-books in science teaching (Figure 5.1).

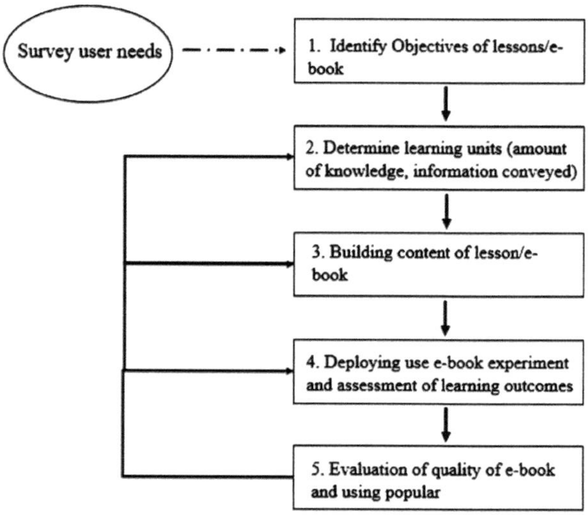

FIGURE 5.1 Model design and use of an e-book.

The first task in this process is to have surveys of learners' needs. E-book's content will be presented differently, which depends on different needs. For example, if users who are teachers, e-book's content will be intensive, scholarly, and suitable for reference or teaching purposes. Conversely, if users are students, the content should be presented in a vivid, easy to understand, no difficult to understand, and academic words, aiming to increase motivation and attractiveness through image channels and illustrated videos. To design an e-book, the following steps should be followed:

1. *Identify the objectives of the lessons:* This step needs to identify the knowledge and experience objectives that the learner can get after studying. The objective determination should be based on the standard of knowledge and skill of the basic curriculum framework, concurrently, developing the common and special competences of the learner on the direction of education innovation. For example, e-book used in teaching biology of Grade 10 is designed to follow the below objectives:
   a. Students can understand the basic knowledge of cell biology, microbial biology.
   b. Students can understand the structure, phenomena through experiment, pattern.
   c. Students are able to apply their knowledge to the practice.
2. *Determine learning units (amount of knowledge transferred):* Each learning unit corresponding to the amount of knowledge transferred, student will solve the requirement of this amount of knowledge in learning. For example, the e-book of biology of Grade 10 is divided into chapters; each of them contains theoretical knowledge and exercise which apply the theories.
3. *Building content of lesson:* Based on teaching techniques, designers/teachers put in the content and integrate with multimedia technology in order to present vividly and intuitively. For example, each theoretical content will be described as text, images, models, experiments/virtual experiments which clearly show the content of the e-book.
4. *Test deployment and evaluation of learning outcomes*: Deploying experiments in face-to-face learning or instructing students to use for the purpose of learning support. Assessment of learning outcomes may include:
   a. Assessing the performance through the small tests at the end of each lesson;

   b. Assessing the learning outcomes of the entire course after using e-books in teaching. Based on each type of these assessments, the designer/teacher can make questions and application exercises with the useful features of an e-book. Besides, it is possible to check and evaluate the effectiveness between groups using and not using e-books in teaching.

5. *Evaluation of e-book quality and wide deployment*: A small survey directed to the users after their experience of using e-books such as their interface, contents, features related to e-books, or the users' satisfaction level. Then, the designer/teacher can edit the negative points combined with improving e-book efficiency, from which an e-book can be widely deployed.

5.3.2 APPLICATION OF AN E-BOOK IN EDUCATION

5.3.2.1 PROCESS OF DESIGNING AN E-BOOK

In Vietnam, there is little information on the designing process of an e-book used in teaching in high school (Tran et al. 2008, 2015). Therefore, this research has described a detailed five-step process, as shown in Figure 5.2. The e-book structure of biology of Grade 10 is illustrated in the figure, which makes desiging e-books easier for teachers themselves, for the subjects they teach in high schools.

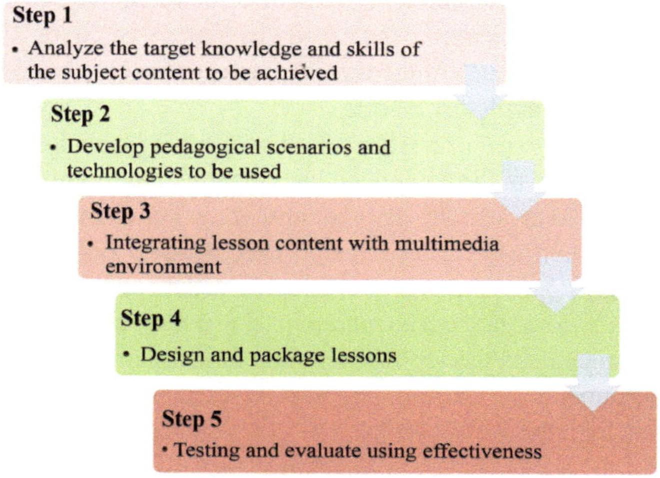

FIGURE 5.2 Process of designing and building an e-book.

Step 1: Analyze the target knowledge and skills of the subject content to be achieved.

When developing the lesson content, the teacher points out the goals of knowledge, skills, attitudes, and formed competence of students

- Determining the title: identify the chapter and the specific lesson name.
- Determining the objectives of the lesson: basic contents and practical meanings of biological programs.
  + Knowledge: Presentation of the contents of cell biology (composition and the structure of cells), microbiology (bacteria and viruses)
  + Skills: May be utilizing IT in searching, collecting, and analyzing the obtained information order to solve the given problems.
  + Attitude: Raising awareness and responsibility in protecting the body.

Step 2: Develop pedagogical scenarios and technologies to be used.

Need to build a framework for readers to easily visualize and having a consistent view of the structure, using techniques to implement the lesson. The details are as follows:

- Purpose of the lesson (What is this lesson for?)
- Content of the lesson (What is the core content? What is the expanded content?)
- Implementation method (What should the teachers do? What should the learners do?)
- Modes of implementation
- General characteristics of the user—object
- Feasibility of technological factors when transferring the content.

When developing the scenario and e-book framework, it is necessary to choose the designing technology and multimedia tools, consistent with presenting the effective content, user-friendly interface, and pedagogical intent.

Step 3: Integrating lesson content with the multimedia environment.

The designer needs to select, classify, and organize the lesson content-related material resource according to the characters for the core content or supplementary content. After that, there is a need to select and coordinate the appropriate technology tools to design the e-book's content, such as follows:

- Selecting software/utilities for digitizing documents and figures
- Selecting software/utilities for processing graphics, sound, images, clips
- Selecting software/tools for supporting learning, assessment, and testing.

Step 4: Design and package lessons.

Based on the designed framework in the above steps, the designer will develop the content in a specific way using the text format, picture, audio, video in accordance with the general format of the e-book and easy to export format with the completed file.

Step 5: Testing and evaluate using effectiveness.

Teachers control the e-book usage of students. At the same time, they must support and answer students' questions during the use of the e-book. After completing the lesson, the teacher can get the student's attention about the interface and lesson content (such as the knowledge is too difficult, or the amount of homework is too much/little, or the images are not vivid yet), so that there are improvement options for the e-book.

5.3.2.2 INSTRUCTION OF AN E-BOOK

After determining the needs of e-book users and the educational goals and IT using trends in teaching, the structure of an e-book will have a common interface, as shown in Figure 5.3.

- Name of the e-book: The title of the book
- Chapter/topics: The content of chapters/topics, material resources are presented systematically and logically.
- Theoretical content: Content of the e-book is systematized in each chapter/small topic. Each lesson will provide information about knowledge for reader/student
  + Introduction: Presenting contents or activities that promote the learning motivation and knowledge creation for students.
  + Content: This is the central part of the lesson, with the knowledge content with models or example order to help students understand and master the knowledge.
  + Extension knowledge (Do you know?): Providing more practical information and expand knowledge from the designed lesson that

helps students further understand and like science and biology subject.

- Questions and exercises: Including exercises in textbooks and additional exercises systematized and performed in the interaction form with learners.
- Instruction: The specific instructions for using the e-book that support users studying quickly.
- Online teaching: The e-book is designed with the purpose of improving the interaction between users and authors. The Question and Answer section allows users to look up academic words or documents. The Classroom Link section helps users interact with the authors, as well as with other users, by creating a connecting space through applications such as Edmodo/ClassDojo/Google Classroom.

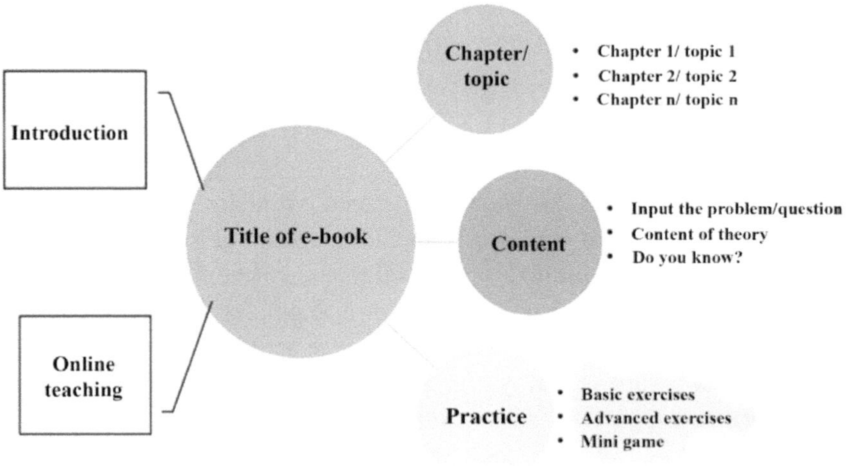

FIGURE 5.3 Structure of an e-book.

5.3.2.3 SOFTWARE FOR DESIGNING AN E-BOOK

Kotobee Author

This is an e-book design tool suitable for education, training, and publishing. Kotobee Author with a friendly interface, easy to use with integrated multimedia tools such as video, images, and audio, and helps users easily create an e-book with their own style.

- *Advantages*
  + Users can easily create interactive e-books with video, audio, 3D, widgets, questions, and exercises.
  + Users can edit the interface in any format they want.
  + E-books can be exported to over 10 formats to be used on web applications or electronic devices.
  + Users can download the Kotobee software (for phones or computers) to the app or computer. The version of the software can be downloaded and installed from the website.
- *Disadvantages*
  + Some users claim that some interactions of the trial version are limited. They can only use the single function without using the insert another one.
  + Using latex in Kotobee is quite complicated, so designing the mathematic e-book is limited.

iBooks Author

If you love the exclusivity and are an intimate user of Apple, you will not be a stranger to this software.

- *Advantages*
+ iBooks Author is integrated with many interactive features; a variety of fonts, widgets, and charts are automated to help users.
+ Software is not limited to interaction and is free for Apple users.
+ iBooks allows its users to publish and get profit from their books when accepted by Apple or posted anywhere.
+ This software is always improving and increasing interactivity as well as supporting latex or mathematical typing, and it can be said that this is a worthy tool for designing e-book in the mathematic field.
- *Disadvantages*

It can be said that Apple always gives their users the best and high-security product, but this is the minus point of this application.

+ iBooks only allows exporting files in two standard formats: EPUB or PDF.
+ This is proprietary software and is only available for Macs and can be downloaded free at the Mac App Store. Thus, it is difficult for most users who use PC with Windows platforms.

Book Creator

This is an application built on the basis of the Google Chrome browser with an easy-to-use interface that helps the beginner to design an e-book.

- *Advantages:*
  + The easy-to-use interface can run on the Google Chrome browser and can be used on a mobile phone.
  + Allows the exploiter adding features such as sound, image, and videos with very simple manipulations.
  + Allows publication quickly of an e-book and can share with everyone or can print in the PDF format.
- *Disadvantages*:

This is an interesting online application. However, it has the following shortcomings:

+ The app does not support creating homework exercise.
+ Users cannot share or create an e-book without the Internet.
+ An application that allows us to create the first e-book and to the continuous one will require a fee for the upgrade.
+ The application has not many interactive elements to create transition effects for the e-book's interface.

5.4 APPLICATION E-BOOK IN TEACHING

5.4.1 IN BIOLOGICAL TEACHING

Based on the purpose and object of using an e-book, we have used Kotobee Author software to design and introduce an illustrative product that applied in teaching biology of Grade 10. Kotobee Author software can insert files in the form of photos, videos, moving images, and linkages that help students easily learn more additional knowledge. Besides, one advantage of Kotobee Author software is that it can design multiple choice questions conveniently. Students can check and evaluate the level of comprehension quickly through these tests.

This e-book can be used as a reference material for teachers in a way that supports student's self-study competence. In order to exploit the lesson knowledge fully and develop this ability, the teacher needs to:

- Understand the content of the lesson and needs to relate to expansion and practical knowledge.
- It understands the characteristics and features of the used software.
- It identifies the right competencies that need to be formed and developed through the selected content.

Figure 5.4 is a screen interface that introduces the knowledge content in topic 1 in biology of Grade 10. In order to increase enjoyment when students study himself at home, the teacher should build an interesting e-book with more pictures and videos. Thus, in this topic, we designed videos of experiments related to the theory (Figure 5.5). Students can check and evaluate the level of comprehension quickly through tests in e-book and (Figure 5.6) or can join in an online class or forum through the link inserted in an e-book. In the classroom, teachers may not have enough time to expand the knowledge content, but e-book will support teachers in the teaching process and help students in the study at home. With the goal of helping students expand the lesson-related knowledge, we added interesting information to the section. Do you know (Figure 5.7).

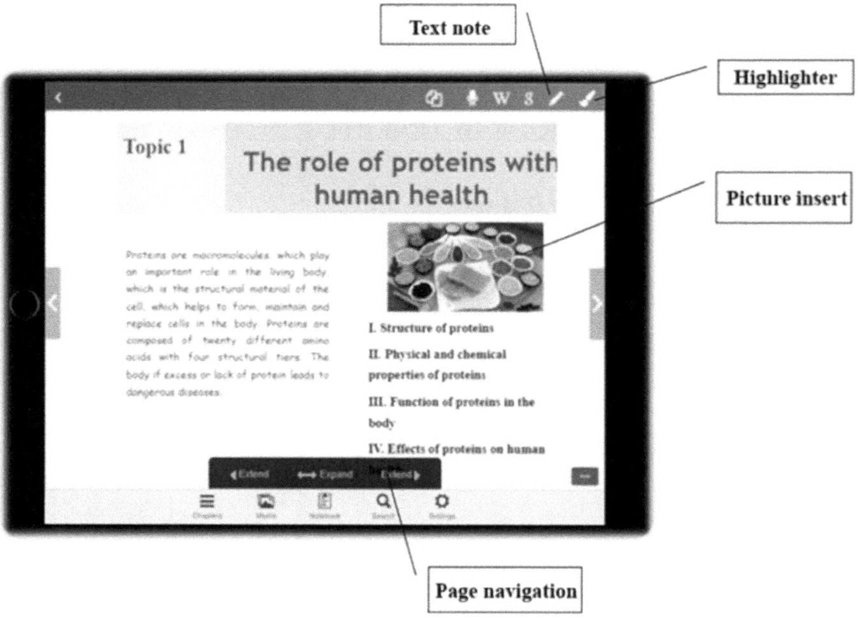

FIGURE 5.4 Interfaces of the knowledge content in topic 1.

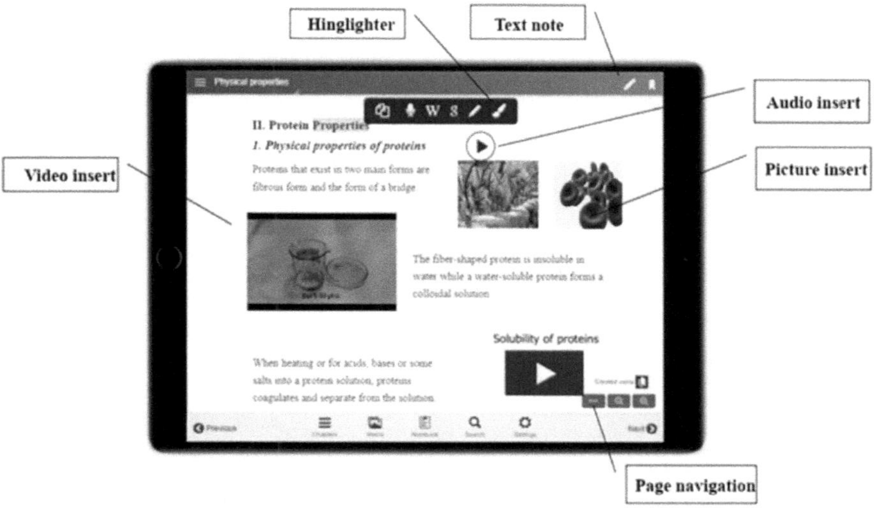

FIGURE 5.5 Interfaces of lession content.

FIGURE 5.6 Interfaces of a workout section.

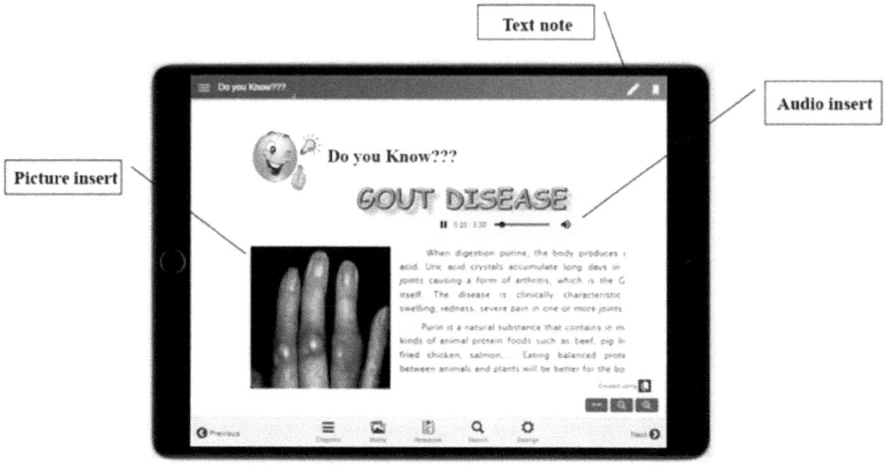

FIGURE 5.7 Interfaces of extended knowledge.

5.4.2 IN CHEMISTRY TEACHING

This is an illustration product applied in 10th-grade chemistry teaching and designed based on Kotobee Author software that we would like to recommend. This e-book can be used as a reference for teachers in a way that supports student's self-study. With the multimedia utilities integrated in Kotobee Author software, teachers can help to give 3D images of atomic, molecular, compound structures, etc. illustrate chemical reactions in the classroom that do not have time to do or are not safe enough for students to do.

Figure 5.8a is a screen interface that introduces the content of knowledge in a subject in 10th-grade chemistry. To increase interest and illustrate student reactions not done in class, teachers should build an electronic book with lots of pictures, videos, and virtual experiments. Therefore, in this topic, we have designed videos of experiments related to theory (Figure 5.8b). The assessment is given in the form of a small experiment so that students can better understand the theory presented in the lesson (Figure 5.8c). With the goal of helping students expand their knowledge, understand the life phenomena that are relevant to the lesson, we have added interesting information to the Did You Know section (Figure 5.8d).

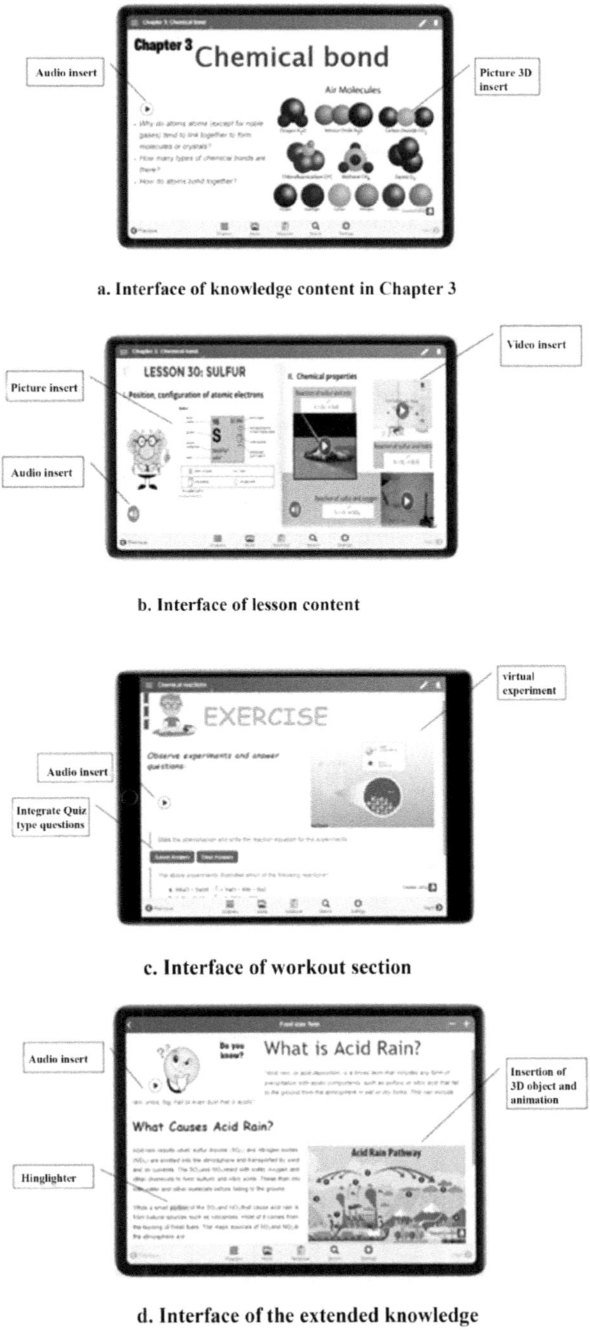

FIGURE 5.8 Interfaces of an e-book designed by Kotobee Author sorfware for chemistry.

5.4.3 IN PHYSICAL TEACHING

In 10th-grade physics teaching, we introduce an illustration product design based on Kotobee Author software. This e-book will be a back-up for teachers to use as a reference tool to develop student's self-study capacity. By describing phenomena, motions, models, and structures transmitted through multimedia utilities integrated in the Kotobee Author software, teachers will be able to stimulate learning excitement and actively learn. When assigning a task to a student studying at home.

Figure 5.9a is a screen interface that introduces knowledge content in a subject in 10th-grade physics, often incorporating small questions related to life's phenomena to stimulate curiosity. In order to increase the interest and help students continue to learn the problem, teachers should build an electronic book with lots of images, videos, models, and virtual experiments. Therefore, in this topic, we have designed and linked the repository of simulation physics experiments related to theory (Figure 5.9b). In order not to be boring when students take teacher assessments, students can ask questions based on real-life situations in order for students to see the practicality and better understand the theory in the lesson (Figure 5.9c). In order to expand our knowledge and understand the phenomena of life related to lessons for students, we have added interesting information to the Did You Know section (Figure 5.9d) to help students see the practicality of the subject.

a. Interface of knowledge content in Topic 1

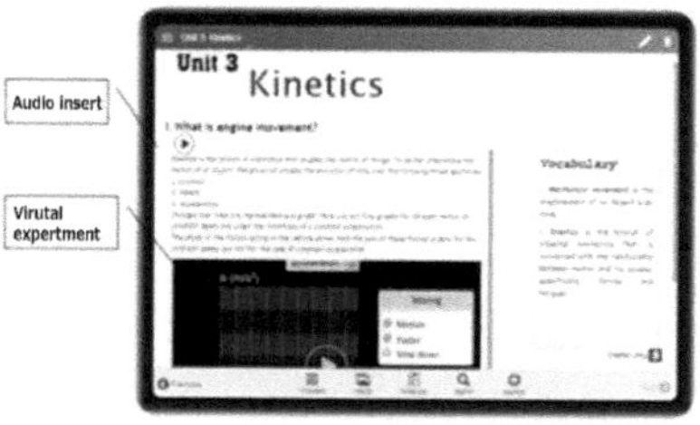

b. Interface of lesson content

FIGURE 5.9 *(Continued)*

c. Interface of workout section

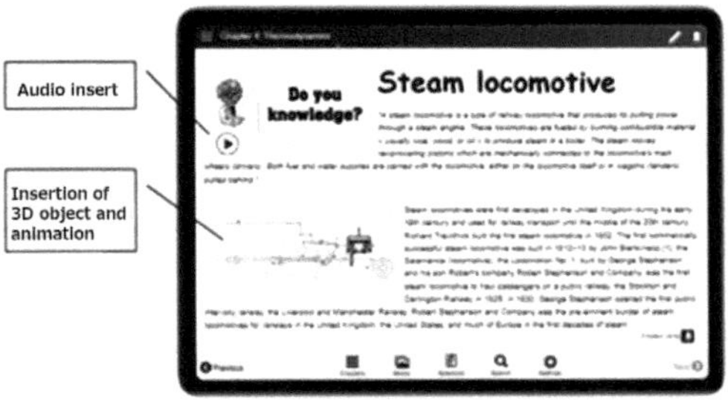

d. Interface of the extended knowledge

FIGURE 5.9 Interfaces of an e-book designed by Kotobee Author software for chemistry.

5.5 CONCLUSION

From this study, the high awareness of biology teachers in Hanoi in using e-books for teaching and learning has been identified, but not really effective. The use of an e-book is quite different between teachers with different seniority levels; most teachers have only worked with e-books as PDF versions or reference documents, and have not yet interacted between the learner and the teacher through an e-book. Teachers have different interests in e-book features. Thus, in order to be proactive in teaching and designing e-books according to personal wishes with the best lessons, in this study, a process of designing, using, and evaluating e-books was developed to help readers have an overview of e-books. We also illustrated an example of e-book designing in teaching biology of grade 10th, partly to help teachers visualize how to use Kotobee Author software in designing their lesson and e-book.

In order to promote fully lifelong learning as well as improve self-study capacity for students in Vietnam, there should be solutions to some outstanding issues like raising awareness of using e-books for learning. Besides, needing the development strategy of e-book detail, such as it is necessary to organize training sessions in high school in Hanoi and other regions, we can be sure that in the future, using e-book will develop in Vietnam.

ACKNOWLEDGMENT

This research is funded by VNU-University of Education under research project number QS.20.01.

KEYWORDS

- **biology**
- **Kotobee author**
- **e-book**
- **high school**
- **science teaching**

REFERENCES

Grzybowski, M. "Educational technologies in South Korea." *General and Professional Education*, Vol. 1 (2013), pp. 3–9.

Jaime, J., Pellicer, E., and Yepes, V. "Teaching and learning using a case study: application to a master degree in construction management." *Procedia Social and Behavioral Sciences*, Vol. 15 (2011), pp. 696–702.

Kisnieriene, V., Kisnieriene, B., Sakalauskas, V., and Daktariūnas, A. "Information technologies for biology education: Computerized electrophysiology of plant cells." *Informatics in Education*, Vol. 7, No. 1 (2008), pp. 91–104.

Luaran, J. E., Rom, K. B. Md., Nadzri, F. A., and Jain, J. "E-Book usage in teaching science: The perspectives of science teachers in secondary schools in Selangor." *Malaysian Journal of Distance Education*, Vol. 18, No. 1 (2016), pp. 13–27.

Mutalib, E. A., Noor, A. M., Ali, R. M. M., Abu Bakar, Z., and Amin, A. R. M. "Teachers' perceptions of the use of e-books as textbooks in the classroom." *World Academy of Science, Engineering and Technology*, Vol. 70 (2012), pp. 580–586.

Nishat, N. and Islam, Y. M. "A proposed e-book model for engaging peer interaction." *Daffodil International University Journal of Business and Economics*, Vol. 9, No. 2 (2015), pp. 117–126.

Orora, W., Keraro, F. N., and Wachanga, S. M. "Using cooperative e-learning teaching strategy to enhance students' creativity in secondary school biology: a study of selected schools in nakuru county, Kenya." *International Journal of Education and Practice*, Vol. 2, No. 6 (2014), pp. 137–146.

Rhodes, A. E. and Rozell, T. G. "A constructivist approach to e-text design for use in undergraduate physiology courses." *Advances in Physiology Education*, Vol. 39, No. 1 (2015), pp. 172–180.

Suleman, Q., Danial, A. H., Shakeel, S., Shakir, M. M. N., Shabbir, F., and Hussain, I. "Effectiveness of educational technology in teaching chemistry to secondary school students in Khyber Pukhtunkhwa (Pakistan)." *American Journal of Scientific Research.* No. 41 (2011), pp. 115–131.

Taizan, Y., Bhang, S., Kurokami, H., and Kwon, S. "A comparison of functions and the effect of digital textbook in Japan and Korea." *International Journal for Educational Media and Technology*, Vol. 6, No. 1 (2012), pp.85–93.

Thai, H. M. and Oanh, D. T. "Developing e-books to support pedagogical students to practice information technology application capacity in teaching chemistry at high schools." *Journal of Science of HNUE*, Vol. 61, No. 6A (2016), pp. 42–53.

Tran, T. N., Mai, N. T. A., and Nga, N. T. "Designing the E-book to improve the effectiveness of teaching chemistry in high school." *Journal of science of hnue education alsci.* Vol. 53, No. 4 (2008), pp. 24–31.

Tran, T. N., Thanh, N. T. A., and Tu, L. T. C. "Using E-book to support self-studying chemistry 10 to improve the effectiveness of lessons using teaching methods from the perspective of constructivist theory." *EducationalSci*, Vol. 60, (2015), pp.75–81.

CHAPTER 6

Flipped Classroom: Google Classroom Application on Teaching Philology

PHAM THI THANH PHUONG[1*], NGUYEN THI HAO[2], and TRAN THI HONG HOA[3]

[1]*Faculty of Educational Technology, University of Education, Vietnam National University, Hanoi, Vietnam*

[2]*Center for Higher Education Studies, The Vietnam National Institute of Educational Sciences, Hanoi, Vietnam*

[3]*Faculty of Publishing, Academy of Journalism and Communication, Hanoi, Vietnam*

[*]*Corresponding author. E-mail: phuongptt@vnu.edu.vn*

ABSTRACT

By outstanding advantages, the application of a flipped classroom model in teaching Philology is a right and satisfactory way to meet the new Philology curriculum's requirements. With the enhancement of information technology, the applications of Google Classroom to build online classrooms have supported and promoted many outstanding advantages of the flipped classroom. The integrated tool of Google services (Gmail, Google Drive, and Google Docs) is completely free, very convenient, and effective in creating and managing online classes. This chapter would like to suggest the main contents when designing an online classroom and a process of teaching Philology according to the flipped learning model with the support of Google Classroom, thereby improving the quality of teaching Philology to meet the requirements of the new Philology curriculum in Vietnam in the era of industrial revolution 4.0.

6.1 INTRODUCTION

Along with the world's transition to the fourth industrial revolution, the education of nations also entered a new phase called education 4.0. It is an education with the basic characteristics of being open, lifelong learning, interacting, personalizing, training people for integration, innovation, and creativity. With radical and comprehensive innovation, Vietnamese education is also gradually integrating with the development trend of world education. The official enactment of the Literature General Education Program in December 2018 (along with the master program and other subjects programs) was a positive movement in the development step.

To meet the requirements of the new literature program in context 4.0, many modern models and teaching methods have been proposed, including the flipped classroom (F.C.) model. This has been a new model of teaching in the United States for more than 10 years, widely used in schools in many countries around the world, and has been tested in Vietnam in recent years. With the development of information technology (IT), there is now a lot of useful software available to support the organization of teaching in a flipped learning model.

This chapter introduces the application of Google Classroom (G.C.) in teaching literature according to the F.C. model at high schools in Vietnam.

6.2 LITERATURE REVIEW

6.2.1 FLIPPED CLASSROOM

6.2.1.1 THE DEFINITION AND CHARACTERISTICS OF FLIPPED CLASSROOM

In 1993, Alison King emphasized the importance of using time in class effectively instead of one-way transmission from instructors to learners (King, 1993). Although Alison King did not call this model, "flipped classroom," he had mentioned the issues of the model, and this was the content of F.C. This concept of "the flipped classroom, or flipped learning, dates back to the turn of the millennium (e.g., Lage et al., 2000; Bergman and Sams, 2012; Hamdan et al., 2013)." In 2007, the first time, Jonathan Bergmann and Aaron Sams practiced F.C. at Woodland High School in America. They wrote their direct lessons and uploaded them on a website for learners who were in no condition to take part in a classroom directly. Since this time, online lessons have developed larger, and learners have used time in the classroom to do exercises and approached concepts. After that, many concepts of F.C. have been published by many authors.

Most recently, Rainne A. M. Bouwmeester showed some highlights of F.C., and they are as follows. Students' workload during F.C. was similar, and exam preparation was significantly less time-consuming; teachers in F.C. spent less time on factual knowledge as opposed to in-depth information; the in-class observations suggest more interactivity in F.C., but this was not important; students in F.C. perceive higher self-efficacy those in traditional ones; retention of knowledge and self-efficacy scores showed no difference between them after 10 months later (Bouwmeester et al., 2019). The methodological guidelines underpinning F.C. are far from new. Reading or studying the course material before a lesson is by no means a novel concept; students are usually required to read certain materials in order to prepare themselves for the next lesson. Nevertheless, the traditional approach to lessons is changing with the progressive advance of technology (Kurtz et al., 2014). Four keys that have been posited to learn this model well are flexible environments, professional educators, intentional content, and a shift in the learning culture (Hamdan et al., 2013).

To sum up, an F.C. model is to use digital technologies to record lectures and demonstrations for their learners, and learners could access and download all digital lectures whenever and wherever they need them.

F.C. has many advantages for instructors and learners. First, for instructors, F.C.:

1. Help them use time in the classroom effectively and interact with their learners more and more (Bergmann and Sams, 2012).
2. May record and narrate screenshots of work which instructors do on their computer screens, creative videos off by themselves during teaching, or curate video lessons from Internet sites such as TED-Ed and Khan Academy.
3. Can check the level of understanding of each student, help their learners enhance learners' procedural fluency (Greenberg et al., 2011; Hamdan et al., 2013).

Second, learners

1. Can learn whenever and wherever they want, and study online more and more.
2. Expand their knowledge (Hamdan et al., 2013; Bergmann and Sams, 2012).

3. Can prepare for class and plan in autonomy and in an active way (Bouwmeester et al., 2019).

We chose this model because there are the difficulties between F.C. and traditional classroom as shown by many authors. The main contents are as follows:

Changing the teaching approach is a from of education to impart knowledge to students (McLaughlin et al., 2014).

Passive learning way while the world requires information processing more actively and analyze the effectiveness and necessity of F.C. model (Khan, 2012). This is the model that shows the difference between a traditional model and Bloom's taxonomy in the F.C. model:

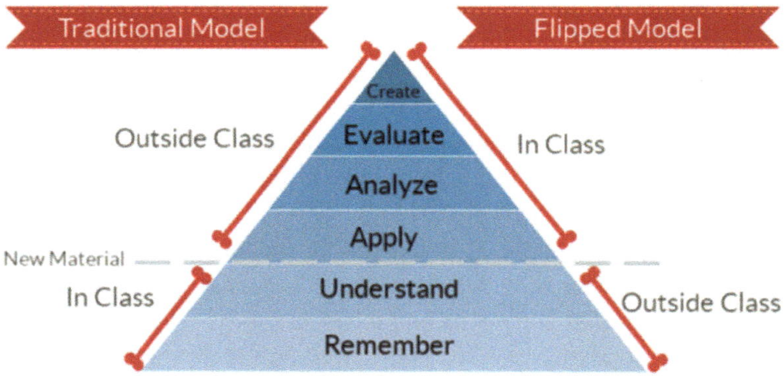

FIGURE 6.1 Bloom's taxonomy (Reprinted from SEE Team, 2019).

6.2.1.2 PREVIOUS RESEARCH APPLY FOR TEACHING PHILOLOGY IN GENERAL AND TEACHING PHILOLOGY IN PARTICULAR

Researchers present contents, ways, and notes of applying F.C. in some areas/some subjects/some terms.

First, this model can be applied in the following three periods: before going to the classroom, in the classroom, and after going to the classroom. For example, applying the F.C. model to teaching Vietnamese language students at the university, Diu Nguyen Thi proposed three periods to organize this term: choose contents, write lesson plans for video recording, and organize teaching. Through experiment teaching, this author affirms that FC is the most suitable model to apply for organizing some important contents of this module (Thi, 2016, pp. 83–85).

Next, Nguyen (2018) and Sokolova et al. (2019) point out the advantages and disadvantages of applying the model. Some advantages are as follows:

1. Learners become more active and legitimate inside and outside the classroom.
2. They can control learning time and reach the highest level: apply knowledge, comprehensive consolidating 21st-century skill.

Some disadvantages are as follows:

1. Waste of time
2. Efforts needed for designing lesson plans and recording videos, dependence on the Internet system, cooperation, the ability to use the technology of instructors. So applying this model is not easy.

Then, researchers also put forward some notes when applying F.C. in teaching in order to develop the effectiveness, such as

1. Organize all-stage favorable conditions for auditory, visual, and kinesthetic learners (Sokolova et al., 2019).
2. Note the subject characteristics, for example, learning foreign languages, are built on the modular principle, so teachers should better provide their students teaching materials. It is a favorable condition for teaching based on F.C. (Nguyen, 2018; Sokolova et al., 2019; Thi, 2016).

Finally, researchers mentioned about applying F.C. in many different subjects: mathematics, citizen education, foreign language, mother language, etc. (Chick, 2014; Sokolova et al., 2019). In Philology, works have focused on language teaching (including foreign language and mother language teaching), and there have been very little researches. They mention the common contents as we present and emphasize the characteristics of language teaching and suitable F.C. application.

In short, the F.C. model is researched and applied for teaching a large number of students and brings many benefits. The authors research on theoretical foundation, contents, methods, and point disadvantages and advantages when applying this model in teaching. Moreover, we recognize that the application of the flipped model in teaching Philology in individual and teaching all subjects in general have still 4 exists:

1. Many difficulties have not been solved when applying this model in teaching, especially in developing countries like Vietnam.
2. Although this model has been more widely applied, it is not yet widespread, even that many instructors have never heard about this model.
3. The number of researcher works in the world is large but very few in Vietnam.
4. In particular, in Philology, research on F.C. application in teaching is very rate, and the number of instructors who apply this model to teaching is still limited.

In the Vietnam context, recently IT has developed rapidly; the world is getting flatter. Vietnam's new high school programs have just been issued. In Vietnam, new textbooks have been published recently (for some grades), and the rest are expected to be published in the next couple of years (for higher grades). With the learners' level in Vietnam getting higher and higher, Philology teaching methods will change more and more. Hence, the research practice on this model application is necessary.

6.2.2 GOOGLE CLASSROOM

6.2.2.1 A BRIEF INTRODUCTION OF GOOGLE CLASSROOM

The definition and features

G.C. is a tool that integrates Google services (Gmail, Google Drive, and Google Docs) to create a perfect application for the education industry. It was common on August 12, 2014 (Wikipedia). Distributed through the free Google Apps for Education suite, G.C. is a technology application not only for creating and managing online classes but also for supporting online teaching and learning effectively.

Basically, G.C. helps organize a classroom through the support of three important features: communication, assignment, and storage. It is useful for both teachers and students. For example, it allows teachers to create classes and invite students to participate; provide learning materials; build class schedules, create notices, questions, assignments for students; and check the progress of students' assignments, mark papers, and give feedback on their work. For learners, the online classroom will create suggestions about the deadline for submission, helping students self-study anytime and anywhere.

Documents and tests that have been taken are saved for the whole course, and the students can look up if they want to know their progress.

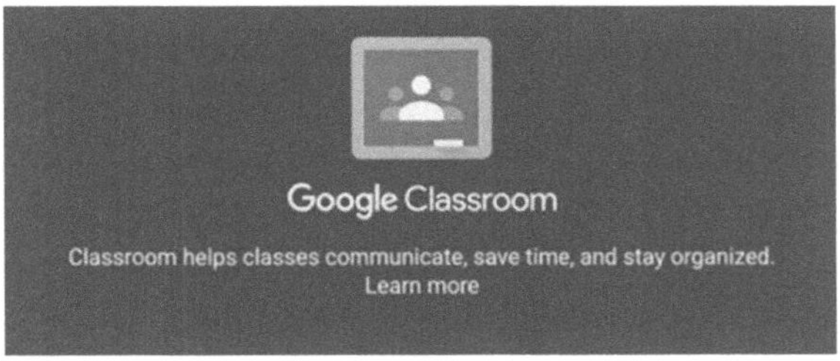

FIGURE 6.2 Google Classroom icon (Onlinehub, 2019).

With G.C., exchanging lessons between teachers and students and discussing among students become more convenient and faster. This application has opened up a new learning method for students on mobile devices such as smartphones, tablets, and notebooks. Moreover, it is even suitable for people who have trouble coming to the school directly.

The basic use

To create and manage online classes on G.C., teachers need to perform the following three main actions:

1. *Create a class*

- First of all, we need a computer and access to the Internet.
- Second, go to https://classroom.google.com and log in with your Gmail account. Or go to https://www.google.com and click on the applications of Google, and then click on the G.C. icon.
- Third, when the G.C. interface appears, click on the "+" sign in the upper-right corner, select "Create class."

FIGURE 6.3 Create a class.

- Then, add students to the classroom: send the class code (top left of the main classroom interface) to students.

2. Create exercises, provide learning materials

Select "Classwork" above the class interface → click "Create" → select the type of data (Assignment/Quiz assignment/Question/Material…) that you want to create in the selection box → fill in the necessary information in the menu that appears (can attach documents in computer/on Google Drive/on YouTube/ link) → click the button "Assign" (for Assignment/Quiz assignment)/ "Ask" (for Question)/"Post" (for Material) … to complete.

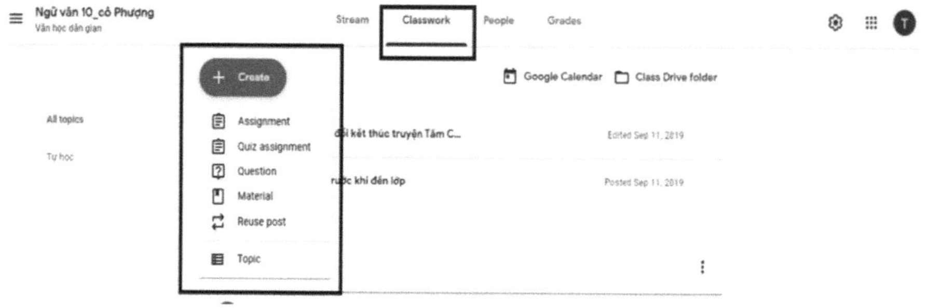

FIGURE 6.4 Create exercises/provide learning materials.

3. Grade and return

To grade the students' works, choose the assigned work → click on "Turned in" → click on the name of the student whom you want to grade → enter a comment or feedback, and a mark on the student's assignment → click the "Return" button to save the mark and notify the student of the work result.

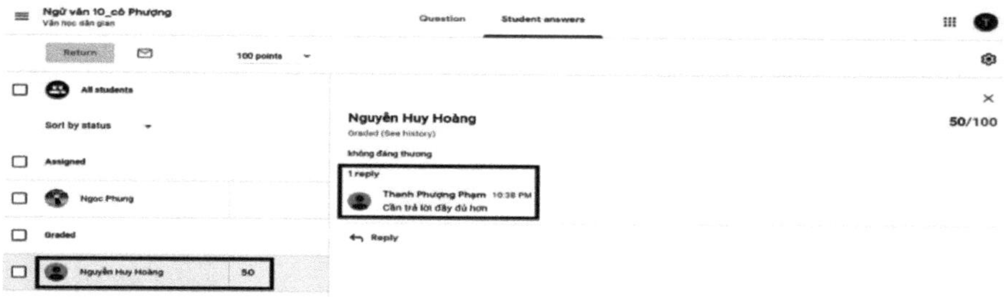

FIGURE 6.5 Grade and return.

6.2.2.2 *PREVIOUS RESEARCH ON GOOGLE CLASSROOM*

As we mentioned above, G.C. was officially launched in August 2014. Since then, it has become the subject of numerous articles, books, and dissertations. Summarizing the materials, we classify them into two main groups: general assessment of G.C. in all aspects and evaluation of G.C. through application into a specific educational aspect.

Group 1: General assessment of G.C. in respect of users' acceptance (teachers and students), effectiveness, limitations, and factors affecting the use.

When a new technology tool is born, a big question is always: What is its feature applicability? G.C. has been used in many institutes, especially universities in Asia; therefore, there have been many studies evaluating its effectiveness as a virtual interactive teaching environment or positive teaching tool. First of all, G.C. is rated to the high acceptance rates of students and teachers, with modern educational theories/models, such as Roger's theory of diffusion of innovation (Iftakhar, 2016), the technology acceptance model (Al-Maroof and Al-Emran, 2018; Mostafa, 2018), and the unified theory of acceptance and use of technology model (Manan and Hanafi, 2019), integrated with technology. Going into the effect of this tool, the authors pointed out and assessed specifically its outstanding advantages: developed students to take part in learning and reformed classroom dynamics (Heggart and Yoo, 2018), "extremely useful in the assignment and collaborative learning" (Ventayen et al., 2018), teachers could interact with the students even outside the classroom which facilitated the overall teacher–student interaction (Azahar and Iqbal, 2018). At the same time, the limitations of G.C. are also pointed out: it is not easy for teachers to manage teaching materials; some of G.C.'s buttons can make a new user confused or spend more time dealing with it. An overview of the advantages and disadvantages of G.C. can be found in the article by Pappas (2015).

In addition, the researchers showed the factors affecting the use of G.C.: speed of Internet, teachers' information and communication technoligy competence, positive attitude (of both teachers and learners) toward technology for instructional purposes (Iftakhar, 2016), perceived ease of use and usefulness (Wijaya, 2016), performance expectancy, hedonic motivation, and habit (Kumar and Bervell, 2019). Users' awareness is very important. That is the reason the authors propose directions for using G.C. more effectively.

Group 2: Assessing G.C. through applying it to a specific educational aspect (a teaching model/subject/method or teaching skill).

Not just looking at G.C. in general, a second direction is that the authors have studied its effect when applying it to the deployment of a specific field of education. Firstly, G.C. is assessed by teaching specific subjects/specialties: history (Fitriningtiyas et al., 2018), economic specialties (Stavytskyi and Urazgaliyeva, 2018), computer science (Kopeyev et al., 2020), especially English (Duc, 2018; Islam, 2018; Sukmawati and Nensia, 2019), even specific skills of English such as listening skill, phonetic ability, writing, and reading. All authors asserted that the use of G.C. tools positively influences the competence of students, stimulates their interest in learning, and improves students' learning outcomes. Besides, researchers appreciate the effectiveness of G.C. when it is applied to implement positive and modern teaching methods such as problem-based learning (Bayarmaa and Lee, 2018), argumentation skill (Pritasari and Jumadi, 2018), interactive communication (Subandi et al., 2018).

6.2.2.3 APPLICATIONS OF GOOGLE CLASSROOM IN THE FLIPPED CLASSROOM

This is a content which we are very interested in because it is directly related to our topic. However, most researchers have only appreciated the role of G.C. as a useful tool in online or blended teaching (Beaumont, 2018; Bhat et al., 2018; Brown, 2018), or just mentioned it when they name the learning management systems in F.C. (Ferreira, 2014; Ozdamli and Asiksoy, 2016). So far, for the best knowledge of the researcher, there was only Puarungroj (2015), who specifically mentioned how to implement content and process when using G.C. in teaching a computer subject at Loei Rajabhat University (Thailand) according to the F.C. model.

In summary, although there have been a large number of studies of the effects of G.C. in general and in specific aspects of education, in-depth research on G.C.'s application in F.C. is very limited, and no studies have investigated this matter in Philology teaching. Moreover, according to the classroom, the F.C. model consists of two main parts: lectures and instructions for students to study individually outside the classroom (at two times: before and after class) and interactive learning activities in the classroom. Therefore, F.C. is one of the forms of blended-learning. Over time, lectures and instruction of lecturers outside the classroom have become more modern, and interactivity and efficiency have increased because of the IT's development (software, web platforms). In the past, lectures were stored on CDs, files on computers and sending mails or printing handouts, and give them to students, they are nowadays saved online, with much more diverse and convenient forms of interaction and management. In this study, we want to mention building F.C. models with the support of web platform 2.0 (G.C.).

6.2.3 APPLYING INFORMATION TECHNOLOGY TO TEACHING PHILOLOGY IN VIETNAM TODAY

6.2.3.1 TEACHING PHILOLOGY IN VIETNAM IN THE CONTEXT OF GLOBAL INNOVATION

In Vietnam, learning Philology is one of the main learning activities and develops the ability to master Vietnamese for leaners. It contributes to creating confident, rich-minded, and responsible citizens in our society. Through studying Philology, people can gain additional knowledge not only inside literature but also outside literature. For example, it helps learners promoting imagination, critical thinking, extending experience, self-enrich, enriching emotions before the beauty in art, and developing read interest. With such an important role, the curriculum of Philology at all levels has always been concerned and adjusted to meet the needs of the time and the educational objectives of each period.

In response to the need for global innovation in the 4.0 era, on December 26, 2018, the Vietnam Ministry of Education and Training officially issued the "General Philology education curriculum." With the transition from education which emphasizes the importance of knowledge to the educational development of qualities and competencies for learners, the new Philology curriculum is compiled in an open way, meaning not specified in detail in the content but only raising the basic requirements; emphasizing some key knowledge, selecting some required text, the remaining encourages teachers to actively and flexibly select learning materials and organize the lesson to suitable for learners. In this spirit, Philology to help students develops two specific capacities: the capacity for language and literacy and contributes to helping students develop the general capacity such as the ability of autonomy and self-study, ability to communicate and collaborate, problem-solving ability, and creativity. To develop the above-mentioned capabilities, the general education program will be based on the following methods: differentiated teaching, integrated teaching, and activity-based teaching giving autonomy for teachers. Besides, the assessment and evaluation process in Philology will change, toward the goal of providing accurate, timely and valuable information on the level of meeting the required quality, the competence and progress of students throughout the course of coursework, to guide learning activities, adjust teaching activities, manage and develop programs, and ensure the progress of each student and improve the quality of education. (Vietnam Ministry of Education and Training, p. 85)

It can be seen that the application of IT in teaching Philology is not only a suitable direction for the digital age but also an urgent requirement to respond to the requirements of the new curriculum.

6.2.3.2 CURRENT SITUATION OF APPLICATION OF INFORMATION TECHNOLOGY AND FLIPPED CLASSROOM IN TEACHING PHILOLOGY

To investigate the current situation of using IT and F.C. in teaching literature today, we conducted a survey of 23 literature teachers and 435 students at four high schools in Hanoi (Vietnam): Nhan Chinh, Phan Dinh Phung, Tay Ho, and Education Sciences. The results are as follows.

1. *On the teacher side*
    a. Teachers' evaluation of the necessity of using IT in teaching literature: According to the survey results, 30% of teachers said that applying IT to teaching is very necessary, 57% said it is necessary, and 13% considered it quite necessary. This proves that most teachers feel that using IT in teaching literature is appropriate and has certain benefits. According to the statistics, most teachers said that using technology in teaching literature helps students to take more interest in the lesson (61% agreed completely), creates a greater ability for student-centered and active learning (65% agreed completely), and technology provides teachers and students with a large knowledge resource (65% agreed completely). This shows that most teachers recognize the great benefits that IT brings to teachers and learners.

The advantages and effects of IT in teaching literature have been recorded by practical teaching such as *vivid image illustrations* help the lecture become interesting so that learners must mobilize both hearing and vision into the process of receiving and discovering knowledge, increasing comprehension efficiency; *essential diagrams* help teachers save time in the classroom, increase the visualization of the content of the subject, help learners understand quickly and deeply, increase the ability to synthesize problems, and absorb the work easier; rich treasure of knowledge from *open sources* for optimal teaching and learning; *interdisciplinary knowledge* between literature and film, music, theater, painting, etc. makes the lecture come lively and increases interest for learners; *the exchange of information via email* allows teachers and learners to be interactive quickly and easily, and promptly answer questions in the teaching process.

Of course, besides these advantages of the use of IT in teaching literature, remaining shortcomings need to be overcome, such as many teachers abuse IT, use slides in place of the lecture, changing the role from "listening to copy" to "looking to copy"; asthetics in education are also not high because many teachers select gaudy and glitzy slides; complex effects make student tired and fatigue; many use abusive video and audio to illustrate the lecture, thus diluting the lecture, and also the focus is not emphasized. These disadvantages are a result from the sense and ability to handle lectures and technology applications of lecturers. Many lecturers do not really understand the role of IT in teaching, applying technology in a mechanical, superficial, lack of investment, and irresponsibility with lectures. Obviously, this is a challenge for teachers to design vivid and engaging lessons using technology, and they have to understand the way to use technology for serving the lesson without making the technology factor overwhelm the literary nature of the lesson.

b. Teacher's evaluation of F.C. in teaching literature: All teachers have heard about F.C.. There are 22% of teachers who know this model quite well, 43% of teachers know a little bit about F.C., and 26% listen to but do not understand it much. Although 22% of teachers know F.C. quite well, the level of use of this model by teachers is modest: 4% of teachers use it regularly, and 9% do not use it. Thus, the FC is quite a strange model for general teachers. Not only a few teachers operate it but also a very few teachers have ever heard about and understood it. Most of them agree with the view that this model of class helps students become the center of the classroom (61%) and helps students to be self-conscious and in self-study (78 %). Besides, teachers also think that it is not easy to operate F.C. because this model requires thoughtful preparation of teachers' IT expertise and skills (65%). In addition, some teachers do not really believes in the advantages of F.C., especially, the idea that this model helps parents better supervise and support their children to study (17% disagreed).

2. *On the student side*

    a. Regarding the necessity of using IT in teaching literature: 94% of students think that using IT in teaching is necessary. However, there are a few students who are not ready to accept and adapt to this teaching method (6%), even if they do not want to change the traditional teaching method. In our opinion, the challenge for teachers is how to apply technology to teaching and to help students feel this new method.
    b. The students' views on F.C.: most of them agree with the view that through the lecture video at home, they can watch the lecture many times, which is appropriate for the speed and the level of their comprehension (51%), and the viewpoint of this classroom model requires children to have self-discipline and high self-study spirit (47%). Otherwise, similar to the teachers' thinking, students also wonder if F.C. helps parents to supervise and support their children's learning or F.C. will help students take the initial study and interact more with teachers and friends or not.

Thus, through the above survey results, we realize that the majority of teachers and students recognize some of the advantages of F.C. Besides, both teachers and students have concerns and doubts about the effectiveness of this model. This is reasonable and apparent because F.C. is little tested and operated at the school, and both teachers and students do not have much time to experience, so there are not many grounds for evaluating and recognizing the benefits that F.C. brings.

6.3 APPLYING GOOGLE CLASSROOMS IN TEACHING PHILOLOGY ACCORDING TO THE FC MODEL

We suggest building F.C. consists of two main parts: the online learning activities outside the classroom and the interactive learning activities directly in the classroom. So in this section, we will discuss two main content: online classroom design and teaching process design.

6.3.1 ONLINE CLASSROOM DESIGN

First, teachers need to create an online class by visiting the https://classroom.google.com website and logging in with their Gmail account, clicking the

"+" sign in the upper-right corner, selecting "Create class." Then they invite students to join the class by sending the classroom code (left above the main interface of the classroom) to their students.

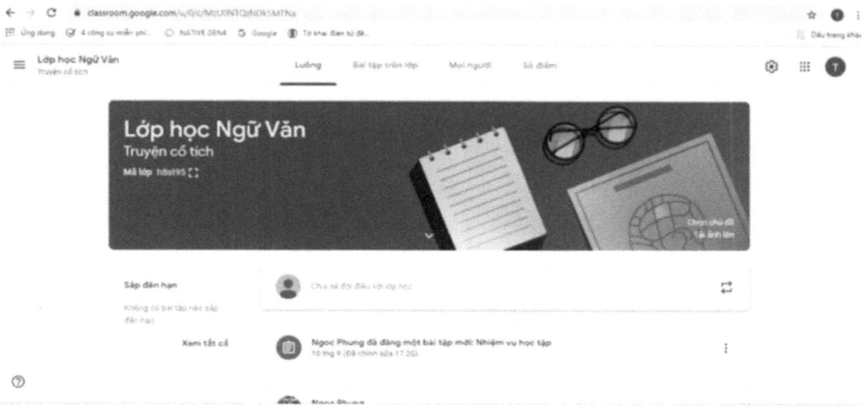

FIGURE 6.6 The main interface of an online classroom in Google Classroom.

In each classroom, we design the following five main contents (here are also five learning contents that students will experience during the online self-study before class):

1. Instruction-learning tasks,
2. Video lectures,
3. Documents,
4. Exams, and
5. Exchanging discussions.

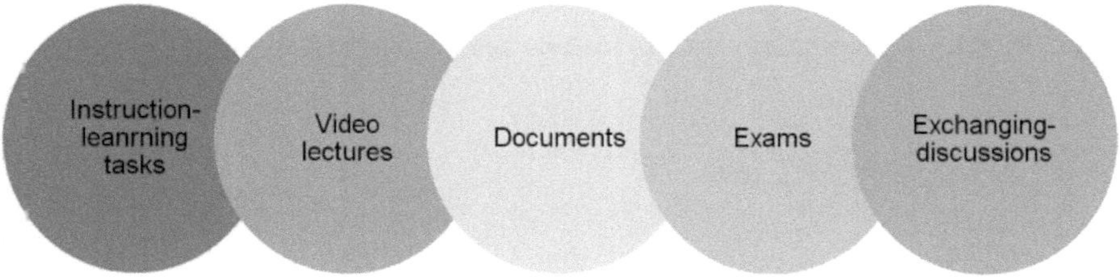

FIGURE 6.7 The main contents in a Google Classroom.

6.3.1.1 INSTRUCTION–LEARNING TASKS

These are instructions on order, how to perform, and the goals to be achieved in each learning task. Students usually do four obligatory tasks before going to class: studying lessons in textbooks, watching video lectures, taking multiple-choice tests, and giving questions and exchanging ideas on topics "Discussion."

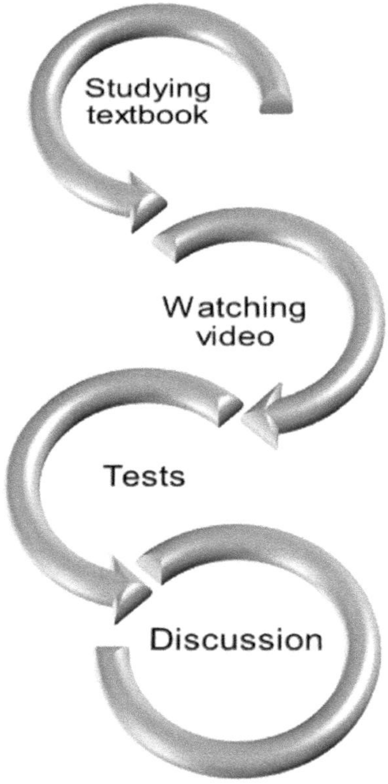

FIGURE 6.8 Four obligatory tasks of students before going to class.

6.3.1.2 VIDEO LECTURE

This is the most important tool that teachers have to invest when they design for students to self-study at home. To create quality videos, ensure the aesthetic-educational criteria we have designed based on the following criteria:

- About capacity: A maximum of 10 min, enough to maintain concentration and attract students.
- Regarding the content: This is the most basic and essential knowledge of lectures so that after watching the video, students can solve at least 70% of the tasks stated in the learning card. According to the theory of F.C., students will learn first at home low-level knowledge on Bloom's scale (low memory, comprehension, and manipulation level). Therefore, before designing the video (as well as the multiple-choice exercises), teachers need to build a target matrix table based on Bloom's cognitive scale, so that there is a basis to select the content to build in the lecture video.

Regarding the form: This is a multimedia lecture to attract learners. To do this, there must be a combination of clips, images, sound, higher than the 3D and 4D character simulation technologies, and the image of the teacher standing in front of the board and lecturing throughout the video. In our products, we have used software such as I movie or Powtoon to design lecture videos, with a combination of diverse images, sounds, and clips, based on the voice of teachers. You can refer to a video of our lecture on "Folk about groan, loving and gratitude" (Literature 10, Volume 1) at https://www.youtube.com/watch?v=hS9S79i0CVo.

6.3.1.3 DOCUMENT

These are articles, texts, and videos related to the content of the lecture and PowerPoint developed by teachers for students to research for themselves. There are two types of documents: compulsory documents and references. The first type is the documents with basic and practical content, directly serving the contents of insurance, and students are required to learn before going to class. References are the documents whose content expands and enhance knowledge intended for students who want to learn more about the lesson. These resources not only help students master, consolidate, and expand their knowledge about the lesson, but also create a habit for students to refer more and read more during the self-study process.

As mentioned above, the Philology program only stipulates a number of compulsory documents. Teachers can be proactive and flexible in choosing other materials as long as they meet the educational requirements. The inclusion of additional texts of the same genre, the same topic, along with the suggestions and instructions of the teachers on the "Extended documents," will help develop students' reading skills and self-study ability. In addition, the development of optional materials in online classrooms will help teachers make teaching differentiated and personalized. Students who are gifted have a career orientation in literature or have conditions to explore their strengths and areas.

FIGURE 6.9 The document that we design for students.

6.3.1.4 *EXAMS*

These are all online tests to quickly check the knowledge that students have learned, are built in the form of objective tests, are set to limit the work time and deadline-completion notification process, if, over time, students will not be able to access this section to work. There are two types of exercises with two points of time to do: before going to class, mainly check basic knowledge students learn by themselves through textbooks and video lectures; and after finishing the lessons, total check matching knowledge with basic to advanced levels that students receive after learning in the classroom and at home (including the self-reading of additional reference materials that teachers provide online). When students attempt for the test, the timer on the system runs, and the system calculates points and indicates the wrong sentence as soon as students complete the test. In an online classroom, teachers can view the results of each child's work and statistics of the results of the whole class in the form of a chart, about the score, or the number of students answering right and wrong. In addition, the system also stores all the test's scores of learners in the process. These are databases to help teachers and students evaluate the learning outcomes of learners; from these evaluations, the teachers have reasonable teaching adjustment plans.

FIGURE 6.10 Exams that we design for students.

6.3.1.5 DISCUSS

This is the section for students to share the questions and opinion about themselves on a content's insurance for teachers and other students publicly to exchange and discuss. There are two periods for students to post their ideas.

- Before class time: The teacher will create the topic "Questions/questions not understood when preparing the lesson," and students will put their questions/problems in that forum. These questions/problems will be answered and discussed in class time.
- After class in class: Students will actively post unknown questions/issues or their own topics on which they want to learn. Teachers and students in the class will respond directly to these questions/issues. The closing of knowledge will be selected and summarized by the students who "post the forum."

The online classroom forum (in addition to the results of the objective tests) before class is a basis for teachers to select the teaching content in the class that is suitable for students' levels, needs, and characteristics. On the other hand, teachers also check awareness of homeschooled students - Forum more questions, comments, and hardly posed as demonstrated students were very attentive study, deepen lesson. From there, the students can learn from each other and contribute to improving the quality of self-study in general and self-study at home in F.C. in particular.

6.3.2 DESIGNING THE TEACHING PROCESS OF LITERATURE IN THE FLIPPED CLASSROOM MODEL WITH THE SUPPORT OF GOOGLE CLASSROOM

After building an online classroom, we have built the teaching process that consists of three stages: before going to class, in class, and after going to class.

6.3.2.1 BEFORE GOING TO CLASS

This is the period of the home study of students in online classes. It is important for teachers in this period to have specific, reasonable self-study instructions, stimulate students' interest in self-study, and manage self-study of students. Whether F.C. succeeds or not depends greatly on this period, so teachers need to follow these steps:

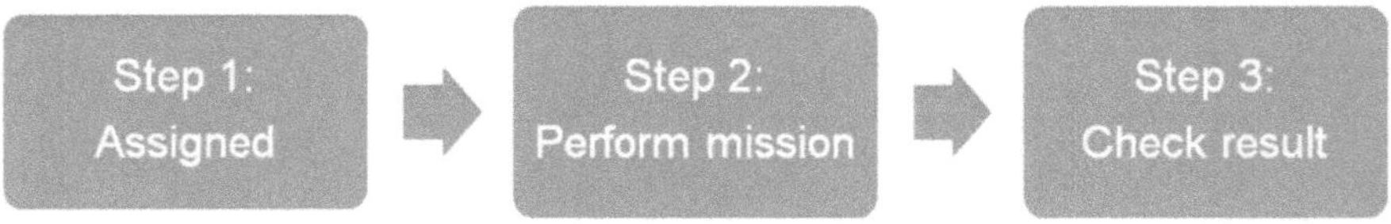

FIGURE 6.11 Three steps before going to class.

Step 1: Teachers assign specific tasks to students to go home to learn and research before insurance. Teachers will give students exercises at the previous session so that they grasp tasks and have time to prepare for the new post. Each student is given a separate learning account, so that teachers can manage their online learning. Typically, students will have four main tasks: research lessons in textbooks, learning lectures via video in the classroom learning, take the tests of objective knowledge of basic self-study, ask questions in the process of self-study on the class forum.

Each task will have very specific instructions in the "Tutorials-learning tasks" in the online classroom, as we have mentioned above.

Step 2: Students perform self-study. With learning tools such as textbooks, video lectures, and reference materials uploaded by teachers to online classes, students will implement learning activities according to the instructions and in accordance with the time, space, and condition and facilities (learning via computer/iPad or smartphone with an Internet connection). Self-study before the class ends with students posting questions on the topic "Discussion" created by teachers.

Step 3: The teacher checks the results and reminds (if necessary) the self-study of the students before going to the class.

G.C. application will help teachers check the following issues:

- Staining quality and test results of student assignments.
- The problem questions posted on forums and students' issues still not fully understood through the responses ignore or wrong in private sector posts.

If by the deadline students do not complete the self-study, teachers can remind via contact channels (email/group class on facebook/chat box on online classroom). With questions from students on the board, teachers will record an overview of the problem to the class discussion during school hours in the classroom session the following day. The capture of issues that students are unclear through the forum, the test results on the web will help teachers have time to better prepare the content of teaching in the class. On the other hand, when students see their classmates' questions on the forum, they also have time to think and prepare for group discussions the next day.

6.3.2.2 AT CLASS

As mentioned above, it is teaching hours in class in F.C. toward higher levels of Bloom's cognitive scale (analysis, evaluation, and creativity) with multidimensional interaction: teachers with students and students with students; positive methods are used thoroughly. To do so, teaching in class should follow five steps:

FIGURE 6.12 Five steps of teaching in a class.

Step 1: Warm up the lesson. Teachers create a mood in lesson and comment on the process of preparing homework for students. Similar to a traditional school, teachers need to design an operational startup or a fast test for students who entered the lecture. Besides, the teacher also makes a comment about the situation of self-study, the process of preparing homework for the whole class, so that students can promote their advantages and learn from the experience for the following self-study times.

Step 2: Systematize knowledge. The teacher systematizes the basic knowledge of insurance. Before going to the class, students process the basic self-learning knowledge of lessons in an online classroom. However, knowledge of students cannot be connected and systematic in a logical and profound way. On the other hand, there may be students who have not completed the tasks assigned by the teachers or have not actively studied. Therefore, before organizing the discussion activities, teachers must systematize the basic knowledge of the lesson, with the support of PowerPoint and technological equipment, using tables, thinking diagrams. In the process of systematizing basic knowledge, on the one hand, teachers have to pay attention to the knowledge units that students often fail in objective tests, and, on the other hand, they can test using the question and answer form to test real learning of them. Although this is not a central part of the class time, it will be a cornerstone of follow-up discussions.

Step 3: Organize discussion. At this stage, the focus is on classroom hours applying the F.C. model. Teachers express teaching perspectives and teaching is student-centered. The topics given by the teachers form the content of the discussion. Also, issues posted by students on the forum can form the content of the discussion. Teachers often give a multidimensional perspective on an issue. They either require a comparative comparison, explain with knowledge, personal experience, or creative thinking, or engage with an action plan. All aims to develop higher order thinking in the Bloom scale. In terms of form, teachers can organize whole-class discussions in small groups or in two large groups with opposing views. Many modern and positive teaching methods can be used at this step (project teaching, playing the role of an expert, etc.). Thus, students receive training; develop high-level thinking, presentation skills, teamwork, and critical listening; and receive suggestions from others. These issues, which are discussed, will answer questions in the process of self-study at home by students and make the lecture more deepen, expand, improve, inspire, motivate learning because it is "challenging" and "useful" for students.

Step 4: Finalize knowledge. Teachers finalize knowledge, and students draw some individual conclusions. Discussion activities may "explode" a lot of different ideas, or students cannot give final explanations on a problem. So after Step 3, teachers need to summarize and finalize the important issues, providing the most appropriate analysis to guide students' thinking. On the other hand, to help students to draw the most meaningful values and recognize the relationship with the reality of lesson, teachers should end the lesson with questions associated with the personal experience of students or real life today. This activity has created an opportunity to end lecture, explored the thoughts of all members of the class, and made students realize the usefulness of each class of literature.

Step 5: Assigned the tasks home. Class hours will end with the assignments to students. The requirements and detailed instructions in the instruction-learning tasks section are posted on the learning website. In class, the teacher only reminds the main tasks and the deadline for completion. There are usually two types of tasks: mandatory and incentive. Mandatory tasks are aimed at strengthening the knowledge of insurance while encouraging the tasks of expanding and improving in-depth research knowledge (usually for gifted and knowledgeable students in the subject). Besides, students will have to start performing the task of self-study before the next new lesson (if the teacher continues teaching this lesson by applying F.C.).

6.3.2.3 AFTER GOING TO CLASS

Students will continue self-study to consolidate and expand knowledge about insurance quality in online classrooms, with the same steps as before class (read the instruction of learning tasks → view necessary documents → do test article → ask questions to the forum, if any). Students can review the PowerPoint (teacher uploads it to after-class materials in class) and related documents to do the self-assessment test after class (this test is required). Moreover, students who are interested in literature can view the references. In addition, the test section also provides banks with a variety of exam questions about the lesson (which may include suggestions or reference articles) for students to refer more. In the process of self-study, if there are problems or questions which students cannot resolve by themselves, they can post their queries on the forum, and then teachers and other classmates will respond directly to that topic within the allowed framework. Along with reinforcing old lessons, students will repeat the self-study process before the class with new lessons.

Thus, with online classes in F.C., students will develop self-study capacity. Through a self-study process under the guidance of teachers, in the future, they can continue to read and understand texts of the same category outside the textbooks. Abundant learning resources with multidimensional interaction on G.C. have greatly supported the success of F.C.

6.4 PEDAGOGICAL EXPERIMENT AND DISCUSSION

6.4.1 PEDAGOGICAL EXPERIMENT

In order to test the feasibility, suitability, and effectiveness of the teaching proposals, we implemented the F.C. with the support of G.C. for the lesson *Tam Cam* (a folk tale in the Grade 10 Philology curriculum) at the class 10A8 (with 35 students) in High School of Education Sciences (HES), Vietnam National University, Hanoi, in May 2019. The pedagogical experiment process is conducted as follows.

Step 1: Building online classrooms with the G.C. tool.

1. First, the teacher relied on Bloom's taxonomy to select students' suitable goals (usually low-level goals, related to memory, understand, and so on), thereby building the corresponding teaching content on G.C.
2. Second, the teacher prepared lesson plans, designs, lecture videos, PowerPoint, and multiple-choice questions for *Tam Cam*. Next, the teacher asked all students to create a Gmail account and send the list to the teacher. After receiving the list, the teacher sent the password to the class through Google support tools to invite students to access the classroom. This was completed one week before the practical class.
3. Third, teachers assigned specific tasks for students to learn and research before the *Tam Cam* lesson:
    a. Task 1: Studying the text of *Tam Cam* in textbooks.
    b. Task 2: Learning the video lecture of *Tam Cam* (https://www.youtube.com/watch?v=DDnk2eZoHi4&t=303s).
    c. Task 3: Take the basic knowledge test on *Tam Cam* on the G.C. (https://drive.google.com/file/d/1CnEDc_DwZrxCx_YOwtVz13bPc8UsRblf/view?usp=sharing).
    d. Task 4: Ask questions during the process of learning *Tam Cam* on the class forum.

FIGURE 6.13 The online classroom on G.C.

FIGURE 6.14 Students study on the G.C. before going to the class.

Step 2: Teaching directly in class 10A8 (periods 4–5) on May 22, 2019.
The teaching process in the class included the following activities:

1. *Activity 1*: The teacher warmed up by the game told the names of the fairy tales and called who tells the most fairy tales. The students bought into the lessons with the bonding of fairy tales with human childhood.
2. *Activity 2*: The teacher commented on the process of performing the tasks in the student's home. Based on the access to the lecture video, test results on the G.C., and questions posed in the forum, the teacher commented on the process of preparing homework for the whole class.
3. *Activity 3*: Teacher systematized knowledge by the mind map on PowerPoint, combining explanation to suggest the knowledge base on G.C. lesson for students.
4. *Activity 4*: The teacher held a discussion about the issues surrounding the lessons on the basis that students have mastered the basic knowledge on their G.C. The teacher organized the class for students to discuss the content and art issues of the lesson (in groups or pairs) such as:
   a. What are the similarities between the incarnations of Tam? Why does it have to be the fourth incarnation (the last incarnation) for Tam to be fully happy?

b. Could it be that if there were no Buddha, Tam would not be happy?
c. What do you think about the end of the story? Please create a new ending for the fairy tale *Tam Cam*.
d. We are approaching the *Tam Cam* fairy tale in the form of poetic genres, but now there is a textual approach by attaching all the ancient events in the story into the behavior and thinking of modern society and draw lessons on life skills. What do you think about this approach?

FIGURE 6.15 Students present their opinions in discussion activities.

FIGURE 6.16 The student wrote down their own lesson from the story on a small piece of paper.

5. *Activity 5*: The teacher commented on the student's ideas, answered the questions asked by students, extended, and finalized knowledge for students.
6. *Activity 6*: Students drew their own useful conclusions on lessons, and the teacher assigned tasks to students.
   a. At the end of the hour, each student wrote down their own lessons from the story on a small piece of paper and returned to the teacher.
   b. The home task continued to be performed in the G.C., with the differentiation:
      i. All students reviewed the lectures through PowerPoint the teacher had uploaded in the classroom to reinforce their lessons and did the basic exercises (analyze a problem in the text).
      ii. For students who were quite good or had a career orientation related to literature: they could do more advanced exercises (optional) → read versions and works of the same topic, a module with *Tam Cam* in other countries to compare and conclude.

Step 3: Got feedback from teachers and students to evaluate the effectiveness of F.C. implementation.

Initially, we received very positive feedback about the *Tam Cam* class in F.C. with the support of the G.C. tool.

1. About test results after self-study with lecture videos: Figure 6.17 shows that, after study with video lectures and other materials at home, the majority of students were aware of the background knowledge of the lesson. Scores of the class were distributed between 5 and 10 points, of which up to 68% were good, 29% were good, and only 3% were average; none of them had weak points (3–4 points). With the above score, we think that the task of assigning students to self-study first at home had a good signal; the majority of students seriously performed their tasks before the class.
2. About students' interest in class time: Looking at Figure 6.18, we see that 100% of the surveyed students took interest in the lesson. None of them felt normal and boring. They also felt that class time was exciting (94%). The atmosphere had been fun and not boring. Hundred percent of students assessed the content of the lesson to reach their needs and desires for the lesson of "Tam Cam." When

conducting experiments, in addition to ensuring the basic knowledge of the lesson, we oriented and spent most of the time discussing the difficult issues in the story, so the majority of students in the class were excited to discuss and debate.

According to the statistics from the questionnaire, a majority of students said that the teacher created conditions for students to practice and be more active (91%), and the method of the teacher was attractive and hence attracted their attention (88%). This proved that during the lesson, students were more active. Instead of just sitting notes, they spent most of the time to discuss competitive presentation.

FIGURE 6.17 The results research self-study with lecture videos.

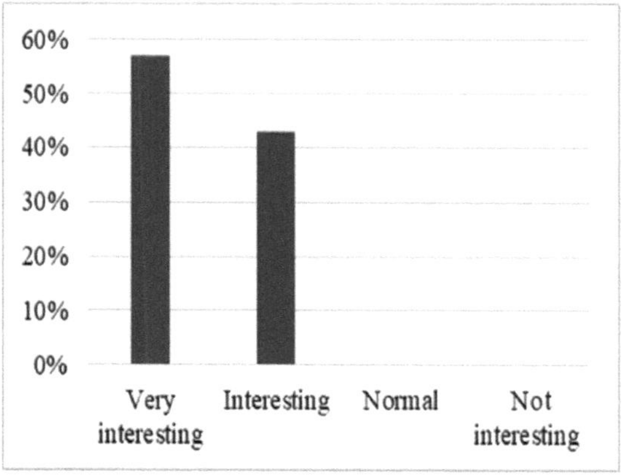

FIGURE 6.18 The results of students' interest in the class.

When interviewing fast for teachers of literature class 10A8 for HES School, Ms. Nguyen Thi Tuyet Nhung (Vice Principal of the school) after the lesson, we received very positive feedback. In general, she acknowledged the outstanding advantages of F.C. with the support of G.C. compared to traditional classes and hoped that such experiments would continue to be implemented in other lessons, other classes throughout the school.

Through experiments in class in the reverse classroom model, we also found that teachers had more time to interact, listen to students' opinions, and comment and correct students not only about the lesson knowledge, but also about the posture, style, and presentation style in front of the crowd. Therefore, we believe that, if this model is implemented regularly, students will greatly improve their soft skills in addition to subject knowledge.

6.4.2 DISCUSSION

1. *First of all, through the experiment process, we realize that the application of G.C. along with F.C. in teaching Philology has many advantages in Vietnam.*

   The most joyful thing is teaching contents and forms of literary work are vividly intriguing. Teachers and students can easily learn through pictures and videos. Students also experience a variety of learning forms and more flexible learning. In addition, they have plenty of time discussing G.C. in class. It helps students understand better. Thus, they do more practice. Furthermore, students can develop many soft skills during learning, such as working group, presentation, and adaptation. These are essential skills in the twenty-first century and also the comprehensive educational goal in the renovation of Vietnam's education nowadays.

   Clearly, this form of learning contributes to the development of self-learning ability. This is a big change from the traditional model. It helps students can become more proactive and active in the learning process. This means that they will actively acquire knowledge, formulate skills, and apply them in their life. Students will also develop better competence. An important thing with this form of learning is that teachers can differentiate each student's competence. Each teacher will have individual lessons, not copy and paste. Both teachers and students have to access IT with a quite simple operation, easy to use, and interact if they have basic knowledge and skills.

2. *Second, during the experiment we realize that if applied more, we have to face some difficulties and put forward solutions to solve each difficulty.*
   The most obvious thing with the application of the F.C. model is that teachers must spend a lot of time to prepare lessons because they have to prepare many things: pictures, videos, exercises, etc. Before going to the class, they also guide, check, and interact with students regularly. Hence, although teachers want to apply this model in teaching, they do not have time to do so. Thus, managers need to facilitate much time for teachers to prepare their lessons.

   Besides, if students do not fulfill the teacher's requirement and tasks on G.C. before going to the class, teachers and students will face difficulties. Teachers had better consider the plan of additional support for these cases. Moreover, it takes a lot of time in class. Therefore, teachers must also consider the plan of separation and assigning an individual task to each group according to the level of G.C. preparation before going to the classroom.

   The fact shows that the teaching process, according to F.C., depends on teaching aids, especially on the Internet and computers. But then in remote areas, they do not have any computers and the Internet. Moreover, the teachers' and students' computer using ability also affects the effectiveness of the teaching and learning process. Some teachers and students cannot use them well or do not use the computer. So we prepare and develop good facilities such as the Internet, computers, classroom (large enough), and teaching aids.

   In particular, in Vietnam, at present, many teachers and students are still familiar with the traditional way of teaching. So, changing traditional thinking of teachers and students is not easy. In addition, despite time conditions and curriculum frameworks as well as fixed lessons, we only can test this model and teaching tools in some suitable lessons. In order to increase its applications, we need to have a specific plan and roadmap. This is a long process and needs perseverance, step-by-step implementation, and full preparation of the guaranteed conditions. We implement further research on applying F.C. in teaching Philology in each level, grade/level education. Discussion is organized in professional groups to find teaching issues. Each teacher and student will have an individual experience. Therefore, sharing will help them develop their strengths and overcome weaknesses. This teaching model not only enhances sharing experiences but also makes students more connected, so there can be many interesting learning activities. We had better organize training courses for teachers: G.C. using course, F.C. using course, and technology course.

A major difficulty in the application of G.C. and F.C. is finance. Normally, teachers can use G.C. for free. But when using free, teachers can use only some functions for teaching. These free applications only apply within certain limits. If teachers want to use more applications, they have to pay for them. So we also had better consider who pays it? They may be teachers, schools, or schools or others? If funding is made by teachers, then it is very difficult to implement this model because they will not get enough to paid or do not want to pay.

3. *Eventually, we discuss the continuing researches trend of the F.C. model and G.C. in the future.*
In order to apply G.C. in the F.C. model when teaching Philology, conducting in-depth and extensive research is essential. These continuing research trends of applying F.C. in teaching Philology are effective. We can apply F.C. in teaching a specific genre of literature. There are many genres that can provide teaching according to this model: legend, fairy tales, short stories, etc. Each genre will have different characteristics, so researchers can compare similarities and differences when teaching categories. On that basis, they draw notes for each genre in order to make teaching more effective.

Another research trend, we can research on, is designing the way to apply G.C. in F.C. in teaching Philology for lessons of some genre in order to use references sample lectures for teachers. This is a specific study that can be published in a book for lager publications. Teachers can use teaching materials without much research. Even students can easily use such materials in the learning process. In the context that this model is rather new, this is really a necessary study.

Next, research on applying F.C. in teaching Philology in different grade/level education should be a continuing research trend in the future. Each grade/level will have advantages and disadvantages. Moreover, new textbooks are about to be published; applying new methods of teaching is imperative. If researched at all grades, this will be a significant contribution to the development of the quality of Philology teaching in the new curriculum. However, this is a very difficult task and requires a lot of effort, money, and time.

Finally, although we chose the trend, the most important thing is passion for conducting serious research.

6.5 CONCLUSION

F.C., with the support of the G.C., will open up flexible, positive, effective, personalized learning opportunities for learners. Applying this model in teaching Philology will help students experience a variety of forms of learning, increase interest, and develop common and specific competencies. At the same time, it also requires the teacher to be "omnipotent" more to make the most of the methods, means, and techniques in teaching. Although teachers will spend more effort and time in designing learning materials and teaching scripts, if implemented widely, this will be a teaching model that fully meets the requirements of the new Philology program, in line with today's digital age.

However, in order to implement widely, we need to ensure the conditions: human, material, and financial. The essential requirements are the Internet and computers which are indispensable. If we do not know enough about these conditions, we cannot apply this model. In the context that the world has witnessed many changes, for example, epidemics, natural disasters, this is a suitable choice. In fact, students and teachers may not always interact directly in class, but they can discuss online without any restrictions on space. Therefore, although this model is still strange to many teachers, it is essential to research and implement it, and hence its larger applications to teachers and students are necessary.

ACKNOWLEDGMENTS

We would like to thank the National Project "Research and propose the model managing electronic education and teaching-learning electronic education in K12 school of Vietnam" CODE: KHGD/16–20.ĐT.042.

KEYWORDS

- **flipped classroom**
- **teaching philology**
- **Google Classroom**
- **online classroom**
- **active learning**
- **the new philology curriculum**

REFERENCES

Al-Maroof, R. A. S., Al-Emran, M. "Students acceptance of Google Classroom: An exploratory study using PLS-SEM approach." *International Journal of Emerging Technologies in Learning*, Vol. 13, No. 6, 112–123, 2018.

Azahar, K. A., Iqbal, N. "Effectiveness of Google Classroom: Teachers' perceptions." *Prizren Social Science Journal*, Vol. 2, No. 2, 52–66, 2018.

Bayarmaa, N., Lee, K. "A study on the application of Google Classroom for problem-based learning." *Journal of the Korea Academiai-Industrial Cooperation Society*, Vol. 19, No. 7, 81–87, 2018.

Beaumont, K. "Google Classroom: An online learning environment to support blended learning." *Journal of Learning and Teaching*, Vol. 11, No. 2, 2018. Retrieved from https://pdfs.semanticscholar.org/1537/78fb6948d9a4f5fc798a359d754f09c09781.pdf

Bergmann, J., Sams, A. *Flip your classroom: Reach every student in every class every day.* Washington, DC: International Society for Technology in Education, 2012.

Bhat, S., Raju, R., Bikramjit, A., D'Souza, R. "Leveraging e-learning through Google Classroom: A usability study." *Journal of Engineering Education Transformations*, Vol. 31 , No. 3, 129–135, 2018.

Bouwmeester, R. A. M., et al. "Flipping the medical classroom: Effect on workload, interactivity, motivation and retention of knowledge." *Computer and Education*, Vol. 139, 118–128, October 2019.

Brown, M. C. "Google Classroom for the online classroom: An assessment." *Distance Learning*, Vol. 15, No. 3, 51–56, 2018.

Chick, H. "Flipping the Classroom: A case study of a mathematics methods class." *Eric No. ED572650*. Accessed 2014. https://files.eric.ed.gov/fulltext/ED572650.pdf

Duc, T. P. "Google Classroom and Google Forms in the EFL classroom". *In Proceedings of the the 26th Korea TESOL International Conference*, Korea, September 15, 2018, pp. 58–60.

Ferreira, J. M. M. "Flipped classrooms: From concept to reality using Google Apps." *In Proceedings of the 11th International Conference on Remote Engineering and Virtual Instrumentation (REV)*, Porto, Portugal, 2014, pp. 1–5.

Fitriningtiyas, D. A., Umamah, N., Sumardi. "Google classroom: As a media of learning history." *In Proceedings of the IOP Conference Series: Earth and Environmental Science*, Indonesia, November 17–18, 2018, pp. 1–8.

Greenberg, B., Medlock, L., Stephens, D. "Blend my learning: Lessons from a blended learning pilot." Accessed 2011. Retrieved from http://blendmylearning.files.wordpress.com/2011/12/lessons-learned-from-ablended-learning-pilot4.pdf

Hamdan, N., McKnight, P., McKnight, K., Arfstrom, K. "A review of flipped learning." Flipped Learning Network. Accessed March 19, 2014. http://flippedlearning.org/cms/lib07/VA01923112/Centricity/Domain/41/LitReview_FlippedLearning.pdf

Heggart, K. R., Yoo, J. "Getting the most from Google Classroom: A pedagogical framework for tertiary educators." *Australian Journal of Teacher Education*, Vol. 43, No. 3, 140–153, 2018.

Iftakhar, S. "Google Classroom: What works and How?" *Journal of Education and Social Sciences*, Vol. 3, 12–18, February 2016.

Islam, M. S. "Bangladeshi university students' perception on using Google Classroom for teaching English." *International Journal of Psycho-Educational Sciences*, Vol. 8, No. 2, 57–65, 2018.

Khan, S. *The one world schoolhouse: Education reimagined of knowledge*. London: Hodder & Stoughton, 2012.

King, A. "From sage on the stage to guide on the side." *College Teaching*, Vol. 41, No. 1, 30–35, 1993.

Kopeyev, Z., Mubarakov, A., Kultan, J., Aimicheva, G., Tuyakov, Y. "Using a personalized learning style and google classroom technology to bridge the knowledge gap on computer science." *International Journal of Emerging Technologies in Learning*, Vol. 15, No. 2, 218–229, 2020.

Kumar, J. A., Bervell, B. "Google Classroom for mobile learning in higher education: Modelling the initial perceptions of students." *Education and Information Technologies*, Vol. 24, 1793–1817, 2019.

Kurtz, G., et al. "The Flipped classroom approach to future learning." European Journal of Open, Distance e-Learning, Vol 17., No. 2, 171–181, 2014.

Lage, M. J., Platt, G. J., Treglia, M. Inverting the classroom: A gateway to creating an inclusive learning environment. *Journal of Economic Education*, Vol. 31, 30–43, 2000.

Manan, N. Z. A., Hanafi, H. F. "Google Classroom: Student's acceptance using UTAUT model." *Journal of Applied Arts*, Vol. 1, No. 1, 64–72, 2019.

McLaughlin, J. E., et al. "The flipped classroom: A course redesign to foster learning and engagement in a health professions school." *Academic Medicine*, Vol. 89, No. 2, 236–243, 2014.

Nguyen, T. P. "About the flipped classroom and applying this model at university amphitheater." *Language and Life Magazines*, Vol. 4, No. 271, 39–42, 2018.

Onlinehub. "How to Use Google Classroom Tool in Classrooms". Accessed 2019. Retrieved from https://onlinehubng.com/how-to-use-google-classroom-tool-in-classroom/

Ozdamli, F., Asiksoy, G., "Flipped classroom approach." *World Journal on Educational Technology*, Vol. 8, No. 2, 98–105, 2016.

Pappas, C. "Google Classroom review: Pros and cons of using Google Classroom in eLearning." *eLearningIndustry*, August 20, 2015. Accessed 2015. Retrieved from https://elearningindustry.com/google-classroom-review-pros-and-cons-of-using-google-classroom-in-elearning

Pritasari, A., Jumadi, J. "Development of science learning tool based on problem based learning with Google Classroom to improve argumentation skill." *Journal of Biology and Biology Education*, Vol. 10, No. 2, 348–355, 2018.

Puarungroj, W. "Inverting a computer programming class with the flipped classroom." *In Proceedings of the Twelfth International Conference on eLearning for Knowledge-Based Society*, Thailand, December 11–12, 2015, pp. 40.1–40.7.

SEE Team. "Flipped classroom—How this model can enhance the educational". *SOFIA Education Experts*. Accessed 2019. Retrieved from https://www.sofiaeducationexperts.com/post/flipped-classroom-how-this-model-can-enhance-the-educational-procedure-flip-your-classroom

Sokolova, O. L., Kagan, E. B., Shadrina, Y. Y. "Innovative 'Flipped Classroom' technology in teaching foreign languages." *In Proceedings of the 1st International Scientific Conference: "Modern Management Trends and the Digital Economy from Regional Development to Global Economic Growth,"* 2019, pp. 811–813. Retrieved from

Stavytskyi, O., Urazgaliyeva, M., "Using Google Classroom tools in teaching students of economic specialities." *Journal of Advanced Education*, No. 10, 76–81, 2018. https://doi.org/10.20535/2410-8286.149361

Subandi, S., Choirudin, C., Mahmudi, M., Nizaruddin, N., Hermanita, H. "Building interactive communication with Google Classroom." *International Journal of Engineering and Technology*, Vol. 7, No. 2.13, 460–463, 2018.

Sukmawati, S., Nensia, N. "The role of Google Classroom in ELT." *International Journal for Educational and Vocational Studies*, Vol. 1, No. 2, 142–145, 2019.

Thi, D. N. "Apply the flipped classroom model into teaching Vietnamese language teaching methodology term for primary education's learners." *Science Magazine*, No. 21, Tay Nguyen University, December 2016, pp. 21–26.

Thi, K. N. "Applying a flipped classroom in teaching citizen subject education in upper secondary school." *Magazine of Science and Technology*, Vol. 179, No. 3, 109–113, 2018.

Ventayen, R., Estira, K., Guzman, M., Cabaluna, C., Espinosa, N. "Usability evaluation of Google Classroom: Basis for the adaptation of GSuite E-learning." *Asia Pacific Journal of Education, Arts and Sciences*, Vol. 5, No. 1, 47–51, 2018.

Vietnam Ministry of Education and Training. *General Philology Education Program*, Hanoi, December 2018.

Wijaya, A. "Analysis of factors affecting the use of Google Classroom to support lectures." *In Proceedings of the 5th International Conference on Information Technology and Engineering Application*, Palembang-Indonesia, February 19–20, 2016, pp. 61–68. Retrieved from https://en.wikipedia.org/wiki/Google_Classroom

CHAPTER 7

Teaching Education and the Digital Process of Assessment, Testing, and Evaluation

PUVVADA GEORGE RAJA KUMAR

Department of Education, Central University of Odisha, Odisha, India

Corresponding author. E-mail: kingsonpuvvada@rediffmail.com

ABSTRACT

Teacher education institutions are powerhouses of a country which produce excellent (influential) teachers so that responsible and good citizens are developed in the country. If a teacher makes a mistake, this affects the entire generation. The instructors, curriculum, and learning environment in a teacher education institution play significant roles in framing the student teachers. Today's student teachers are tomorrow's teachers. Good teachers are the builders of an excellent nation. The economic, psychological, sociological, esthetical, cultural, moral, and philosophical developments of the citizens of a country ultimately depend on the educational system of that country and the shoulders of the teachers. Along with the curriculum, the evaluation procedures determine the quality of the student teachers' achievement. Excellent curriculum, adequate teaching and learning environment, and appropriate evaluation techniques can make perfect teachers, in return, perfectly molded citizens. If we neglect the procedures, techniques, and tools of the evaluation system, teacher education institutions become paralyzed and after that unproductive. Keeping the importance of evaluation procedures in mind, the author of this chapter clearly explains every term that comes across the evaluation system, types, steps, and course of action in the evaluation procedure with keen interest to strengthen the teacher education institutions and specifically the student teachers by equipping them with knowledge about the evaluation procedures.

7.1 INTRODUCTION

Teacher education is an interdisciplinary gathering for innovative practices and research in instructor training in teaching. The advanced teaching method has its starting points in separation realizing, which appeared as correspondence courses conveyed via mail during the mid-twentieth century. The University of London offered the main separation learning degrees with its External Program of 1858. Charles Dickens broadly said that the External Program was the "People groups' University," as it permitted more extensive access to advanced education. With the improvement of correspondence advances, it was accepted that the TV and radio would start to change the course of training. In the United States, instructive TV projects and radio stations were made by establishments, for example, the University of Louisville, with the participation of the National Broadcasting Company. Telecourses saw a resurgence in the mid-1970s with the spread of junior colleges. These separation courses were prepared for online instructions, which came to pervasiveness during the advancement of the Internet, starting toward the end of the twentieth century.

During the 1980s and 1990s, the Association for Computers and the Humanities offered workshops and gatherings on showing PCs, and thus the focus on humanities, similar to computerized humanities, increased. In any case, the teaching method was not a concentration for gatherings in computerized humanities between 2001 and the 2010. From 2011 onward, symposia and gatherings concentrating on computerized teaching methods are becoming progressively normal day by day.

Instructor training plans to create information and aptitudes in student educators for the proper use and incorporation of the right innovation in a reasonable manner. It is the need of time that each educator should know the best possible utilization of innovation, instructional method, and substance of a specific subject to make their everyday study hall showing powerful and productive. The computerized instructional method or techno-teaching method comprises three subject matters, that is, substance, instructional method, and innovation. Content (C) is the topic that is to be taught. The instructional method (P) describes the gathered strategies, practices, techniques, forms, methodologies, and approaches of educating the instructors in teaching. It incorporates information about the points of directing, assessment, and student learning. Technology (T) includes most recent advances, for example, PC, Internet, mobiles, computerized video, digital books, open instructive assets, and typical advances, including overhead projectors, boards, and books. Contingent on the nature and extent of substance, and level of students, based on the individual contrasts of the students' suitable innovation incorporation, must be searched for. Innovation as an associate improves the way toward learning and aides in the accomplishment of more elevated level destinations, similar to examination, amalgamation, and assessment (Figure 7.1).

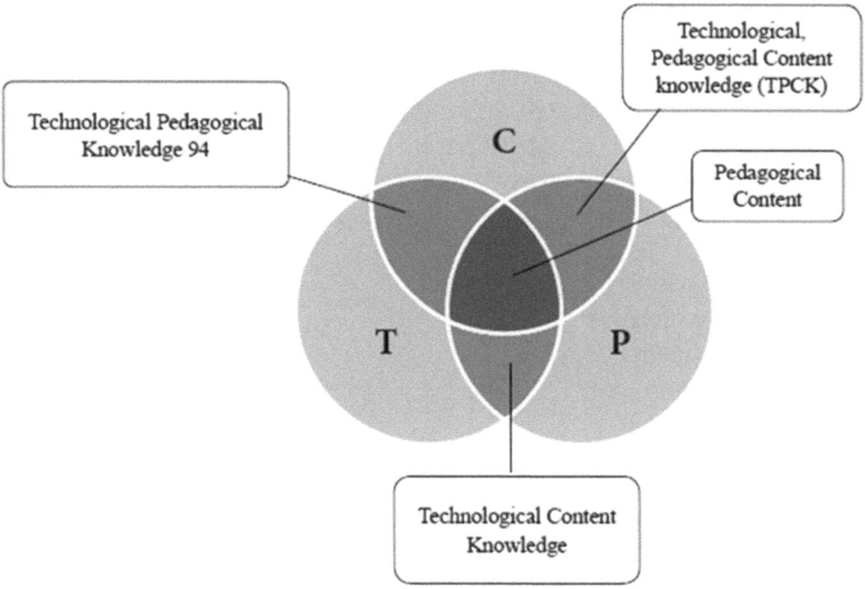

FIGURE 7.1 TPCK framework (Koehler and Mishra, 2005).

7.1.1 NEED FOR MEASUREMENT AND EVALUATION IN EDUCATION

Teachers have always endeavored to measure the results of their teaching efforts, as indicated by the progress of their pupils toward desired educational goals. Many have been equally concerned about the need to diagnose and remedy revealed defects in instruction. However, only recently a large degree of accuracy has been injected into their methods of measurement and diagnosis. Measurement of progress, evaluation of the efficiency of instruction, and the accompanying attempts at diagnosis were largely a matter of personal observation and judgment on the part of the individual teacher as recently as two decades ago. Actually, the recent development of modern educational instruments of measurement and evaluation may be regarded as an extension and improvement of an old practice. The modern educational measuring instrument presents a surprisingly accurate picture of the course objectives as well as an analysis of the underlying skills, knowledge, concepts, understandings, and other outcomes upon which accomplishment in different subject matter fields depends. It points out weaknesses in learning and learning procedures. It permits the establishment of specific and objective goals of achievement that is based upon the actual attainments of children under typical school conditions. Educational tests and the information resulting from their use in the classroom have come to be almost universally identified with good teaching practice. Today, the professionally equipped teacher is expected to be well versed in their construction, selection, and use in the classroom. First, let us know about the definitions of the frequently used terms in measurement and evaluation:

1. Test
2. Examination
3. Measurement
4. Assessment
5. Evaluation.

7.2 CONCEPT OF TEST, EXAMINATION, MEASUREMENT, ASSESSMENT, AND EVALUATION

7.2.1 TEST

7.2.1.1 DEFINITIONS OF TEST

The test can be defined as follows:

1. Definition 1: An educational and psychological test is defined as a systematic method to get information by qualitative and/or quantitative means of a student.
2. Definition 2: A test is a formal, systematic procedure to gather information about a student's behavior or performance.
3. Definition 3: A test is a task or series of tasks or questions that students must answer or perform.

7.2.1.2 TYPES OF TESTS

Tests can be classified in many ways. To classify tests, we use the following criteria:

1. Kind of item
2. Nature of item scoring
3. Standardization
4. Conditions of administration
5. Emphasis of the responses in line with the language
6. Time to respond to items
7. The attribute (behavior) being measured.

7.2.1.2.1 By Kind of Item. The questions or exercises or tasks appearing on a test are called items. According to this, there are two types of tests:

1. Selection-type item tests.
   Ex: True–false tests, multiple-choice tests, matching tests
2. Supply-type item tests.
   Ex: Short answer tests, completion tests, essay, or descriptive tests.

7.2.1.2.2 By Nature of Item Scoring. Based on the way we score test items, we can classify tests into two types:

1. Objective tests: These are types of tests having a plain and clear-cut scoring standard. That is, different scorers can give exactly the same scores, or they can agree on the number of points students can receive.
   Ex: Selection-type items tests, completion item tests, and short answer item tests are objective-type tests.
2. Subjective tests: The scores of subjective tests are not constant but different.

7.2.1.2.3 By Degree of Standardization. This can be defined as the application of fixed procedures in all aspects of the test. According to this, we can classify tests into two types.

1. Standardized tests: Standardized tests are constructed by specialists working with curriculum similar to those of experts and teachers. To compare the performance of one school, students with other schools of the same grade use standardized tests.
   Ex: Secondary School Certificate Examination.
2. Nonstandardized test: These are administered through teachers any time they wish to get information about their student's progress. Nonstandardized tests are prepared by teachers to use the results in their classrooms. Nonstandardized tests cannot be used to compare students of one school with other schools.

7.2.1.2.4 By Administrative Conditions of Tests. According to this, we can classify tests into two types:

1. Individual tests: These tests allow the maximum amount of interaction between the examiner and the examinee. These are rich in opportunities for getting information by observing the student when taking the test. In these tests, the examiner can also ask questions that follow up on an examinee's response to clarify it and to understand it more completely. But it requires much training, time, money, and experience. It means that there are more complications.
2. Group test:
   a. Administered to a group at the same place and time
   b. All students are given the same amount of time to complete tasks
   c. Less complicated.

7.2.1.2.5 By the Language Emphasis of Response. Based on the nature of responses required by the items, we can classify tests into three types.

1. Verbal tests
   a. These tests highlight the use of verbal communication as the main means of responding to test items.

b. They use written language.
Ex: Tests in high schools.

2. Nonverbal tests
   a. These tests deemphasize the role of reading or written language in Figure 7.2.

Ex:-

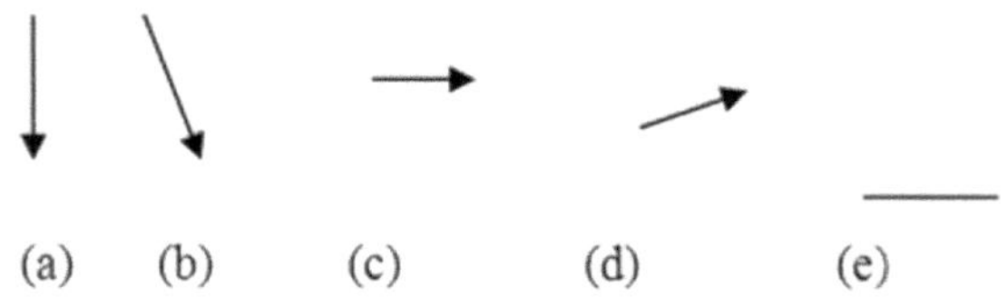

FIGURE 7.2 Example of test questions.

What will be in "e"?

This type of test is important and useful for children and illiterates.

3. Performance tests
   a. These tests require a student to carry out an assignment other than answering items.
   b. Performance tests are often administered individually, so that the examiner can observe the behavior of the examinee while performing the task.

7.2.1.2.6 Emphasis on Time to Respond to Items. Another way to classify tests is by speed. Based on this criterion, we classify tests into two categories:

1. Speed tests: These types of tests required students to complete the items as quickly as possible within a very short time.
   Ex: Typing tests.
2. Power tests: In these types of tests, students are assessed for the amount of knowledge, comprehension, or understanding they possess. Power tests have liberal time boundaries so that most students will be complete each item.

In comparison to speed tests, power tests are difficult.

7.2.1.2.7 Score-Referencing Scheme. There are two types of score referencing schemes.

1. Norm-referenced test:
   These are:
   a. Designed to measure individual differences in the students relating to achievement, interests, attitudes, and intelligence among the students.
   b. Involve the comparison of a student's performance to the performances of other group students.
   c. Measures such as the height of a student, how much it rained, and how fast he/she was driving.
   d. It rained less than it usually does; I have never driven this fast and so on. These are all norm-referenced interpretations.
   e. This type of interpretation of test scores was necessary to limit the number of persons who could be admitted into a college or hired for a job.
2. Criterion-reference test:
   a. In this type of test, relate a student's score on an achievement test to a certain criterion or domain of knowledge rather than to another student's score.

   In this type of test, the comparison takes place between one domain score of one student with another in the same domain.
   Ex: These are used in elementary and secondary schools.

7.2.1.2.8 The Attributes being Measured. These tests are used to get information about a variety of human attributes (behavior) or characteristics.

These are of two types:

1. Cognitive tests:
   a. Designed to measure the intelligence, reasoning ability, and academic achievements of students.
   b. Have correct or best answers.
   c. Academic achievement tests measure what students have learned in the classrooms.
2. Noncognitive tests

a. Also called nonacademic tests or affective domain tests.
a. Do not have correct or incorrect answers.
b. Measure human attributes such as personality, interests, attitudes, and motivation through questionnaires asking the examinee's the degree of agreement with various statements.

7.2.1.3 CLASSIFICATION OF TESTS USING DIFFERENT CRITERIA

7.2.1.3.1 By kind of item

1. Selection type
    a. True–False item tests
    b. Multiple-choice items tests
    c. Matching item tests.
2. Supply type
    a. Short answer item tests
    b. Completion item tests
    c. Essay item tests

7.2.1.3.2 By the Nature of Scoring

1. Objective tests
2. Subjective tests

7.2.1.3.3 By Degree of Standardization

1. Standardized tests
2. Nonstandardized tests

7.2.1.3.4 By Administrative Conditions

1. Individual tests
2. Group tests

7.2.1.3.5 By the Language Emphasis of Response

1. Verbal
2. Nonverbal
3. Performance tests

7.2.1.3.6 By Emphasis on Time

1. Power tests
2. Speed tests

7.2.1.3.7 By Score-References Scheme

1. Criterion-referenced tests
2. Nonreferenced tests

7.2.1.3.8 By Attributes Being Measured

1. Cognitive tests
    a. Achievement tests
    b. Aptitude tests
        i. General aptitude tests
        ii. Specific aptitude tests
2. Noncognitive tests
    a. Personality tests, interest tests, aptitude tests, and motivation tests.

7.2.2 EXAMINATION

The examination can be defined as the collection of a series of tests to award a degree or a diploma.

Ex: If there were four subjects in the semester, after four tests in four subjects, they award you the first semester.

7.2.3 MEASUREMENT

Definition:

1. It is a procedure of evaluating or doling out a number to execution as per express guidelines.
2. It is worried about precise assortment, capability, and requesting of data.
3. Measurement is the procedure that allows numbers to qualities or attributes of people, items, or occasions as per elucidate definitions or rules.

Scales of measurements: It is commonly concurred that there are four distinct levels or scales of measurements. These scales are as follows:

1. Nominal scale: Nominal scales are the simplest form of measurement. Scale involving the classification of objectives, persons, or events into discrete (small) categories.
2. Ordinal scale: These scales involve ranking of objects, persons, traits, or abilities with regard to equality of differences.
3. Interval scale: This scale is having equal differences between successive categories.
4. Ratio scale: Scale having an absolute zero and equal intervals.

Most complex scales: Ratio scale and interval scale.
Least complex scales: Ordinal scale and nominal scale.

- The ordinal scale of measurement is most frequently used in psychology.
- Psychologists want to treat their assessment data as "internal" when those data would be better described as "ordinal" because data in the interval scale can be manipulated statistically.

7.2.4 ASSESSMENT

As indicated by Airasian (1996), assessment is a procedure of gathering, orchestrating, and interpreting data to aid in decision-making.

It centers around the idea of the student in addition to what is to be realized and how (Phye, 1997).

Evaluation is a nonexclusive word implied for a lot of systems that figure the aftereffect of understudies' learning, the extent that data acquired, an understanding made, and aptitudes or limits got.

Evaluation is an orderly assortment of proof of understudy learning. It is a procedure for consistent improvement of understudies dependent on explicit understudy learning results and a device for improving understudy learning. It is likewise an instrument to improve showing techniques, archive best practices, distinguish open doors for fundamental enhancements, and give proof to institutional adequacy. Evaluation is not only a regulatory action based on understudy reviews as it were. It is not something directed for accreditation, workforce execution assessment, and does not mean it speaks to pointless numbers (*Oklahoma City Assessment Handbook*, 2011).

As shown by Brown and Knight (1994), evaluation is an orderly explanation behind creating determination about the learning and improvement of understudies. It is the path toward portraying, picking, organizing, gathering, analyzing, disentangling, and using the information to extend understudies' learning and headway. From this definition, one can perceive that appraisal is a human movement that includes connection, targets looking to comprehend what the students have accomplished, and it is clear in nature as opposed to critical. Its job is to build understudies' learning and advancement by helping students to analyze their issues and to improve the nature of their resulting learning.

An appraisal can be formal or casual.

7.2.4.1 MOTIVATION BEHIND ASSESSMENT

Appraisal fills a few gaps. It gives the premise to choices on whether an understudy is prepared to continue to fit the bill for honor or has shown the skill to rehearse. It additionally empowers understudies to get input on their learning, their qualities, and shortcomings, and makes a difference in them improve their presentation. Also, it gives the methods by which understudies are evaluated as pass or come up short. The presentation is judged with the accomplishment of the points and proposed learning results of the particular seminar on which the understudy is selected. Moreover, it empowers staff to assess the viability of their education.

Based on its purpose—designed to collect information—assessment is categorized as follows.

1. Diagnostic assessment: It is planned to offer a meter of a student's capacity and attentiveness for a program of study and see probable learning problems.
2. Developmental appraisal: It counsels understudies' headway, likewise inform progress. It is an evaluation that is led to give criticism before the fruition of the exhibition (program), which empowers the understudy to alter and improve the understudy execution (program).
3. Summative assessment: It is considered to offer a compute of attainment made with respect to a student's presentation against the planned learning results of the module and/or program of study.

Any instrument in this can and often does engage more than one of the elements.

1. Analytic appraisal: It is intended to give a marker of an understudy's fitness and readiness for a program of study and recognize conceivable learning issues.
2. Developmental evaluation: It is intended to give understudy's input on advance and educate improvement. It is an appraisal that is directed to provide feedback before finishing the presentation (program). It empowers understudy to alter the understudy execution (plan).
3. Summative evaluation: It is intended to give a proportion of accomplishment made in regard to an understudy's presentation against the planned learning results of the unit and additionally program of learning.

An appraisal gadget frequently involves more than one of these mechanisms. In this way, for instance, much coursework is developmental in that it gives a chance to understudies to get criticism on their degree of fulfillment, yet in addition frequently checks a summative proclamation of accomplishment. A finish of module or end of assessment is planned principally to make a summative judgment fair and square of accomplishment which the understudy has reached. Both developmental and summative evaluation can have the asymptomatic capacity. Evaluation principally focused on finding is inherently developmental despite the fact that it may infrequently add to a summative judgment.

When the assessor recognizes understudies' presentation on a particular learning angle or course utilizing the various sorts of tests or estimation systems, the outcomes can be utilized for making judgments about the status of understudies or somebody who is being surveyed. These decisions or assessments can fulfill a few needs.

7.2.4.1.1 Here Is a Portion of the Motivations Behind the Assessment

1. Placement of understudies, which includes bringing understudies properly in the learning succession and order or gushing of understudies as indicated by capacity or subjects.
2. Selecting understudies for courses: general, proficient, specialized, business, and so on.
3. Certification: This assists with confirming that an understudy has accomplished a specific degree of execution.
4. Stimulating getting the hang of giving input and proposing reasonable practice can be an inspiration for the understudies or educators.
5. Improving education by assisting with checking on the viability of instructing courses of action.
6. For look into purposes.
7. For direction and guiding administrations.
8. For the adjustment of the educational program purposes.
9. For the reason for choosing understudies for business.
10. For the adjustment of instructing strategies.
11. For the reasons for the advancement of the understudy.
12. For revealing understudies' progress to their folks.
13. For the honors of grant and legitimacy grants.
14. For the affirmation of understudies into instructive foundations.

7.2.4.2 KEY POINTS ABOUT STRATEGIC ASSESSMENT

Evaluation should be lined up with the learning results and the picking up of the educating approach received. It is imperative to embrace an increasingly key way to deal with an appraisal by posing a progression of inquiries.

7.2.4.2.1 For What Reason Would We Say We Are Evaluating Understudies?

The fundamental reasons for existing are right off the bat to help understudies' learning through drawing in them in learning exercises and giving criticism, besides to gauge learning against the expressed learning results lastly to guarantee the measures of grants. While the subsequent reason may concentrate on imprints and foundations accentuate the need to guarantee measures, our primary spotlight ought to be on how well appraisal advances powerful understudy learning.

7.2.4.2.2 What Are We Evaluating?

Progressively, abilities are viewed as being significant for understudies' future employability. Understudies' mentalities, values, capacity to act in an expert setting, to perceive their need to obtain new information and abilities and to see adapting comprehensively should be evaluated.

7.2.4.2.3 When Are We Going to Survey?

Experience has frequently demonstrated that when an appraisal is led toward the finish of a course, understudies will invest a large portion of the class energy attempting to spot prompts with respect to what they will be surveyed on and, ideally, the appropriate response that the teacher needs. An assortment of constant evaluation strategies is important to use to survey all encompassing parts of discoveries that are arranged as learning results.

7.2.4.2.4 Who Is Going to Complete the Appraisal?

It is critical to consider who will be associated with the appraisal procedure as an assessor previously directing the evaluation. For example, instructors may plan to attribute to the evaluation process of peer appraisal and/or appraisal by manager other than the educator, or instructors themselves may direct the evaluation procedure.

7.2.4.2.5 Where Will the Evaluation Happen?

Evaluation can happen at any place where the understudies are learning: at home, workplace, and in a study hall. The central matter here is that evaluation serves a number of capacities, and in the event that it is adding to understudy learning, at that point, it can occur in various settings and not simply in informal study settings.

7.2.4.2.6 How Are We Going to Review It?

An appraisal can be utilized for assessment reasons, or it may very well be utilized to distinguish what understudies have been learning and on which zone they need further help without utilizing it for assessment or making a decision about the presentation of the understudy.

7.2.4.2.7 What Input Will Understudies Get?

Over and over again, understudies get criticism either past the point where it is possible to assist them with improving their learning later on or in an arrangement that does not assist them with making the enhancements. Note that inputs cause understudies to learn better by evaluating themselves against the standard.

In view of the objective of evaluation concerning the choices made subsequent to distinguishing understudies learning results, the appraisal can be ordered as evaluation of learning and appraisal of learning.

7.2.5 EVALUATION

Definition: According to Airasian 1996,

> Evaluation is the process of making a judgment about a student's performance, instruction, or classroom climate. It occurs after assessment (assessment is a process of collecting, synthesizing, and interpreting information to aid in decision making) because this is when the teacher is in a position to make informed judgments.
> … Evaluation includes both quantitative and qualitative images of student performance plus worth judgment concerning the appeal of that performance.

For understudies to be effectively occupied with making their own understanding, they should figure out how to be basic assessors who comprehend data, relate it to earlier information, and use it for new learning. This is the administrative procedure in meta-perception. That is, understudies got skills of specifically checking what they are realizing and use what they find from their observations to make alterations, adjustments, and even significant changes in their reasoning. Evaluation as learning is situated in examine about how learning occurs and is described by understudies pondering students' comprehension.

Assessment of learning implies strategies planned to assert what understudies know and show whether they have met instructive arrangement results or the destinations of their individualized ventures.

7.2.5.1 THE NECESSITY OF EVALUATION IN TEACHING

In general, evaluation plays a vital part in the process of education and in schools particular. An evaluation provides information about:

1. *Instructional management decisions*. These involve:
   a. Planning instructional activities (deciding what to teach to students)
   b. Placing students into learning sequences
   c. Monitoring student's progress
   d. Diagnosing student's learning difficulties
   e. Motivating students for learning
   f. Providing feedback about achievement to students and their parents
   g. Assigning grades to student
   h. Evaluating teaching effectiveness.
2. *Selection decisions*. These are the decisions made on the selected candidates by an institution.
3. *Placement decisions*. These are made after selecting applicants. The purpose is to assign them to different categories.
4. *Counseling and guidance decisions*. These are made by the students about their careers after a series of tests are administered.
   Ex: Series of tests are interest inventory, various aptitude tests, personality questionnaires, and an achievement battery.
5. *Classification decisions*. These are made to place candidates in the best program to increase the possibility of success.
6. *Credentialing and certification decisions*. These decisions are concerned with assuring that a student has attained certain standards of learning.
7. *Educational diagnostic and remedial decisions*. These are made to determine a person's capacities in order to pick up performance.

7.2.5.2 TYPES OF EVALUATION

According to Airasian 1996. There are four types of evaluation:

1. *Placement or preliminary evaluation.* Placement evaluation occurs during the first days of school and provides a base for expectations throughout the school year. The purpose of this evaluation is to determine the entry behavior of students.
2. *Diagnostic evaluation.* It occurs before instruction. This evaluation is concerned with skills and other characteristics that are prerequisites to the current instruction.
3. *Formative evaluation.* Formative evaluation occurs during instruction by letting the teacher know if students are meeting instructional objectives. The purpose here is to determine what adjustments should be made to instruction.
4. *Summative evaluation.* Summative evaluation occurs at the end of instruction, a unit, a chapter, or a course. This evaluation is used to:
   a. Certify students' achievements and assign end-of-term grades or marks
   b. Promote students from one grade level to the next
   c. Group students into different categories
   d. Determine whether teaching procedures should be changed before the next school year.

7.2.5.3 TYPES OF EVALUATION PROCEDURE

An evaluation procedure is broadly classified as quantitative and qualitative. In addition to this, there are two major ways of classifying an evaluation producer. They are as follows:

1. Classification by aspects of behavior
2. Classification by evaluation method.

7.2.5.3.1 Classification by Aspects of Behavior. It can be subdivided into two general categories. They are used to:

1. Determine a person's abilities: This type is concerned with how well an individual performs or whether he/she is motivated to put forth his/her best effort. In short, the evaluation results indicate "what an individual can do." A person's learning abilities involve:

1. Aptitude and achievement tests are included in this category.
2. An aptitude test is primarily designed to predict success in some future learning activities. If measure what a student will learn in the future.
3. An achievement test is designed to indicate a degree of success in some past learning activities. If measures student's ability to remember what they have learned in the classroom.

2. Reflect a person's typical behavior: Evaluation of typical behavior falls in the general area of personality appraisal. Methods designed to evaluate interest, attitudes, and various aspects of personal social adjustment are included in this category.

1. This type of evaluation is concerned with answering the question, "How does the individual usually behave in 'normal' or routine situations?"
2. Results in this area tend to indicate "What an individual will do rather than what he/she can do."

7.2.5.3.2 Classification by Evaluation Method

Based on this aspect, the evaluation procedure can be classified into:

1. *Testing procedure*. A test is a progression of errands which is utilized to quantify an example of an individual's conduct at a given time. The most widely recognized tests utilized in schools are accomplishment tests.
2. *Self-report techniques*. Every individual has a wealth of information about him/herself. The information may be one's interests, personal problems, how he/she feels about certain situations, etc. This information can only be obtained directly from the individual through interviews or different types of questionnaires.
3. *Observational techniques*. Reliable information about an individual's typical or usual behavior is best obtained from persons who have observed him/her, in a variety of situations. There are various

observational techniques: Anecdotal records, checklists, rating scales, socio-metric techniques.

The socio-metric procedure is a technique for assessing the social connections existing in a gathering.

7.2.5.4 PROCEDURE OF EVALUATION

There are five steps in the evaluation process. They are as follows:

1. Formulation of objectives
2. Providing learning experiences for the attainment of objectives
3. Selections and use of suitable evaluation tools to assess the expected objectives
4. Analysis and interpretation of data
5. Improvement of instruction using the results of the evaluation.

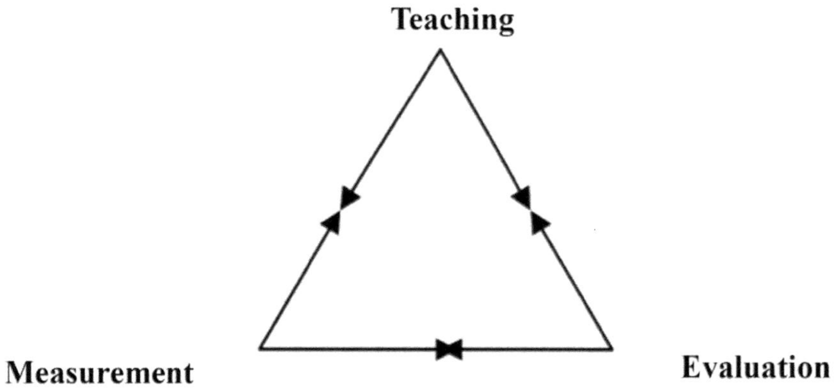

FIGURE 7.3 Relationship among teaching, measurement, and evaluation.

There is a triangular relationship between teaching, measurement, and evaluation, as shown in Figure 7.3. Based on evaluation, if a teacher does not attain the teaching objectives of the lesson, he/she has to modify the strategies of teaching, and also measurement techniques. If teaching strategies and measurement techniques are well enough, he/she needs to change the tools of evaluation. If teaching strategies and evaluation techniques are well enough, he/she needs to change the scale of measurement.

7.3 CONCLUSION

Plainly no single assessment configuration will be fitting for each computerized trade program worldwide. I suggest, nonetheless, that all projects base both structure and assessment on understudy learning results. I found that diverse learning results require distinctive estimation draws near and, thus, a successful assessment requires various techniques for estimation.

This is particularly significant in the qualification among worldwide and general learning results. I likewise found that the utilization of numerous strategies is especially significant for understudies of ages 10–13 who are toward the starting phases of creating worldwide skills, and who may show development in various manners and at various occasions. Hence, I likewise suggest that assessments of advanced worldwide trade center around development in understudy learning results, as opposed to on end-capability. Evaluators can expand on existing examination by adjusting measurements created in related fields.

The utility of these understudy learning results for worldwide computerized trade stretches out past program assessment. They ought to advise all angles regarding the program structure, paying little mind to varieties in the model. In light of my survey of worldwide training programs and the writing, just as profound information on the Global Scholars model, I distinguished the general qualities of an investigative model that makes this structure available to other people.

KEYWORDS

- **teacher education**
- **test**
- **examination**
- **measurement**
- **assessment**
- **evaluation**

REFERENCES

Airasian, P.W. *Assessment in the Classroom*. New York: McGraw-Hill, 1996.

Brown, S. and Knight, P. *Assessing Learners in Higher Education*. London: Kogan Page, 1994.

http://www.okhighered.org/state-system/policy-procedures/2011/AA%20Procedures%20Handbook%20December%202011.pdf"www.okhighered.org/state-system/policy-procedures/2011/AA%20Procedures%20Handbook%20December%202011.pdf.

CHAPTER 8

Voice of the Future Classroom: Virtual and Augmented Reality

BIKRAMJIT SARKAR*, DWAIPAYAN SAHA, and AINDRILA DAS

JIS College of Engineering, Kalyani, Nadia, West Bengal, India

**Corresponding author. E-mail: sarkar.bikramjt@gmail.com*

ABSTRACT

Information and communication technology (ICT) plays a vital role in today's education system. Recent research shows that the pedagogies as well as pedagogical reasoning of teachers highly inspire their use of ICT, thereby impacting the achievement of their students. In this chapter, the use of virtual reality and augmented reality in digital pedagogy has been discussed. First, the basic difference between these two has been explained, and then their applications in both face-to-face teachinglearning processes and online mode of education have been illustrated. Along with the advantages of these technologies, the disadvantages are also mentioned. Also, the facilitators and the barriers to adopting these applications are elaborated.

8.1 INTRODUCTION

Experimental research in the domain of virtual reality (VR) and augmented reality (AR) has increased in the recent past, so has the investment and the interest among the stakeholders of the education industry. Indubitably, there are certain areas in the education system that have been empowered by the current state of the applications of VR and AR.

AR technology is preordained to impact every industry and multiple societies. AR applications utilize the collection of software, sensors, and devices to display information on a device simultaneously with the practical world in a sort of hallucination digitally. In AR, the actual world is utilized, and it is finalized with information that is digital. Fundamentally, AR elevates the quantity of detailed taken by a person amidst the surroundings. Essentially, in real time, processing and production are done in everything. Among many differences, the VR is one where artificially created environments are used. Actually, AR and VR are absolutely different technologies that use different R&D paths and practices.

Fundamentally, VR embraces the action for inducing a specific feature in an organism by the use of sensory stimulation that is artificial, unless there is an awareness by the organism, interfering. The computer simulation, which is interactive, conveys information that is sensory to users, and it is perceived by them while being substituted. The definition of VR can be stated, therefore, as an environment that a computer system has designed that imitate an actual scenario.

8.1.1 AUGMENTED REALITY VERSUS VIRTUAL REALITY

"Virtual reality" and "augmented reality" are widely discussed terms these days. Although they sound similar, there are individual differences between these technologies. Table 8.1 illustrates the difference between the AR and VR at a glance.

TABLE 8.1 Augmented Reality Versus Virtual Reality

Augmented Reality		Virtual Reality	
1.	The virtual elements are used by the AR for the enhancement of the actual world that a user experiences.	1.	A computer-generated environment is required by VR to run. Everything that is taken, touched, or interacted by the user is absolutely virtual.
2.	Through the AR, the full connection is maintained with the physical world by the user. Full awareness of their environment is experienced by the user where perceiving, touching, as well as interacting with the actual world can be experienced with the help of all the information provided digitally by the application.	2.	Through VR, users are fully disengaged from the actual world, where the level of involvement of VR is 100%.

3.	AR offers interesting services with the help of media such as cell phones or tablets. Most importantly, AR is only not dedicated for devices having highly demanded Microsoft, HoloLens, or Meta 2, but also for different applications of AR, which are used in mobile phones with all the features.	3.	Powerful processors are needed by VR. Mobile phone processors are also launching new applications, but they come with limitations. The quality substantially differs from Oculus Rift or HTC Vive and other dedicated devices.
4.	Finally, AR is real for 75% and virtual for 25% (general approximations are on the basis of current market applications).	4.	VR is real for 10% and virtual for 90% (general approximations are on the basis of current market applications).

8.2 AUGMENTED REALITY IN EDUCATION

Since the beginning, a lot of development has been done in the field of VR once compared to AR. MetaVisión is such a startup known for the raising of plenty of investment capital. They have launched applications regarding education on the basis of different domain knowledge, such as engineering or health, the verifications of which are still pending (Villarán et al., 2015). There are other domains, like design, where the AR is furnished as the underlying tool to create representations that are digitally generated on the basis of the reality where the manufacturing of the physical elements is supposed to take place. Within the academic processes, the students will be allowed to work using AR, which may increase the creative mindset without the fear of costs and risks of manufacturing (Di Serio et al., 2013). The challenges of having a little experience that remarkably elevated the intricacy to accomplish projects having implications of all kinds could substantially be decreased. With the technology of this type, an image of the ultimate result over the real space can be displayed by a student, with no requirement to finish the process of manufacturing, which is quite physical.

In a similar way, sessions on engineering and health domain (Boletsis & McCallum, 2013) will provide allowance to the instructor to share knowledge among the students by the usage of images having superimposition of the real scenarios in a classroom. The human body modelized in a digital fashion can also be shown in a three-dimensional area, where any type of information can be accessed by a teacher about its elements, where each of the parts is separated in order to exhibit the details, as well as having the interaction between the model and the student to help for the development of any sort of activity. Proceeding with having this initiative mindset, in the engineering domain, a model of an engine made digitally would be with a teacher—circuits that are printed or structure that is architectural (Figure 8.1). By all these models students are allowed to interact, but the social factor would also be taken into account where there would be the sharing of experience with real individuals in real time, that is, their fellow mates (Ibáñez et al., 2014).

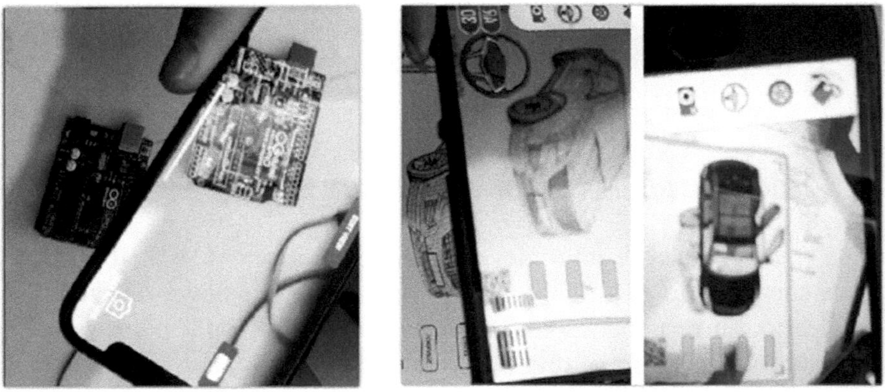

FIGURE 8.1 AR application.

There are numerous examples in the market that frame how powerful this technology is. For example, users are helped by Amikasa (www.amikasa.com) in styling the rooms and evaluating the layout that could be appropriate before procuring the particular piece of furniture, thus making this a great opportunity for students of design. AR Liver Viewer from ISO-FORM (www.iso-form.com/apps/ARLiver/) is another interesting example. This is a real-time 3D tool dedicated to medical education and patient communication, which features anatomical models having high detailing. Another impressive solution is 4D Anatomy (www.4danatomy.com/), where AR is used in the field of health. Aeronautical engineering is such another area where more interesting examples are found. Users are allowed for real flight visualization by HoloFlight (http://www.valorem.com/) through holograms data, which is three dimensional. Finally, HoloStudy (www.holo.study/) is another example, where offers are given in a sequence of lessons in chemistry, physics, geology, and biology. Unequivocally, these applications are established for the disruption of the paradigm of education.

Possibilities are huge there. Recently, identifying the technology suitable for each subject area is a huge challenge. This process is under development and is yet to be illustrated or contrasted. Definitely, the balance in these initial moments will be refined by the applications that are available on the market, but it is necessary for exposing the experiential/experimental results which would provide support to different proposals that the developers make.

TABLE 8.2 List of Some Augmented Reality Systems for Education

Subject Area	Features of the AR Design	Accuracy and Authenticity	Impact on Learning
Art and history education	HMDs	High	Various enhanced exhibits and how the verbal improvement is more useful for the academic purpose than the visual enhancement.
Art and history education	Mobile devices	High	Effect of the AR for motivating the students.
Augmented, virtual and mixed reality	HMDs, interfaces, mobile devices, and tracking types	Not applicable	Development of AR, VR, or mixed reality projects for cultural heritages well as different technologies and their applications.
Chemistry education	Authoring tools and mobile devices	High	Testing of the usability/enjoyment. Students found the system easy to use and really enjoyed it.
Computer education/authoring AR tools	Authoring tools and mobile devices	Not applicable	Assessing an AR system upon toward the effectiveness in education.
Computer education	Not applicable	Not applicable	Attitudes toward the serious games and how the enjoyment impacted the learners' motivation.
General education	Not applicable	Not applicable	Different categories of motivation and how they impact the process/quality of learning.
General education	Not applicable	Not applicable	Gamified learning and how to device it within the classroom.
General education	Not applicable	Not applicable	The impact of motivation on learning.

TABLE 8.2 *(Continued)*

Subject Area	Features of the AR Design	Accuracy and Authenticity	Impact on Learning
General education	Not applicable	Not applicable	Technical apprehensions of the distractions of gameplay or intellectual overload.
General education (various disciplines)	AR learning environments	Not applicable	Binding academic contents to long-term memory.
History education	Not applicable	Not applicable	Historical perspective illustrating with historical figures.
History education	Handheld devices and marker-based AR	High	Historical perspective illustrating with historical figures. Involving the experts to comprehend the consequences of the historical decisions.
History education	Digital tours and handheld devices	High	Sense of a place within the learning environment and the usefulness of an AR system within it.
History education	Mobile devices	High	Not applicable
History education	Mobile devices and object recognition	High	Location-based learning process, historical variances, and changes over the periods of time.
History education	Mobile devices	High	Not applicable
History education	Not applicable	Not applicable	Historical perspective illustrating with historical figures.
History education	Mobile devices	High	Not applicable
History education	HMDs and interactable content	High	Pleasure of interacting with an AR-based museum exhibit.
History education	HMDs	High	The comfort of the users when wearing a HoloLens HMD and their ability to intermingle with the control set.
Holocaust/history education	Not applicable	Not applicable	Historical empathy.

TABLE 8.2 *(Continued)*

Subject Area	Features of the AR Design	Accuracy and Authenticity	Impact on Learning
Holocaust/history education	Not applicable	High	Comprehensive virtual replication of life within the Auschwitz Holocaust Camp and historical empathy.
Holocaust/history education	Not applicable	Low	An accurate depiction of a demolished Holocaust site, design based on historical evidence.
Holocaust/history education	Mixed reality and natural language processing	High	Interactions of natural languages and historical empathy.
Holocaust/history education	Mobile devices, GPS-based markerless augmented reality	High	Historical understanding, a sense of a place and the location-based learning process.
Holocaust/history education	Not applicable (a project proposed for markerless AR)	Not applicable	Location-based learning, historical empathy.
Holocaust/history education	Mobile devices and object recognition	High	History of Holocaust victims.
Mathematics education	Not applicable	Not applicable	Educational autonomy.
Mathematics, language, arts and scientific literacy	GPS-based markerless AR and handheld devices	Not applicable	Engagement and motivation of students. Hazards of intellectual overload.
Medical education	Projected data of computerized tomography and simulated training environments	High	Virtual 3D training environments and simulated saneness.
Multimedia learning theory	Not applicable	Not applicable	When correctly applied, the motivation of the students increased.

TABLE 8.2 *(Continued)*

Subject Area	Features of the AR Design	Accuracy and Authenticity	Impact on Learning
Pathology/medical education	Handheld devices and markerless AR	High	Location-based study and its impact on learning.
Science education	Object recognition	High	Development of the intellectual behavior and its impact on the users' motivation.
Storytelling/physics education	Augmented books and mobile devices	High	Students were able to retain more of the learning contents when taught with the help of AR than with the traditional means.

Source: Challenor and Ma (2019).

Notes. AR, augmented reality, VR, virtual reality.

8.3 VIRTUAL REALITY IN EDUCATION

A different aspect in education is found by VR, which displays the full potentiality of it. Every element in education, which this technology brings through, is not only the value but also it becomes an extremely differential value (Kumar, 2017). The approaches that are in the learning stage and greatly impact the current educational systems confront with the students with a real-life scenario that needs to be solved by means of the knowledge in theory that has been acquired or by enhancing capacities of the students until the moments become underdeveloped or nonexistent. Till now, the description of the circumstances has been using text, or there may be cases where audio with or without a video is used. Using the technology of VR, keeping in mind, on which the student can act, the programming of a particular situation can be done using environments and several variables. Applications are completely customizable subject-wise, in the knowledgeable area, in the segment of the population, or in the field of geography. Transferring the message among all the students will be possible, with the case matching the messages as described (Falloon, 2010). With these technologies, democracy will be maintained while accessing knowledge. The students, struggling to attain some goals while learning with a poor success rate, will now be able to attain the goals with a higher rate of success. There is a logic behind this claim. There will be a contribution by these technologies to make many tangible concepts that are abstract such that it should be built by these students within their minds. As, not every student is blessed with these skills, this exercise will be supported by these technologies, thus improving the success rate.

Another important area where a greater substantial value in the representation of the abstract notions is provided by the VR (Curcio et al., 2016), for example, the applications which enable the representation functions of complex mathematics in space: allowance that is provided by the solutions using the resources that are digitalized for the representation of works that are artistic in any one of the branches which are able to define them (painting, music, or sculpting), and products were moving across the structures that are architectural and are allowed to facilitate for accessing all the layers which help in structural composition (wiring, conduits, or every other type of existing material).

A significant opportunity is offered as well by VR in the simulating area. This technology simulated laboratories where students can be able to interact with the devices (Figure 8.2) (Hoffmann et al., 2014). It is evident that measuring devices that could be updated on the environment of a new version is definitely a benefit. It will be a great opportunity for the students to work with cutting-edge technology without involving the physical elements, which is of a higher investment to the institutions. Analyzing, even more, conservation of the cost in spaces would be noticeably big. The spaces that are underutilized within the centers would have a significant reduction. Replacement would be done by "multilaboratory" rooms depending on the subject, and access could be thereof one or another laboratory (Lindgren et al., 2016). The availability of these products is noticed on the market. These products are developed depending on the promises made for them by the traditional education sector.

FIGURE 8.2 VR application.

The applications having the potentiality in the field of education spread as much as imagination or ability is allowed for adapting materials in the traditional format. In today's market, there are numerous examples like ThingLink (www.thinglink.com/) which have a bunch of videos and pictures that are much interactive and cover many topics such as language, science, and arts. University (unimersiv.com) is another choice having a bunch of VR apps related to education, which includes three experiences of education at launch:

1. International Space Station discovery,
2. Anatomy of a human, and
3. Traveling 5000 years ago to places like Stonehenge in Wiltshire, England.

From the university, MoleculE VR, yet another choice, is an app on VR, which introduced some basic concepts on how communication can be done in a cell. Another one is InMind (luden.io/inmind/), which is a VR-based scientific game. InMind allows a player to experience a journey into the patient's brain for searching the neurons, which are the cause of the mental disorder. Through VR, thus, we can get a variety of applications in the field of education.

The actual problem falls in the sector of education itself. It is such an area where it is usual to have institutions individually consisting of small geography, whereas the manufacturers who are actually in need of these technologies have a scope of business that is globally geographic. Global partners are needed in the academic field with the help of which they can try solution testing before products are launched into an international market. Sadly, the options are very less. This unequivocally harms the development, even though it is not eliminated; it just decelerates it. When this opportunity is given, small companies are established by integrating solutions of technology from big manufacturing companies and sign agreements with an applicable number of small institutions to be flexible enough, which is a large multinational lack. It is already on the verge of an outstanding result. It is necessary for emphasizing the fact that it is a great opportunity to explore when there is an agreement between a global educational partner and a large manufacturer. The following table shows some of the advantages and disadvantages of certain existing VR technologies [27]:

TABLE 8.3 Advantages and Disadvantages of a Few Existing VR Systems

Technology	Advantages	Disadvantages
Google Cardboard	• The weight is almost like a standard smartphone. • Very much portable. • Very cheap. The price is around ₹500.	• Dependent on smartphones. • Very low battery life. • Field of view is ~90°. • No positional tracking is available.
HTC Vive Pro	• Resolution is 1440 × 1600 pixels per eye (2880 × 1600 pixels combined) • Field of view is around 110. • External virtual controllers available. • Tracking area is quite reasonable (10 m × 10 m).	• Weight is a bit higher (~550 g). • Costlier (~₹90,000) • Accessories have to be purchased separately (~₹20,000). • Setup time is reasonably higher.
Oculus Quest	• The resolution is 1440 × 1600 pixels per eye. • External virtual controllers are available. • Very much portable.	• Weight is ~600 g. • Price is reasonably higher (~₹30,000). • The field of view is ~90°.
Samsung Gear VR	• The resolution is 1480 × 1440 pixels per eye. • Very much portable. • Weight around 350 g. • Price is comparatively reasonable (~₹8000).	• The field of view is ~100°. • Dependent on smartphones. • Very low battery life. • No positional tracking is available.

8.3.1 ADVANTAGES AND DISADVANTAGES

An investigation is going on regarding the benefits of VR. They should, without a doubt, focus on the learning outcomes of students who are improving. In recent years, the first experiments introduced the first experience, which is virtual in traditional teaching. Contributions are important for understanding the concepts that are abstract, and in training as well in actual situations and environments. Students are allowed to have learning experiences that use all of their senses in made-to-measure environments by the immersion. As trivial may it seem, but experience like such becomes crucial while understanding the concepts which an individual is unaware of without question (Huang et al., 2010). Spatially representing a complex mathematical function or navigation to the nooks and corner of the human body are some examples of the applications which are the early-stage success story of the application of VR in the field of education. If we consider going back to the period when VR was originally invented, there is an evolution to the phrase "a picture is worth a thousand words" to a level where experience has become something that is beyond the image. Rephrasing as "experience is worth more than a thousand images." So, the prime advantage of VR is to utilize it as a piece of equipment for improving the approach to understanding the concepts that are complex or abstract (Hwang & Hu, 2013).

On the contrary to this, using this technology has a set of risks as well as negative features in the field of education. Using this sort of application in an abusive way can call for the isolation of individuals from their peers. The use of VR can also be in environments that are collaborative, but this is absolutely true that currently they have become quite common than those where there is individual participation. Human DNA is planted with personal interaction, and with such interaction imitation, exchange of experiences and collaboration is used for learning.

The actual risk of VR within the field of education is on the misinterpretation of the technology as the end rather than a means to the end, improving the outcomes of the students in the process of learning.

8.4 LANDSCAPE OF CURRENT TECHNOLOGY

Recently, the division of the state of the art of both the technologies, that is, VR and the AR, is clearly between the prototypes that the major advertising campaigns have announced and the major manufacturers which launched the devices that look forward to capturing the first prescribers.

Regarding VR, there has been progress from the devices that only provided the information to allow transfer in one sense, that is, those that only served for observation. Nowadays, the version of technology is in advancement in the area of work, which has been intense in incrementing the interaction between environments and the user. This gives a provision of the addition of value that is very much important, since it gives empowerment to a person for feeling like he/she is acting in the situation, rather than just being a trifle spectator. For using this technology, the packages created are formed mainly by audio device and visualizing, to which there are the addition of elements use to enable the using hands, thus giving the allowance for interacting with its surroundings. Mainly two types of visualization devices are found: Oculus Rift and HTC Vive. These devices cost over ₹40,000 and also need a computer having a capacity of a high processor, which brings the complete set of devices that are necessary for the enjoyment of experience through VR to exceed ₹25,0000. However, it is not to be forgotten that the device that Sony offered for exclusive use on its PlayStation platform sold out in the 2016 Christmas campaign for less than ₹30,000 per device.

In contrast, there are devices having less price, which use the processing capacity of smartphones. Different features are there, in mere housings, which are dependent on their price that costs around ₹5000 per device. Wearable glasses and headphones are provided in maximum cases. A small track-pad has been added by some manufacturers, like Samsung, where more interaction is allowed between the user and the environment. Headphones having high sound quality have been added by others. So far, supplementary devices have not been provided by these manufacturers, which have an increment in the interaction, as in the case of dedicated devices. Supplementary devices will be available very soon, with the smartphones having increased processing capacity.

The manufacturer supplied the assemblies that are most common, made up of elements, and also less compatible with other suppliers. In the last year, launching peripherals began in the smaller companies by the offer that is compatible with different packages that the renowned brands marketed. Possibilities for the developers are increased, and prices are lowered with a hike in competition. This offers the users with more options and the development in the capabilities as well. A startup created the haptic gloves, which are a clear case of this type of element (Arnab et al., 2011) in 2015.

Another important feature is the physical mobility of the actors through VR. Until very recently, the connection of the sound devices and display was required to be done with the processor through cables. This made the interaction limited by eliminating comfort from the experience. Currently, the main prototypes are already presented by the manufacturers publicly that subtracted this wiring, making the connections wireless. Mobility is gained by the user, and with it, interaction between the user and the environment is enabled, which is programmed.

In relation to AR, the description can be given of two very clear states of technology. At the start, the first applications that deployed this technology focused on interpreting the symbols or codes and the substitution in visualizing the system by the elements programmed beforehand. The example is the applications that give users the allowance to superimpose on the furniture spaces of a catalog to guess how they would be looking like if they were installed. Another instance which has been cited.that has supposed a huge amount of income for the company that it is put on the market is based on the game Pokémon animations. Using the camera that is in the wide variety of smartphones, animated characters are superimposed by it above the neighboring reality. On the part of its producers, the impact on the market has surpassed any optimistic expectations. They are, therefore, simple applications that overlay the images that are digitally created on images taken in reality with both the images that are exposed on a mobile screen, tablet, or any other similar type of device.

AR glasses are in another order of possibilities. Although there is an increase in the offerings every year, it is the Microsoft HoloLens device that has opened the market, making applications that are physical and tangible, which, until now, were just present in cinematic fiction. The real environment is enabled to map around this gadget and superimpose the viewers with information, getting to perceive the real images by the human eye and those digitally included as the elements of the same reality. Some of the most noteworthy examples have been the design of the interiors and its recreation, and calls between the individuals, having both parties placed in the same area. The effect that is received is that in each of the interlocutors, there is a digital image with fully synchronized size and movements with the other end of the communication, having a simulation of a hologram of the second individual.

The concept of AR has been developed within the last few years in order to reform the learning experiences and the interactions inside the classrooms. Unlike VR, AR provides matchless flexibility owing to the fact that it does not obscure the field of users' vision but to add some more relevant information.

8.4.1 FACILITATORS AND BARRIERS TO ADOPTION

Students have been an integral part of the education sector who have adopted this technology unquestionably. It is a trend which, by and large, is reaching the zenith with a massive multichannel promotion. The entire technology sector has shown a commendable character by pioneering projects, applications, and development platforms. Individual computers or very popular mobile phones have manifolded access prospects. This, undoubtedly, has been one of the great successes in its endeavor. Due to technological standardization, makers have suggested that openings not to have to make supplementary investments. A smartphone has the potential to act as a display device. This attains predictions of the universal access from the high-tech perspective very rapidly (Kerawalla et al., 2006).

The veteran individuals have undoubtedly gambled on a pursuit which focuses on resources rather than gadgets. It is anticipated that huge profits will be made in the field of software, whereas hardware will have small profit, courtesy to the users with advanced needs such as professionals and gamers. Undeniably, this transformation in the education sector is worth praising. The present devices in the institutions are existing digital natives with last-generation gadgets used regularly. The contribution of educational centers will be minimal in the future, as the users will already have their respective and customized smartphones and devices (De Souza e Silva, 2006).

Undoubtedly, the institution and the faculty are the next most important parts of this chain. They are simultaneously the torchbearers and the main barrier. Teachers are untrained in this field of use of technology within their teaching function. Thus, this is one of the obstacles, but not least in this context.

In addition to the first barrier, there are two essential parts of the training of teachers. Technological training is the first key: knowing about VR and AR and its potential and possibilities. The second key, which is equally important, is to know the use within the course of teaching. The obstacle is appropriate as the utilization of this technology as a medium for teaching is not clear to a teacher who wants to use it to the full potential. And humans are reluctant to embrace any change. Being a teacher, one should have the inclination to learn and initiate new tools and methodologies. VR inherited the use and ability to provide the services for interaction but has a lot of complexity while adapting existing curricula and improving them with the help of the gigantic scopes that are present for office workers (Cheng & Tsi, 2013). The stipulation still persists, and the hindrance is there on the bidders.

Next, another hurdle is the unavailability of proper content under this technology. There are several pieces of training proposed by veteran companies who are involved in this development. The ultimate blockade is the content of the training, which is not relevant to the curriculum but is grounded on the acquaintances obtained, which are likely to be interesting to the learners. Now, a couple of options are available: either compatible or nonexclusive. These developments can be used for some segments in the current matter and using them as elements for teaching and modifying the material related to the experience. Parallely, experiences can be modified according to the subject recognized by the regulators. Both of the ideas are important and must complement each other.

The procedure suggested to espouse these technologies as the basic elements with the existing education has six different steps.

The initial juncture is to train educators on these latest technologies. Owing to their freshness, the major portion of the faculty members does not have an adequate understanding of the technologies, and thus, they are unaware of their prospective use linked with teaching. The process will have two distinguishable portions. One is to make the faculty familiar with the basic and essential concepts that make the base of the technology, such as the mobility, ambiance, devices, existing platforms, modes of interaction, and the course that will establish the massive transformations in the technology in the coming future. Simultaneously, it is necessary to complete the initial phase by training using the existing samples. Various solutions will be expected, depending on the need and the ambiance required. Another prime requirement is to know the various ways of communicating with the main stakeholders with proper needs like tools and their own hands, which can perform tasks such as painting, running, holding objects, and catching them. Solutions in different fields of knowledge are also important, such as anatomical atlases, mathematical notations, and mechanical workshops.

After completion of the initial phase, faculties will be able to perceive the usefulness of technology as applied in this sector. This knowledge will enable the teachers to realize and visualize the first applications they will need to inculcate in the courses. This will strengthen the students' understanding in areas in the subject with the proper use of various technologies. This will thus focus on making the subject matter of the course better with experience. Teachers should remember that the available or the upcoming technologies are a means to increase experiences rather than mediums of online or face-to-face communications.

The circumstances that the faculty members want to use in the course of study have to be the first conceptual prototype and the beginning of the second phase. The prototype might need some advancements and suggestions from existing solutions or the direct use of the marketable products. The results will contain both the methodology required to be initiated within the course, the purpose of learning is attained, and the experience gained, and also the functional needs of the as acquainted while using the software.

For this, a detailed study of the composition of the subject is required by the teacher. The integral parts will be contents, activities, and the usability of the fresh technologies readily (the circumstances, communications, ideas, time frames, ample number of collaborators within the experience, and other fundamentals that should be considered with great importance).

Ultimately, for the evolution of this work, it is desirable not to exceed 20 days so that everything is done along with the training of the first block of work. Moreover, this being a new technology and not available to everyone, and there is a chance that important concepts might be forgotten, and thus losing potential in attaining the experience.

The third phase contains activities that require teamwork. The groups will be made involving the subject matter experts. The team also has to have a technical programmer expert who can use the technology to the fullest and challenge the technology with ideas, and eventually an educational architect. This architect will suggest and recommend a teacher a number of better practices for improving the quality of experience gained. Moreover, he/she will be efficient and will be a sound grasp in the field of technology that will help in making the prototype prepared by academic staff, having the real potential of the suitable technology.

This team will obtain the result, which will give the requirements that will be needed for the production if required. Such types of processes can be executed either within the institution internally or through an external vendor. Considering the freshness of technology, the quality of the processes must be meticulous. This is because the new implementation of the technology will mark the success for both the faculty members and the students.

The fourth stage requires the production of the experience, and this will provide the results in two parts. Firstly, a small-scale testing is done within a controlled environment, which will be judged by the students. In this phase, debugging of the program is done, and also some details are incorporated to give an edge to the concept proposed by the teacher. In the second phase, a completely functional version is showcased, which can be used as an example for which it was created within the subject. This new version has to be modified further based on the results and feedback obtained from the teachers and students.

The experience already gathered marks the beginning of the fifth stage. Now teachers have to be taught to implement these solutions in their teaching methodology. After this, customized courses will be created depending on the requirement of this new technology. The training through which the teachers have to be taught should either be done through face-to-face and/or online classes. This training will we based on the basic facilities and aids available in the course as an introduction. Once the elementary knowledge imparted, the next part of the training will be imparted with pedagogical elements like the creation of rubrics, knowing about the learning objectives, instances of group problem, and defining the real problems which will be solved using virtual situations.

Ultimately, the experience gained is now implemented in regular subjects with the students, and this is the sixth and the final stage. Students have to be explained the refinement objective brought through the medium of these latest technologies. It is also elementary to have feedback for the fine-tuning and upgradation of the experience in questions. At this initial point in time, the technical as well as pedagogical support for both the faculty members and the students is very important. The initial applications will determine the success of these new techniques. Irrespective of the solutions, the ambiance around the student experience must be ideal and faultless. The amount of advancement in the outcome of the student will help to continue such innovations in the education sector.

8.4.2 AUGMENTED AND VIRTUAL REALITY IN FACE-TO-FACE EDUCATION

The gradual expansion in the field of technologies related to VR and AR will unquestionably attract the entertainment world. Engines and devices of interaction with higher efficiency will be devised for processing and amplifying the probable human touch. Products involving touch are already present in the market (Inoue et al., 2015). Areas like smell and taste are yet to be explored (Gatto, 2011). Thus, the developing technologies will always be ahead of the readiness of the application.

With regard to the academic field, two distinct characteristics are defined to make the most out of technologies. Firstly, the scope of face-to-face teaching will be present. From the theoretical standpoint, it aims at the upgradation of simulators that creates a scope to clone factual situations within a particular environment. The main agenda is to make the students feel as if the conditions are real. In the near future, we will have platforms that will authorize the user to modify the environment quickly, and it would not require any development from scratch once launched. These, when in practice in real life, would work better in a virtual way. Thus, improving the experience of the students, curtailing the cost and having greater updates.

Simultaneously, AR is connected. In the future, we are going to share images digitally within a face-to-face session in a class, where areas such as architecture, mechanics, or healthcare require such initiatives (Scholz & Smith, 2016). This will enable students to communicate in real time, and not only can they see but can also act actively in the class. An instance is the use of a motor or a human body, and unlike a theoretical class, a student will be able to choose and displace each element of the model as and when required.

They will not only see the model but can also get detailed information about it in a better way.

In the future, technology will have more senses to enhance the experience obtained. Recently, the touch sensitivity is being developed with the aid of intensity. Moreover, the newest experimental developments are concentrating on detecting a point and fixing it virtually using a display device, using this as a reference to interact with nature. Another milestone will be the increase in subject portability within the surroundings. Thus, this will require a lesser use of cables, and increase in the development of platforms that allow the user and give the edge of mobility, and move without moving from his or her present place, and development of interaction peripherals suitable to human biomechanics to enhance and make the movements with ease. And all these work are presently under process in real time (Soo, 2016).

These innovations aiming to expand the opportunities given to a student will require a better processing engine with more power. The academic centers will make investments that will be greater than investments done domestically; thus, students will be able to avail of the opportunities in universities and colleges. Greater and better processing means better and more powerful engines, thus requiring more funds and investments.

8.4.3 AUGMENTED AND VIRTUAL REALITY IN ONLINE EDUCATION

Without an iota of doubt, the online area will be benefitted and will grow rapidly using these technologies. The recent trends of the market show evidence that the distance learning is growing in double digit with respect to other markets. This leads to more users, and subsequently, an increase in the supply is demanded in the light of a potent business. An increase in competition will expand the quality of commodities to seize greater shares of the market, as a reduction in the price has an ill effect, which results in disgrace of the title to be obtained. If the commodities are too inexpensive in this sector of education, they are not able to fulfill their purpose any longer.

It is said that the existing online classes have been explored to the limit of services available through the present platforms (Meggs et al., 2011): incorporation of social networks, joint learning with the help of conversations through audio and video support, virtual tutorials will be very helpful in the systems suited for videoconferencing, links to the forums, multichannel access to global databases of information, etc. These platforms are exhausted. So, its time to upgrade the available present experience within the teaching field, like a student who has the facilities and reaches to texts, videos, or teamwork. Unquestionable, VR and AR will change the voice of the academic sector.

Technically, a quantum leap is taken with the creation of technology that is capable of functioning and penetrating socially 95% higher than a smartphone. The quality of processors and their screen has increased in leaps and bounds. Now, individuals have the processing powers in their pockets, which they earlier had in a high-end PC. But there is no partner who is willing to be the first bet in this endeavor. Money and investment are not just the only problems, but making this a profitable reality is also a challenge. Moreover, the assimilation of different students and participants and making them come together and executing the methodologies properly is difficult now.

Technically, it is predicted that in the next five years, each and every person will have a personalized high-performance display device such as Oculus Rift, HTC Vive, and PlayStation VR at home like each one has a mobile phone, desktop, or laptop. The price is nearly equal and is available for the common man. Any person will be able to purchase this device approximately in a price range of between ₹1500 and ₹5000.

Conceptually, the future VR and AR through online channels will be a quantum jump but will surely happen as we will introduce real experiences using such tools and technologies. This will definitely make distance learning more effective and detailed, and enriched with the inclusion of face-to-face teaching and enjoying the privileges and flexibilities of distance learning. This will make students feel as if they are in a class; the entire ambiance will be of a perfect ideal traditional classroom (Merry, 2016). We can visualize using this that our classrooms will move to any point in this universe without having that feeling of loneliness and isolation, but the feel of a real classroom persists. A lot of benefits are offered when these technologies are used. Without any time constraints, classes can be conducted from home, along with the benefits of having experiences, equipment, companions, and above all, teachers that give the students the same impact as the face-to-face teaching–learning process.

8.4.4 ADVANTAGES OF VIRTUAL REALITY IN EDUCATION AND AUGMENTED REALITY IN THE LEARNING PROCESS

8.4.4.1 DIVERSIFIED TEACHING OPPORTUNITIES

One of the concerns of the present system of education is the restricted potential to recount and epitomize the difficult concepts and discuss them in a way that will better in terms of being practical and easily approachable. The process in which topics are explained presently is like a hit-and-miss scenario, where some students are able to grasp that, whereas some are unable to understand the subject. This problem is comparatively less complex for humanitarian science than physical science, chemical science, and biological science.

The engaging audio-visual effect makes it easier and more convenient to explain the complicated ideas using AR and VR educational apps.

There are projects like augmented education that are taking this technology to new heights. This aid is conceptualized for modeling, presentations, and 3D designs. This is comparatively easier when in use and is better than the present AR tools. It helps the educators and the pupils to try and experiment with constructing different objects and animating them via resourceful libraries of objects, shapes, and actions.

8.4.4.2 INCLUSIVITY: A LIFELINE FOR SPECIAL PEOPLE

Students with various disabilities, and those who are special, are greatly affected and can be included in the class with the implantation of VR, and AR in this is indeed an additional benefit.

For instance, students with visual impairments are highly benefited when Near Sighted VR Augmented Aid is used. There are other applications too, which can add audio effects and commentary, and is also capable of controlling and changing the contrasts and font size. Thus, ensuring the students having impairments and disabilities is also a part of our society.

For the students with impairments, VR can be an ideal ambiance. This not only gives the students the opportunity to experience situations that were considered out of their reach but also provides a practical touch to the process.

For instance, in a VR environment, SignAloud provides the opportunity to convert a sign language into human speech, thus making communication possible.

8.4.4.3 LESSER DIVERSIONS AT VR SCHOOL

One of the challenges that a teacher faces in a classroom is to constantly maintain the attention of students and handling a lot of possible distractions. With the emergence of smartphones, this problem has been aided. However, fighting against technology is not at all required; instead, we should embrace the endless possibilities of technology.

Well-arranged VR experience can purely confine the student's attention inside the classroom and prevent it from going outside. Thus, increasing the concentration level of the students.

With the possibility to interact with the ambiance and environment, things get only better due to immersion. Hence, the probability of development of the VR school environment is very high in the near future.

8.4.4.4 ENHANCED ENGAGEMENT/IMMERSION THROUGH INTERACTIVITY

Honestly, a lot of students think that education is a mundane chore, and teachers have to face this problem of lack of interest of students. This problem is a recurrent one and can certainly never be solved completely. But VR and AR can add interactivity to this problem.

Immersive VR education is very exciting and better than the practice of sitting in a designated area called classroom and listening to the teacher's lecture, which is moderately interesting. Moreover, it aids the pupils to concentrate in a better way on the subject and also makes the students a part of the class. Students find their interests as they are able to do certain interesting and creative things in the class.

VR and AR are so attractive because students can interact with the help of these. The present scenario of learning involves more listening and taking notes. There is a significant gap between practicality and theoretical studies. A lot of imaginations are involved in this. Thus, this is not an appealing experience and quite boring, and one becomes careless about the subject.

VR provides a chance to experience the topics and subjects with a better approach. As a matter of fact, this kind of environment is more appealing and is more effective in delivering knowledge that a conceptual class.

For instance, one can experience the feel of a soldier who was there alongside Napoleon in the Battle of Waterloo or can experience the livelihood of an Egyptian farmer. One can virtually tour numerous countries, rather than wondering about a place and locating them on a map with a small dot. The entire experience will be something new and different and something really very special.

When AR and VR are compared, the VR is capable of giving an experience which will be wholesome. On the other hand, AR can make the educational procedure better and can take up a more supplementary role. Thus, both are very useful.

8.4.5 PROBLEMS WITH AR AND VR

8.4.5.1 THE DOUBT ABOUT COST-EFFECTIVENESS

We know that the education is generally a matter which is funded by the government, and the question which arises now is: How economical will be the implementation of AR and VR technologies (considering that only school education is not taken into consideration but also professional training)?

VR can be very helpful in the process of learning, but certain things are there that have to be kept in mind while calculating the cost:

- Full investments have to be done in the making of the gear to avail the VR solution (glasses or helmets along with the controllers). Fortunately, some good options for the budget are available (e.g., Google Cardboard). A business analyst will be helpful in figuring out what the gear is capable of doing, and then it is up to the user and the investors.
- Another investment has to be done in the development of the content of AR/VR. There are certain blocks that can be arranged according to the need, while there is some special training that requires starting from scratch.
- Another important investment is a time investment, along with money; teachers have to be trained and made completely prepared to teach their students.

When compared with VR, AR solutions are cheaper since gears are not required. The students only need a smartphone. But the content for the apps needs to be developed.

Eventually, the implementation of AR/VR will be very helpful and pay dividends if implemented properly in the form of having some extremely trained professionals who will be ready to teach and guide, and the other benefits are known to us. It will have a tremendous impact on both the system when technology is considered and also for the students who will be able to develop new operating models.

8.4.5.2 THE DOUBT ABOUT CONTENT DEVELOPMENT

One of the biggest hindrances in the way of large-scale implementation of AR and VR technologies is the educational process. AR and VR have to be very practical, so a lot of effort from both the developers' and educators' sides will be crucial, which will eventually make the quality of AR/VR content relevant and will cover all the sectors of the educational programs. There are several reasons for the content being a challenge:

- The basic tools which are present in the present market are not enough for the AR/VR content. They can supplement but not substitute for the present system.
- Teachers cannot improve and improvise on their own, so they need external help or hire someone for support and help.
- The content has to be adjusted, so that the audience gives feedback

Several AR solutions are present like human anatomy atlas, and Sky Map on the market, and the user has to just modify them according to the needs. These are sufficient to understand the areas that are lacking and need further development. Everything will be on the discretion of the user and the board of investors

8.4.5.3 THE QUESTION OF HEALTH

The impact of VR on the mental and physical health of the user has not been studied yet with emphasis. There are two reasons for this: it is either because the technology is new, and on the other hand, there has not been any widespread use of this technology for the development of research.

Many ask, "How can one possibly trust a technology when there has been no research done yet whether they are affecting the physical and mental health or not?" This is a grave issue when adults are considered, and the case worsens when the children are taken into consideration.

To this, the case of Pokémon hunters arise who were engrossed into the game so much, courtesy AR/VR advancements, that there were oblivious to their surroundings and dangers like the cars, roads, and holes. Johnny English addresses this issue perfectly.

KEYWORDS

- **digital**
- **education**
- **virtual reality**
- **augmented reality**

REFERENCES

Arnab, S., Petridis, P., Dunwell, I., and de Freitas, S. "Enhancing learning in distributed virtual worlds through touch: a browser-based architecture for haptic interaction." In *Serious Games and Edutainment Applications,* pp. 149–167, Springer, London, 2011.

Arnab, S., Petridis, P., Dunwell, I., and de Freitas, S. Haptic rendering of virtual shapes with the Novint Falcon in Serious Games and Edutainment Applications, pp. 149–167, 2011.

Blumenfeld, P.C., Soloway, E., Marx, R.W., Krajcik, J.S., Guzdial, M., and Palincsar, A. "Motivating project-based learning: Sustaining the doing, supporting the learning." *Educ. Psychol.* 26 (1991): 369–398.

De Souza e Silva, A. "From cyber to hybrid: Mobile technologies as interfaces of hybrid spaces." *Space and Culture*, 9(3) (2006): 261–278.

Dewey, J. *Experience and Education*, Simon and Schuster: New York, NY, USA, 2007.

Falloon, G., Janson, A., and Janson, R. "Learning objects and the development of students' key competencies: A New Zealand school experience." *Australasian Journal of Educational Technology*, 26(5) (2010).

Gatto, K. A Virtual Reality Scent System that Fools Human Taste, 2011 [Web log post].

Graham, C.R. Blended learning systems. In *The Handbook of Blended Learning*; John Wiley and Sons: Hoboken, NJ, USA, 2006, pp. 3–21.

Halverson, L.R., Spring, K.J., Huyett, S., Henrie, C.R., and Graham, C.R. "Blended learning research in higher education and K-12 settings." In *Learning, Design, and Technology: An International Compendium of Theory, Research, Practice, and Policy*, Springer: Cham, Switzerland, 2017, pp. 1–30.

Heitink, M. C., Van der Kleij, F. M., Veldkamp, B. P., Schildkamp, K., and Kippers, W. B. "A systematic review of prerequisites for implementing assessment for learning in classroom practice." *Educational Research Review*, 17 (2016): 50–62.

Hoffmann, M., Schuster, K., Schilberg, D., and Jeschke, S. Bridging the gap between students and laboratory experiments. In *International Conference on Virtual, Augmented and Mixed Reality* (pp. 39–50), Springer, Cham, 2014.

Huang, H. M., Rauch, U., and Liaw, S. S. "Investigating learners' attitudes toward virtual reality learning environments: Based on a constructivist approach." *Computers & Education*, 55(3) (2010): 1171–1182.

Hwang, W. Y., and Hu, S. S. "Analysis of peer learning behaviors using multiple representations in virtual reality and their impacts on geometry problem solving." *Computers & Education*, 62 (2013): 308–319.

Kerawalla, L., Luckin, R., Seljeflot, S., and Woolard, A. "Making it real": exploring the potential of augmented reality for teaching primary school science. *Virtual Reality*, 10(3–4), (2006): 163–174.

Kumar, C.. A new frontier: How can you profit from augmented and virtual reality. Business. com, 2017.

Lindgren, R., Tscholl, M., Wang, S., and Johnson, E. "Enhancing learning and engagement through embodied interaction within a mixed reality simulation." *Computers & Education*, 95 (2016): 174–187.

Magdalene, R., and Sridharan, D. "Powering e-learning through technology: an overview of recent trends in educational technologies." *Online J. Distance Educ. e-Learn.* 6 (2018): 60.

Meggs, S. M., Greer, A. G., and Collins, S. (2011). "Integrating second life as a pedagogical tool for interactive instruction." *Journal of Online Learning and Teaching*, 7(3), (2011): 380–392.

Miyajima, Y., Sako, Y., Terauchi, T., Inoue, M., Shirai, K., Takai, M. et al. U.S. Patent No. 7,731,656. Washington, DC: U.S. Patent and Trademark Office, 2010.

Richardson, J., and Swan, K. "Examing social presence in online courses in relation to students' perceived learning and satisfaction." *J. Asynchronous Learn. Netw.* 7 (2003): 68–88.

Scholz, J., and Smith, A. N. "Augmented reality: Designing immersive experiences that maximize consumer engagement." *Business Horizons*, 59(2) (2016): 149–161.

Schwarz, S., Aminlou, A., Curcio, I. D. D., Hannuksela, M. M., Moreshini, S., Gama, F. et al. "Tampere pole vaulting sequence for virtual reality video coding." Joint Video Exploration Team (JVET) of ITU-T SG, 16, 2016.

Singh, H. "Building effective blended learning programs." *Educ. Technol.* 43 (2003): 51–54.

Stockwell, B.R., Stockwell, M.S., Cennamo, M., and Jiang, E. "Blended learning improves science education." *Cell* 162 (2015): 933–936.

Villarán, D., Ibáñez, M.B. and Kloos, C.D. Augmented reality-based simulations embedded in problem based learning courses. In *Design for Teaching and Learning in a Networked World.* Springer, Cham, 2015: 540–543

Wittich, C.M., Agrawal, A., Cook, D.A., Halvorsen, A.J., Mandrekar, J.N., Chaudhry, S., et al. "E-learning in graduate medical education: Survey of residency program directors." *BMC Med. Educ.* 17 (2017): 114.

Zheng, S., Rosson, M.B., Shih, P.C., and Carroll, J.M. Understanding student motivation, behaviors and perceptions in MOOCs. In *Proceedings of the 18th ACM Conference on Computer Supported Cooperative Work and Social Computing*, Vancouver, Canada, 14–18 March 2015, pp. 1882–1895.

CHAPTER 9

Using Digital Technology for Evaluating the Quality of Preschools in Vietnam

PHAM QUANG TIEP[1*], DO THI HUONG NHU[2], and NGUYEN THI HUONG[3]

[1]*Faculty of Pedagogy, University of Education, Vietnam National University, Hanoi, Vietnam*

[2]*School of Business and Management, RMIT University Vietnam, Hanoi, Vietnam*

[3]*Faculty of Preschool Education, Hanoi Pedagogical University 2, Vinh Phuc, Vietnam*

[*]*Corresponding author. E-mail: tiep@vnu.edu.vn*

ABSTRACT

Vietnam's preschool education has made great progress toward modernization and standardization. Vietnam has a set of criteria for evaluating Vietnamese preschools. However, it needs improvement to be closer to international evaluation standards. Moreover, it is necessary to take advantage of the current digital age to bring that set of criteria to a wider audience, including parents and schools, helping them to evaluate and improve school quality. We conducted research to solve this challenge in evaluating preschool education quality in Vietnam today.

9.1 INTRODUCTION

In recent years, the quality assessment of Vietnamese educational and, in particular, early years learning institutions in the educational system has received great attention from educational management levels. Its purposes are, primarily, monitoring and managing the quality of education, and secondarily, ranking these institutions based on national standards. In general, the current evaluation process has created a positive effect, pushing early childhood education institutions to change and improve to become better and achieve higher standards to meet regulatory requirements.

However, the quality assessment of Vietnamese early childhood education institutions still revealed many limitations in both scope (what needs to be assessed?) and method (how the assessment should be conducted?). Especially, it has not opened opportunities for early childhood education institutions to make any substantial improvement in terms of quality. The evaluation process and procedures are formal, bureaucratic, and for administrative purposes, rather than focusing on improving education quality. The standards and assessment criteria are still too general and vague for users, especially for beneficiaries of educational services and early childhood educators. The current set of quality criteria does not allow the quantification of the level achieved for each indicator (Pham, 2017a).

The research questions we posed here are as follows:

1. With regard to the *scope of evaluation*: How the current criteria and standards for evaluating early years learning institutions in Vietnam should be improved not only to measure the quality of educational institutions, but also, at the same time, to help assess educational institutions to continuously improve their education quality, instead of merely focusing on dealing with administrative and bureaucratic regulations?
2. With regard to the *subject of evaluation*: Who is the "eligible" subject to conduct the assessment? Is it advisable to allow, besides early childhood education institutions and educational management agencies, other subjects, for example, teachers, parents, and society, to participate in assessing the quality of early years learning institutions, with a view to mobilizing all social resources for the process of evaluating the quality of education in particular, and the process of educating children in general (Quang Tiep, 2017)?
3. With regard to the *method and tools for conducting the evaluation*: The use of information technology in building evaluation tools should be considered as a useful solution, since it makes the evaluation process more convenient for those who are involved. Moreover, if designed properly with appropriate information technology infrastructure, the tools for evaluating early childhood education quality can gather enriched feedback, which can be used not only to rank early childhood education institutions, but also for other

stakeholders to analyze, providing helpful insight. For educational managers, evaluation results are used to monitor and manage the quality of early childhood education institutions. For educational institutions themselves, especially preschool teachers, the evaluation results provide information as to the level of achievement of early years learning institutions in key areas and where the improvements are needed. Even better, evaluation results can assist teachers and schools in how to implement that improvement. For students' parents and society, the evaluation results not only provide an honest view of the quality of each institution, but also suggest appropriate early childhood education service options, suitable to their demand and financial capacity.

9.2 NATURE OF EVALUATING THE QUALITY OF EARLY YEARS LEARNING INSTITUTIONS

9.2.1 THE CONCEPT OF QUALITY EDUCATION

Different authors, depending on perspectives and purpose, have different perceptions about the quality of education (Adams, 1993; Harvey & Green, 1993; Scheerens et al., 2011; Tikly & Barrett, 2013).

According to Adams (1993), there are at least six views on the quality of education. First, education quality is closely attached to the reputation of educational institutions (quality as reputation). This concept is perhaps the most common in higher education evaluation but is not suitable for assessing the education quality at the preschool as well as primary education level. Besides, the reputation of an educational institution sometimes does not correspond to the quality of education, as it could be brought about by other factors, such as material resources and communication.

Second, the quality of education is measured by resources and other inputs; this is a common concept used in professional education quality accreditation and is also widely used by international agencies and organizations on education. Usually, data on financial resources, the number of teachers and qualifications, and conditions of facilities of an educational institution are essential information to indicate the quality of that institution.

Third, quality is considered by looking at the entire educational process. This notion shows that quality is not just about inputs or outputs. It also includes the operation of the institution, the interaction between the student, and the teacher and other members of the educational institution. Today, the gap in education between countries and regions is gradually narrowing, the concept of educational quality as a process has an increasing importance. Sometimes the output is less important, but it is the enthusiasm and enjoyment of both teachers and students in the educational process that count.

Fourth, quality is interpreted as educational content. This concept reflects an overestimation of certain types of knowledge and skills. However, the content of education here is not equivalent to the curriculum. Curriculum can comprise multifaceted processes involving many cognitive, emotional, and social activities to form knowledge, skills, attitudes, and values for learners, whilst educational content focuses on the knowledge and information provided to learners. Therefore, the interpretation of educational quality to educational content is not comprehensive.

Fifth, quality is assimilated to outcomes or outputs. This is one of the most popular concepts today, because educational outputs or outcomes are easy to observe, measure, and evaluate. However, no set of tools to evaluate educational outcomes is good enough to fully measure the quality of an educational institution or a national education. Psychological factors, emotions, and the interest of the learners are difficult to measure in the outcomes, whilst they are changeable, depending heavily on the process of organizing educational activities.

Sixth, quality is understood as a measure of change. According to this concept, the quality of education is considered as "value added" in learners by the effects and impacts of the organization or system. Factors that influence and create change in students from an educational institution include educational curriculum, culture, and organizational standards. Accordingly, education can be understood as "forming and developing capacity in learners" or enlightening the inherent potential of learners. This means, the higher the quality of education, the greater the increase in knowledge, skills, attitudes, and values in learners.

Another famous study by Harvey and Green summarizes the common perception of educators about education quality, conceiving it as a set of different attributes, including (Harvey & Green, 1993):

1. Quality is seen as excellence.
2. Quality is seen as perfection.
3. Quality is seen as fitness for purpose.
4. Quality is seen as value for money.
5. Quality is seen as transformation.

The above analysis proved: Educational quality can be understood in many ways, can reflect many values, and can be explained differently from individual perspectives. Educational quality is often considered multidimensional from the quality of the inputs, the process of organizing educational activities, financial resources, to the outcomes. The quality of education is a dynamic factor, it changes over time and to specific contexts. The quality of education can be measured by quantitative or qualitative metrics. Quality goals and other goals related to effectiveness, equity in education, or similar could sometimes not be aligned. The quality of education has the purpose to create the foundation of the values, culture, and traditions of an educational institution or organization. The quality of education can be unique to a country, region, community, school, or even to individual parents or students.

9.2.2 ASSESSING THE QUALITY OF EARLY YEARS LEARNING INSTITUTIONS

Assessing the education quality is an aspect of the educational planning process; it involves assessments of the achievements of educational institutions via systematic collection, and then analysis of data about the institution's objectives, resources, operational, and educational outputs. Ideally, an educational institution would include an internal (self-assessment) and external assessment of all aspects of the institution and its impact on learning. The evaluation and analysis included in the assessment include a series of inputs, operations, and educational outcomes reflected in factors, for example, personnel, learning resources, curriculum, leadership and management competencies, learning and teaching management, and finally, results that learners achieve after their education (De Grauwe & Naidoo, 2002). From this perspective, assessing schools in general and preschool institutions in particular is not an activity carried out at the end of the educational process; rather, it is the beginning of the process of school quality improvement.

Thus, the evaluation of preschool quality is the process of collecting and processing comprehensive data about the school, based on which making judgments about the extent to which the school meets defined criteria. Conducted through a combination of quantitative and qualitative indicators, quality assessment of preschools aims at three main purposes: quality monitoring, quality assurance, and quality management. There are also two main methods for assessing the quality of preschools: educational institution's self-assessment, also known as internal assessment, and evaluation by an outsider to the institution, also known as external evaluation.

9.2.2.1 SELF-ASSESSMENT, OR INTERNAL EVALUATION OF EARLY YEARS LEARNING INSTITUTIONS

Self-assessment or internal evaluation are terms for assessment activities conducted by the educational institutions themselves, based on defined criteria and standards. The main purpose of self-assessment activities for educational institutions is to identify the status of educational quality, so it can guide the process of improving education quality (De Grauwe & Naidoo, 2002). In recent years, the need for the self-assessment of educational institutions has increased. This is because of the need of schools and teachers themselves to self-assess, and continuously improve their quality. This is also an inevitable trend in the context of increasing autonomy for schools. Besides, practical studies have shown that a schools' self-assessment is a simple but effective method in increasing the quality of education. This provides useful tools and guidelines for teachers and schools to improve preschool quality, in part or in whole. The ultimate purpose of school quality assessment is to improve school quality. To make improvements, schools must know where they are in regard to quality standards; they must be aware of the gaps and have a need to improve these. Self-assessment of the school not only helps them identify strengths and weaknesses, but also ways to change those for the better. Self-assessment activities also help form the motivation and the need for improvement for that school, especially when the assessment is conducted rigorously, and the industry quality benchmark in each locality is publicly announced.

Despite many positive aspects, the self-assessment of educational and training institutions in many countries, especially in Asia, including Vietnam, is still not very popular and has not brought about the quality improvement effect as expected. This situation comes from the following main reasons. First, educational institutions, especially at the preschool level, are under strict administrative management. Autonomy of educational institutions is generally very low. Therefore, the entire time of each school is only used to understand and execute administrative regulations from higher management agencies such as Departments of Education at the district level, Services of Education at the province or city level, and the Ministry of Education and Training. Second, there seems to be a lack of tools for the self-assessment of the quality of an institution, good enough for an institution or an organization to use. The standards and evaluation criteria prescribed by the state are not specific enough to assess the quality of education. In addition, the evaluation criteria focus more on the administrative aspects, instead of helping to clarify the factors that make up the quality of a school. Third, at the end of the self-assessment process, most educational institutions are not sure what to do next to raise their quality. Whilst teachers and schools do not have enough knowledge and capacity to improve the education quality themselves, ironically, there is almost no guidance to help teachers and schools make change post self-assessment. Therefore, when starting a new policy on school self-assessment, attention should be paid to the support and capacity building that comes from the bottom line. Building school capacity to support self-assessment covers aspects such as general understanding of the concept of self-assessment, from the purpose of the assessment to the relationship and balance between internal and external assessments; building a conscious ownership and a strong commitment for quality improvement through self-assessment; developing or clarifying performance indicators, standards, and criteria that will be used to measure school quality. Self-assessment for schools should be conceived and developed as an ongoing and essential mechanism to monitor and improve the education quality in each institution.

9.2.2.2 EXTERNAL ASSESSMENT OF EARLY YEARS LEARNING INSTITUTIONS

This activity is not done by members of the school but by a third party. Third parties of the school include mainly government authorities in charge of education management and independent evaluation organizations. Currently, in Vietnam and many other developing countries, external assessments are carried out primarily by related government authorities. The scope and criteria for assessment are mainly prescribed in legal documents on educational management. The assessment process serves the purpose of supervising and managing educational activities of schools. Therefore, this evaluation is more formal and bureaucratic, rather than directed at evaluating the quality of education.

School evaluation by an independent evaluation organization specializing in the evaluation and accreditation of school quality is gradually becoming the trend in countries around the world. The evaluation and accreditation organization comprises members who are educational experts, with expertise in school assessment, accreditation, and school administration. Their school assessment task is partly to evaluate and recognize whether a school meets or does not meet quality standards, but also aims to provide the school with advice and support in forming a strategy for change and improvement of the quality after each audit (De Grauwe & Naidoo, 2002). In Vietnam, various independent educational accreditation organizations have been set up, but they only provide assessment services to universities, not yet to preschools or K12 schools. The external assessment by independent evaluation and accreditation organization model is gradually replacing that conducted by government authorities. In a foreseeable future, self-assessment and external assessment by independent evaluation and accreditation organizations will be two forms of assessment that exist in parallel, complementary to each other, to help educational institutions to receive both the recognition of quality that they are seeking for and consulting services from independent evaluation and accreditation organizations to help them make continuous progress in their development process.

Third parties in the external evaluation process could also be beneficiaries of educational service. Being customers, they have their own perspectives and approaches. Their feedback is critical in helping education institutions to understand customers' needs and expectations. In Vietnam, this form of evaluation was mentioned in the Resolution 29-NQ/TW (2013) of the Communist Party on fundamental and comprehensive renovation of education and training. However, mechanisms for external evaluation of the quality of education by beneficiaries of educational service have not been legislated.

Here, it is necessary to clarify the concept of educational quality accreditation. The external assessment is in itself the education quality accreditation of an educational institution. Education quality accreditation has developed over many years in America and has been very popular in higher education, but little known in other countries. In the process of decentralization and educational popularisation, educational standards have changed and have been sometimes significantly different among schools in the same country, even in the same locality due to the downgrade in the quality of enrolled students in certain schools where the size increased far greater than the financial resources, or due to external negative factors affecting the school. Furthermore, the education system in developed countries is gradually shifting from state oriented or the school's academic oriented to market oriented. In this context, quality accreditation has become an effective tool to maintain educational quality standards and to continuously improve the quality of education. Quality accreditation is a process of external evaluation conducted by an independent assessment body to decide if a school or an educational program meets the specified standards (Pham, 2005). An assessment not intended to make a quality recognition decision is not an accreditation. Quality accreditation is also used by many countries to periodically review, evaluate, and recognize schools that maintain regulatory standards. The main purpose of quality accreditation is to ensure that certain standards of education are attained, meeting the needs of learners, students' parents, and society. In some places, accreditation is conducted also for the purpose of accountability to society, with authorities or with funding agencies and organizations. A number of organizations and agencies are interested in whether the school has been accredited before deciding to sponsor or not sponsor the school. Students and parents, before selecting a school to enrol in, also consider whether or not the school has been accredited.

Accreditation practices are diverse and complex, but the process is almost unified, and consists of four steps as follows:

1. Step 1: Developing or updating quality evaluation tools
2. Step 2: Self-assessment of schools
3. Step 3: Reviews from third parties
4. Step 4: Recognize or not recognize that school meet quality accreditation standards.

In the first step, quality evaluation tools, including standards, criteria, guiding documents, and especially the regulatory/policy framework for the evaluation process will be developed. Standards are understood as the level of requirements that the school must meet in order to be accredited. These requirements include requirements for quality, results, effectiveness, sustainability, compliance with national laws and regulations, and school stability (Council for Higher Education Accreditation, 2001). This qualitative approach makes quality standards abstract. To make it easier to measure, these requirements are divided into less qualitative subcategories, called criteria. In the second step, the school uses quality evaluation tools to implement school self-assessment. This is a time-consuming and labor-intensive process. Self-assessment is not merely about writing a self-criticism and self-report. Self-assessment is a process of self-study and self-improvement according to the standards issued to be accredited. This process usually lasts from 6 to 18 months. It is the time needed for the school to identify its own shortcomings and strive to overcome them. Self-assessment has the advantage of being directly performed by the school's members. They are the ones who know the school better than anyone else. But self-assessment is often impersonal and performed by nonprofessionals. In contrast, external evaluation is a process that adds value to self-assessment results. External evaluation is carried out by the best experts in the same field of expertise. The external assessment process is intended to shed light on issues that self-assessment has not clarified.

9.3 AN OVERVIEW OF PRESCHOOL EDUCATION AND THE REALITY OF ASSESSING THE QUALITY OF PRESCHOOL EDUCATION IN VIETNAM

9.3.1 VIETNAM EARLY YEARS LEARNING SYSTEM

Like most countries in the world, Vietnam's early years learning sector is a part of the national education system, tasked for providing care and education service for infants to children who are ready to go to primary school. Divided into two stages: nursery and kindergarten, early childhood (preschool) education provides the foundation for a comprehensive, long-term development of children. According to the Education Law of Vietnam, the goal of early years learning is to help children develop holistically, preparing them for school readiness. Thus, it can be seen that the primary aim of early years learning in Vietnam is to develop character and personality, that is healthy, active, with a well-proportioned, and balanced body; loving, caring, giving, helping close people (parents, teachers, friends, ...), honest, polite, bold, innocent; love beauty, know how to preserve beauty and a desire to create beauty; curious, interested in exploring, have some basic skills (observation, comparison, analysis, synthesis, and reasoning) necessary to enter primary school.

From the above goals, early childhood education institutions have the following specific tasks:

1. Organizing pastoral care and learning for children from infancy to six years old according to the preschool education curriculum issued by the Ministry of Education and Training;
2. Mobilizing preschool children to go to school, organize learning environments for children, including those who experience disabilities,

and universalize access to early years learning for all five-year-old children;

3. Self-assessing their achievements of the standards for early years learning universalization for all children over five years old, reporting to competent authorities in writing;
4. Managing officials, teachers, and employees in implementing the task of nurture, care, and education of children;
5. Mobilizing, managing, and using resources to meet the requirements of regulatory and legal bodies; build preschool infrastructure according to standardization and modernization requirements, or at least the minimum requirements for extremely difficult areas;
6. Coordinating with children's families, organizations, and individuals to carry out child nurture, care, and education activities;
7. Organizing managers, teachers, staff, and children in involving in social activities in the community;
8. Conducting quality assessment of the implementation of the task of nurture, care, and education children according to regulations; and
9. Performing other duties and powers as prescribed by law.

According to the 2019 statistics of the Department of Teachers and Education Managers, Ministry of Education and Training, Vietnam has 24,784 preschool institutions, among which 12,445 are public preschool institutions, which means over 50% of preschools nationwide. Nearly 50% of the remaining preschool institutions are community founded and private schools. The ratio between public and private schools is not equal among provinces in Vietnam. In large cities, there are more private schools than in rural provinces.

Based on the conditions of facilities, teachers, class sizes in square meters, and the number of children in each class, Vietnamese early years learning institutions are divided into two types: preschool and childcare groups. A preschool is an early childhood education institution that meets the requirements of state regulations on the area, the number of classrooms and function rooms, utensils, toys and other facilities, the number of classes, the number of children .... Early childhood education institutions that are not eligible to become preschools, but still eligible to organize child nurture, care, and education according to regulations will become childcare groups. Each childcare group is placed under the administrative management of a people's committee at a commune level.

9.3.2 SITUATION OF ASSESSING THE QUALITY OF VIETNAM'S EARLY YEARS LEARNING SYSTEM

In Vietnam, quality assessment of preschool is the assessment conducted by a school's leadership for all aspects of that school's activities, childcare, and education process and results. During the evaluation, related parties will collect, analyze, and interpret information on childcare and education, compared to the school year goals. The school's evaluation results help managers see the current status of educational and teaching activities, creating a basis for making good decisions or adjusting management and direction to optimize their education quality.

The quality of an early childhood education institution is not only considered from the perspective of standards and criteria issued and used nationwide. Each early years learning institution built locally with different economic, cultural, and social development characteristics, serving the children of local families with different traditions, customs, languages, and beliefs. Therefore, besides the general regulations for all early childhood education institutions in the country, the quality of a preschool is also measured by the level of achieving the expectations of teachers working in it and especially of the parents whose children attend the school.

From a practical perspective, the assessment of a preschool has three main levels. First, evaluation is based on the criteria specified in the legal documents of the state and governing bodies, aiming at controlling the quality of the school. The preschool will, based on the standards, criteria, and indicators specified in these documents, conduct a self-assessment to get ready for an external quality evaluation. Standards and criteria for preschool assessment often relate to the following aspects: the legality of the documents giving permission for the establishment and operation of the preschool; the qualification and experience of preschool administrators, teachers, and other school staff; the quality of preschool infrastructure, facilities, equipment, and teaching aids; the development of students in schools; and relationships with stakeholders in child education. Second, the assessment is based on the standards and criteria common to all preschools and specific goals developed by the school itself in consultation with parents and the community and considered as an evaluation for quality assurance. The standards and criteria developed by the school itself may be more extensive or advanced than the general standards and criteria or may include new standards and criteria. Third, the assessment is based on the above two assessment forms, with the consideration of the actual context, the inputs, the implementation process, and the outputs. This also includes the consideration of the satisfaction of parents and the community about preschool education and teaching activities. Such a level of assessment can be considered as an overall quality management review. Which means that the evaluation of the preschool has been considered in all aspects, including the goals that need to be achieved in accordance with the regulations of the state and governing bodies, the expectations of administrators, teachers, and the expected goals of students' parents and the community in the specific local context.

Currently, the standards and criteria for the evaluation of Vietnamese early years learning institutions as set by many legal documents, including the legal basis for the establishment and operation of preschools; standards and criteria for evaluating management staff, teachers, and other staff working in preschools (ethical qualities, training qualifications, work experience, responsibilities, rights, ...); standards and evaluation criteria for the age structure of students in a class, and the ratio of teachers, carers, and support staff to the number of students; relationships with authorities, unions, parents, etc.; standards and criteria for evaluating infrastructure, facilities, equipment, and teaching aids of early childhood education institutions; development standards of five-year-old children; standards and criteria to recognize the schools that meet national standards at levels 1 and 2.

There is missing information and inconsistency in the documents that determine standards and criteria for the evaluation of the quality of early childhood learning institutions, more specifically:

1. Standards and criteria for evaluation are not consistent in some areas. For example, standards and evaluation criteria for principals and teachers, class size, and rooms in an educational institution in the documents regulating each area differ from the corresponding standards for ranking schools' levels 1 and 2.
2. Some standards and criteria for quality evaluation are still general, not considering regional specific factors, so it is difficult for them to be used in the assessment.
3. There have been no guidelines on the formulation, by early childhood education institutions, of specific goals and internal regulations, corresponding with the characteristics of local economic, cultural, and social development.
4. No guidance has been provided on the involvement of parents and community in goal setting and quality assessment.

5. The quality of preschool education is expressed in specific contexts and in all stages from inputs, education process, and outputs. However, the standards and evaluation criteria specified in the current legal documents only focus on outputs.
6. There are no developmental standards for children aged 0–5 years old in Vietnam. Moreover, the set of five-year-old child development standards is not in alignment with developmental goals described in the preschool curriculum and with the standards used in countries with advanced early childhood education, so it still needs to be adjusted.
7. There are no guidelines for the announcement and transparency of the results of preschool quality evaluation for the benefit of all stakeholders.
8. There are no sanctions for parties who are directly involved in preschool education if they do not fulfill their responsibilities (besides the regulations on reward, discipline of school officials, and dissolution of the early years learning institution).

In order for the assessment of preschool quality to make substantial contributions to the quality improvement of early years learning institutions, the following major issues need to be addressed:

1. The system of standards and evaluation criteria in normative documents must be unified, making them simple, understandable, and implementable for early childhood education institutions.
2. Standards and evaluation criteria must be developed, taking into consideration the peculiarities of early childhood education institutions in different areas and regions.
3. Standards and evaluation criteria must be developed in accordance with the current context and have long-term vision based on Vietnam's educational development strategy for the next 5–10 years and subsequent years.
4. Core standards and evaluation criteria need to be studied to be in line with the common standards of early years learning in developed countries and global best practice.
5. Guiding documents for the formulation of preschool development strategies must be elaborated, so schools can base on, and depending on their own situation, together with parents and the community,

build reasonable and feasible development goals for a specific period of time.

6. More thorough standards and evaluation criteria must be developed to cover not only outputs but also inputs and implementation process of preschools.
7. Development standards for preschool children of all ages that are appropriate for Vietnamese children, with reference to the development standards of children in different countries, must be defined, as a basis for assessing children's development and the preschool quality in terms of care and education.
8. Standards and criteria for preschool quality assessment must be communicated to a wider community in the society, making them open and accessible.
9. The quality assessment of preschool institutions should be conducted with the participation of many subjects (education managers, teachers, school officials, parents of children attending schools, and communities).
10. The application of information technology to assess the quality of preschool education should be promoted, making evaluation tools easily accessible to users via popular websites or social networks in Vietnam such as Facebook and Zalo.

Changes in education, especially in the quality assessment of schools, will always entail different reactions from the participants in the educational process. Therefore, careful research and an implementation roadmap for creating a consensus of both the school and wider society about evaluation should be put in place. Assessment is often considered the final stage of the educational process. However, in practice, assessments are carried out at all stages of the education process (diagnostic evaluation, implementation evaluation, and final review). Effective implementation of quality assessment will be a guarantee for improving the quality of early years learning institutions (Nguyen, 2000).

9.4 METRICS TO ASSESS THE QUALITY OF VIETNAMESE EARLY YEARS LEARNING INSTITUTIONS

The criteria for assessing the education quality of early years learning institutions in Vietnam can be found in more than 10 different legal documents, including Education Law; Decision No. 04/VBHN-BGDĐT promulgating the Regulation on kindergartens; Decision 1677/QD-TTg approving the project of development of early years learning for the period 2018–2025; Provisions on universalizing early childhood education for five-year-old children; regulations on organization and operation of public, community founded, and private preschools; documents guiding career payrolls in public preschools; regulations on recognition of preschools meeting national standards; regulations on standards of preschool principals; regulations on standards of preschool teachers; regulations on development standards for five-year-old children. As stated earlier, there are overlaps and inconsistencies between the standards and evaluation criteria in the above-mentioned documents.

The criteria for evaluating preschool institutions in Vietnam are most fully reflected in the Ministry of Education and Training's Circular No. 19/2018/TT-BGDĐT on quality accreditation and recognition of meeting national standards for preschools. Based on this circular, preschools are evaluated and classified according to four quality levels. The standards and criteria for evaluating preschools according to current regulations include:

1. Standard 1: Organizing and managing schools, with 10 criteria, including directions, strategies for school development and growth; school boards (board of directors for private schools) and other councils; presence of Communist Party of Vietnam within a school, unions, and other organizations in the school; principals, vice principals, specialized groups, and admin teams; children's group or class structure; administration, finance and property management; managing officials, teachers, and staff; managing educational activities; implementing democracy regulations at the institutional level; ensuring school security and safety.
2. Standard 2: Managers, teachers, staff, with four types of criteria, including criteria for principals and vice principals, teachers, and staff.
3. Standard 3: Facilities and teaching equipment, with six criteria, including floor space, campus, and garden; nursery's rooms, kindergarten classrooms, and learning support rooms; admin and property management division; dining room; equipment, utensils, and toys; and toilets, water supply, and drainage system.
4. Standard 4: Relationship between school, family, and society, with two related criteria: representative board of children's parents; advice

and collaboration mechanism with Communist Party's executive committees, governing bodies, and organizations and individuals of the school.

5. Standard 5: Activities and results of child nurture, care, and education, with four criteria, including enactment of the early years learning curriculum; organizing educational and nurturing activities; and results of educational and nurturing activities.

It can be seen that some criteria in the set of preschool evaluation criteria above are too formal and do not reflect the true quality of education. For example, the criterion of the presence of "School Council (board of directors for private schools) and other councils"; the criterion of the presence of "the Communist Party of Vietnam within a school, unions, and other organizations in the school" … says nothing about the education quality of schools. Or the indicators relating to the percentage of qualified teachers, the number of years of teacher's experience … do not reflect fully and comprehensively, the teaching quality of that teacher. Meanwhile, a number of important issues that are critical to the quality of education are not considered for evaluation. They are, for example, the classroom arrangement and organization; daily educational activities; teachers' and children's interaction in the educational process. Moreover, the evaluation criteria in the above standards are only suitable for public preschools. They are not relevant to nearly 50% of early years learning institutions countrywide belonging to the private sector, and to many childcare groups that do not meet the standard (mostly not meeting the floor space requirements) for becoming a preschool.

To solve the above-mentioned gaps and shortcomings in the current set of standards for the quality evaluation of preschool education in Vietnam, our research team has built and developed a new set of standards and criteria for the assessment of quality of early childhood education institutions since 2015 under the project "Publicity and transparency in educational quality assessment with SchooListing.Net, an online education quality assessment tool" under the "The Vietnam Anti-Corruption Initiatives Program 2014" (Nhu, 2015). After many tests and improvements, the tool for evaluating preschool education quality now includes 5 standards, with 21 specific criteria as follows:

1. Standard 1: Facilities and teaching equipment, with four criteria, including equipment, utensils, toys; classrooms and space for learning; outdoor play areas; and functional rooms and other areas of the school.

2. Standard 2: Childcare and education activities, with five criteria, including organizing child education activities; organizing childcare activities; implementation of the curriculum; supporting children with disabilities; and child's assessment.
3. Standard 3: Relationship between school, family, and society, with four criteria, including parental involvement in the educational program; parental involvement in the school board; communication between teachers and parents about the curriculum and child development; and interaction between parents and teachers.
4. Standard 4: Managers, teachers, and staff, with three criteria, for managers, teachers, and staff.
5. Standard 5: Organizing and managing schools, with five criteria, including the ratio of teachers to children; enrolment activities; managing educational activities; administrative management; and financial resources.

Each criterion is assessed by a number of different indicators, and each indicator is described at three levels from low to high. For example, the indicators used to evaluate the criteria "Interaction between parents and teachers" in the standard "Relationship between school, family and society" are described in Table 9.1.

TABLE 9.1 Criteria "Interaction Between Parents and Teachers" in School Sting

Interaction between parents and teachers			
Level ①	**Level ②**	**Level ③**	**○ No reviews**
Teachers and parents do not interact and exchange formally on issues related to child education.	Teachers and parents occasionally interact and exchange formally on issues related to child education.	Teachers and parents often communicate informally with each other to update their children's progress (e.g., exchange information at arrival and departure times, send emails, make phone calls, ...).	I do not know what to choose for this item. Ignore.
Level ①	**Level ②**	**Level ③**	**○ No reviews**
The information communicated by teachers to parents about children's development is neutral, unclear.	Some information communicated by teachers to parents about children's development is clear and relevant.	All the information communicated by teachers to parents about children's development is clear and relevant.	I do not know what to choose for this item. Ignore.

TABLE 9.1 *(Continued)*

Level ①	Level ②	Level ③	O No reviews
Teachers do not provide information to advise parents on how to assess a child's needs for special capacity development.	Teachers provide parents with information to assess a child's needs for special capacity development.	Teachers work with parents to identify and evaluate children's special abilities.	I do not know what to choose for this item. Ignore.

To evaluate each criterion, the assessor needs to observe, measure, and collect evidence for each type of indicator. When sufficient information on the indicators is available in each criterion, the level of achievement of the criteria to be assessed will be determined. If the entire evaluation process, through the determination of indicator levels for each criterion, is designed using online tools with the help of calculation software, the evaluation becomes extremely simple. The assessor, whether being an education expert or a teacher, even a parent, can conduct the assessment with ease. The evaluation is now a simple gesture of selecting the level of the indicator that reflects the reality of the school. All collected information will be handled and visualized intuitively by software, helping the assessors and preschool institutions not only see the results but also identify the strengths, weaknesses to overcome and improve school quality. In addition, if the software is good enough, it will draw up the school's quality development roadmap during the long-term improvement process.

9.5 USING INFORMATION TECHNOLOGY IN EVALUATING THE QUALITY OF VIETNAMESE EARLY YEARS LEARNING INSTITUTIONS

9.5.1 USING INFORMATION TECHNOLOGY IN DEVELOPING PRESCHOOL QUALITY ASSESSMENT TOOLS

Information technology has become indispensable in all areas of social life, including education in general and evaluation of education quality in particular. Information technology has changed both the scope and methodology of evaluating educational quality, thanks to a number of outstanding advantages.

First, information technology provides the opportunity for everyone to access information fully and quickly. Preschool administrators, teachers, and parents can access documents about school operation, preschool assessment standards, and teacher assessments. At the same time, information about preschools is also publicly available on the web, making it accessible to governing bodies, parents, and the community. This makes school quality assessment easy and convenient.

Second, information technology, especially computers, Internet, and specialized software, has made evaluation effective. Almost all jobs relating to calculations, visualization, simulation, etc. are supported by information technology. The remaining work in the assessment process that needs to be performed by participants is to observe, find evidence, match, and select indicators (Carlson and Gadio, 2002). This means, with the support of information technology, teachers, parents, and literally everyone can participate in the school evaluation process. This helps turn the evaluation results into greater and more valuable insights for the school to improve education quality, meeting governing bodies' requirements and community's expectations.

Third, according to recent statistics, as of 2017, Vietnam had 120 million mobile phone subscribers, with more than half of mobile phones in Vietnam being smartphones connected to the internet (Cameron et al., 2019), which means that Vietnam has 66% of the population using the Internet. This is comparatively very high against other countries globally and is increasing every year. This is a very positive sign, opening up good opportunities to apply information technology not only in evaluating the quality of education but also in communication, brand promotion, trading activities. An intuitive and convenient online tool for school quality evaluation, which could be a web app or a mobile app, means endless possibilities for it to reach a wider audience. The school's evaluation will not take too much time for the participants, and results will be generated almost instantly.

Fourth, information technology today helps store huge amounts of database storage and can process both quantitative and qualitative data to make sense from it (Williamson, 2017). Therefore, the results of school assessments at each stage by many subjects (managers, teachers, parents, and communities) can be stored, matched, and used to create development graphs and charts of school for a long period of time. This has great value for teachers and schools in continuous quality improvement. In addition, parents and communities will also get a picture of the quality of education of many schools in the same area, so they can compare and choose the right school for their children.

In May 2014, our research team implemented the project "Publicity and transparency in educational quality assessment with SchooListing. Net, an online education quality assessment tool" under the "Vietnam Anti-Corruption Initiative Program (VACI) 2014" co-organized by the Vietnam Government Inspectorate and World Bank. Our key task was to build a set of preschool quality assessment standards and criteria (as mentioned above) and embed them in an online environment (Nhu, 2015). Schooling (domain name SchooListing.Net), an online preschool quality assessment tool, was then built in the form of a website with two user interfaces, with the structure diagram.

Figure shows how easy it is to understand the operation of the online SchooListing system. The system will create accounts for users, namely the preschool institutions (including administrators and preschool teachers) and students' parents. User accounts will be activated when providing information for all required fields, such as the name of the reviewer, name of the assessed preschool and address, contact information. When the account is activated, the user will conduct a preschool assessment by carefully considering each of the criteria in the standards and determining the right level of indicators for each criterion in accordance with the practice of the preschool being assessed.

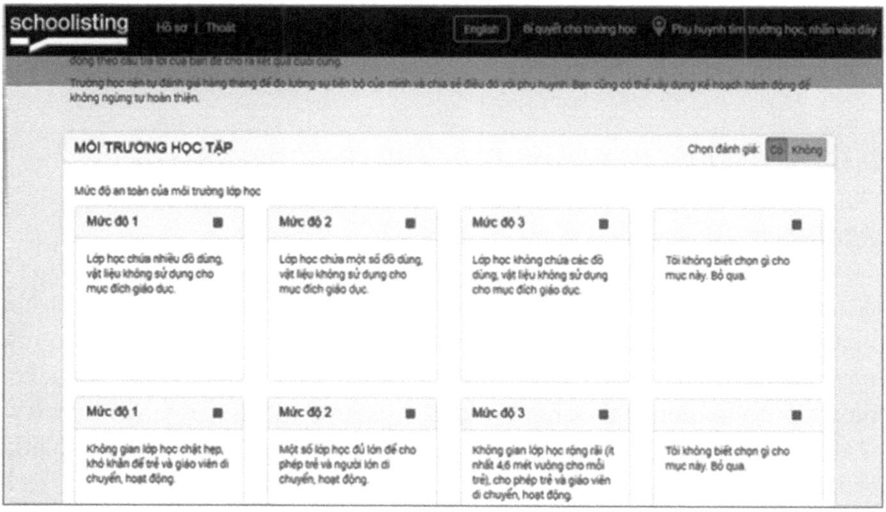

FIGURE 9.1 User interface when using SchooListing to evaluate preschools.

When finishing the selection of indicator levels for all criteria, the assessor completes the school evaluation. Depending on the user attributes, the system will make appropriate choices. If users are administrators and preschool teachers, the system will give detailed feedback on the results achieved by each criterion and the overall results of preschool quality assessment. The members of the preschool will have useful information about whether or not the school is good, where the school's strengths and weaknesses are, to have strategies to address them. All of this feedback from the system is not only the result of the evaluation of individual members of the school but also of everyone involved in the school evaluation, including the students' parents. If the user is a student's parent, the system will give the general evaluation of the preschool and suggest a series of preschools ranked from low to high, in order to provide the parents with an overview of a preschool system in their local area. This will help them in finding and choosing a suitable school for their children.

9.5.2 USE THE ONLINE EVALUATION TOOL TO EVALUATE PRESCHOOL QUALITY

Within the scope of this study, we gathered information, interviewed, and selected 50 preschools across the country to use SchooListing.Net in a pilot program. The tool was later brought to the public. By December 2015, SchooListing had 1713 school users (with 243 users with self-assessment data) and 897 parent users who participated in the evaluation of one or more schools. Here are the main results.

9.5.2.1 OVERALL ASSESSMENT OF THE EFFECTIVENESS OF SCHOOLISTING TO PRESCHOOLS PARTICIPATING IN THE PILOT PROGRAM

After a period of using SchooListing.Net and learning from the preschool quality assessment tool, a number of preschools participating in the pilot program have taken improvement actions to achieve better quality. The school paid more attention to the observations of parents and used them as a basis for making timely adjustments and corrective decisions. The use of evaluation criteria provided by SchooListing.Net also allowed these preschools to be more aware of their strengths, to become more confident in maintaining and promoting them. By regular monthly reviews, all administrators, teachers, and parents were better informed of the school's improvements over time and could identify what needed to be done to further improve the quality of the school.

Schools also proposed the necessary changes to improve the educational quality assessment tool such as organizing regular training workshops, answering questions of schools and parents about the tool; shortening the list of criteria to be assessed by parents to lower the barriers for them to use; promoting the tool via both online and traditional communication campaigns for parents to understand, know and regularly use this useful tool.

9.5.2.2 EVALUATION OF SCHOOLISTING BY PRESCHOOLS AND PARENTS

The research project used SchooListing.Net to evaluate the quality in depth in three preschools, including:

1. Lieu Giai Preschool, No. 6, Doi Nhan Commune, Ba Dinh District, Hanoi, Vietnam
2. KHC Preschool, No. 21–24—No. 3—New Urban Area Dich Vong, Cau Giay District, Hanoi, Vietnam
3. Xitrum Kingdom Kindergarten, Level 5 - V1 - Van Phu Victoria Building, Ha Dong District, Hanoi, Vietnam.

The above preschools believed that the online assessment tool of preschool quality is complete and detailed, convenient to use to comprehensively evaluate an early childhood education institution, help the school confidently promote its strengths, and have a clear orientation on what to do in the development process for the future.

As a preschool quality assessment online tool, SchooListing has been remarkably effective for all three subjects: teachers, school administrators, and parents. More specifically, teachers gradually changed their awareness, recognized their strengths and weaknesses, always tried to overcome difficulties, and made an effort in every activity. School administrators had a long-term perspective on development and in managing all aspects of the activities, from management to childcare, education, and relationships with parents. Parents are becoming more proactive in involving in school activities, contributing to school development, providing more detailed and sophisticated observations when assessing the quality of a school.

However, feedback from preschools also shows that SchooListing's online assessment tool still has some limitations. The preschool quality assessment questionnaire, version for parents, is too lengthy, which takes some parents a long time to evaluate. Parents cannot assess some criteria as they do not have access to the internal information of the school. Some criteria in the preschool quality assessment tool are still high, especially the criteria relating to facilities, which are difficult to achieve for private schools (e.g., the area of the outdoor playground).

9.5.2.3 SOME IMPROVEMENTS TO PRESCHOOLS THROUGH THE USE OF SCHOOLING

The preschools that permanently used SchooListing, all improved their quality after a period of about three months. The most important changes are as follows. The school teachers had a clear shift in awareness and consciousness, using the criteria set as a handbook to adjust activities in their work.

Regarding the learning environment, preschools strive to rearrange learning corners and classroom environments in a more reasonable, more child-friendly way, supporting children's initiatives and creativity, as well as diverse areas of child development. Learning corners are also used more consciously for pedagogical purposes, instead of just being the place to display children's toys as before. In order to enhance the active role of children in learning activities, many preschools have set up a daily plan corner, to record their ideas in designing activities of the day, helping children practice planning. In addition, teachers also pay attention to displaying children's products in the classroom to encourage children to have a positive awareness about themselves, instead of displaying products made or purchased by teachers "beautifully."

In daily activities, teachers spent an additional 10 minutes in the morning to plan with children for that day. While children did the planning, teachers supported and guided them, especially those who did not know how to make an individual plan. The teacher wrote each child's plan on a small piece of colored paper, posted it on the Daily Plan Corner. At the end of the day, the teacher spent 10 minutes reviewing the plans with the child to check what he or she has accomplished, and what had not yet been done. The reforms were small but created a new habit in the children, which is planning, organizing daily activities, and being accountable for the implementation of the plan. Some schools add "sharing the goodness you have done during the week" to the weekend music festival (every Friday), helping children to share their experiences, be more aware of what is good and what is not. Children in nursery groups who do not speak well yet will receive support from teachers with short, easy to understand questions.

The quality of interaction between teachers and children has clearly improved toward giving children more initiatives in selecting activities appropriate to their abilities and needs. Teachers respect children's individuality and will only provide support if they need them; teachers do not act on behalf of or replace children's roles in their tasks; they change the way they design and manage learning activities to create opportunities for children to work in smaller groups; and they successfully support children and their families in the transition from home to the classroom environment, preparing them for school readiness in the next step.

The school is also more aware of the importance of conducting regular child assessments, developing plans to help children make progress. The school wishes to receive training or access to more information about the child's assessment tools, so they can conduct the assessments professionally.

The preschools are all more aware of the importance of professional development activities such as observation/feedback sessions, teacher-training workshops, and daily/weekly teachers' meetings. Many schools make permanent daily observations and meetings with teachers to answer to their questions and support their new initiatives, and others allow weekly peer observations for teachers to learn from each other. Some preschools even built a partnership with bigger schools to send their teachers in for observations, training, internship, or for general professional development.

9.6 CONCLUSION

Assessment of education, in general, and assessment of the education quality of preschool education, in particular, are crucial to developing the entire education system. The most important purpose of preschool quality assessment is not only monitoring and managing the quality, categorizing, and ranking preschools, but, more importantly, also to provide support to schools during their long-term continuous improvement journey. Therefore, it is important for all individuals and organizations involving in the education quality evaluation process, from managers to preschool teachers, parents, and the community, to be fully aware of its meaning. Their participation in the evaluation process will make the results authentic and bring about an effective improvement of school quality as expected.

Standards and criteria for evaluating preschool quality in Vietnam need further research to support policy reform. It is the preschool assessment criteria that will shape the way the whole preschool education system develops. Recently, Vietnam has shown great effort in renovating education, in general, and improving its capacity to assess school quality, in particular. However, there are still too many legal documents on the same matter with overlaps and inconsistencies. Currently, applicable evaluation criteria remain formal and bureaucratic, rather than focusing on those that truly reflect the quality of education. Today with the trend of globalization and integrating Vietnam into the world, regardless of cultural, religious, economic, or scientific differences, the consideration, for a selective succession, of best practice in the world about the standards and criteria for preschool quality assessment will allow Vietnam to learn and develop its education faster and in the right direction.

Information technology today has grown to a universal level for most people around the world. It has crept into every corner of social life and affects every industry, education included. Therefore, the use and application of information technology in education in general, and in assessing school quality in particular, is an inevitable trend and requirement. Information technology makes evaluation simple and easy. Evaluation results are processed, analyzed, and interpreted more comprehensively and convincingly with the help of information technology. More importantly, access by various users, from administrators to teachers, parents, and the community to the school assessment tools, will allow a complete data collection for more accurate and objective evaluation results.

In the near future, with the advancement of science and technology, things such as artificial intelligence, cloud computing, big data, and information technology can be used not only to build an online quality assessment tool for educational institutions, but it also has a great potential for application in assessing all the aspects and all members in a school. Then, a complete suite of online assessment tools for a comprehensive understanding of school quality and performance, yet friendly and convenient for users, will be achievable. Anyone who wants to participate in the evaluation can give their opinions or observations, contributing to the improvement of the education quality. This approach helps to make the education system transparent.

ACKNOWLEDGMENTS

We would like to thank the Vietnam Anti-Corruption Initiative Program (VACI) 2014 co-organized by the Government Inspectorate of Vietnam and the World Bank for funding our project "Publicity and transparency in educational quality assessment with an online education quality assessment tool, SchooListing.Net."

KEYWORDS

- **digital technology**
- **evaluation**
- **quality of preschools**
- **education quality**

REFERENCES

1. Adams, D. K. (1993). *Defining educational quality* [IEQ Publication # 1, Biennial Report in the Improving Educational Quality (IEQ) Project], Institute for International Research, USAID.
2. Cameron, A., Pham, T. H., Atherton, J., Nguyen, D. H., Nguyen, T. P., Tran, S. T., Nguyen, T. N., Trinh, H. Y., & Hajkowicz, S. (2019). *Vietnam's future digital economy: Towards 2030 and 2045*. CSIRO.
3. Carlson, S., & Gadio, C. T. (2002). Teacher professional development in the use of technology. In W. Haddad & A. Drexler (Eds.), *Technologies for education: Potentials, parameters, and prospects*. Academy for Educational Development, UNESCO.
4. Council for Higher Education Accreditation. (2001). Glossary of key terms in quality assurance and accreditation. Retrieved October 28, 2002 from https://www.uv.es/alfa-acro/documentos/documentosinteres/29.htm
5. De Grauwe, A., & Naidoo, J. P. (2002, July 2–4). *School evaluation for quality improvement*. [Meeting: The Asian Network of Training and Research Institutions in Educational Planning], Kuala Lumpur, Malaysia.
6. Harvey, L., & Green, D. (1993). Defining quality. *Assessment and Evaluation in Higher Education, 18*(1), 9–34.
7. Nguyen, D. C. (2000, April 4). *Research and develop a set of criteria for evaluating training quality for Vietnamese universities*. [Paper presentation], Conference of quality assurance in training in Vietnam, Da Lat, Vietnam.
8. Nhu, D. T. H. (2015). Transparency in education quality assessment with Schoolisting. Net: An online education quality assessment tool. Vietnam Anti-Corruption Initiative Program (VACI) 2014 and the World Bank.

9. Pham, D. N. T. (2017a). The problem of education quality in building a learning society in Vietnam. *Journal of Educational Science*, *98*, 1–5.
10. Pham, Q. T. (2017b). Renewing and evaluating the quality of early childhood education institutions. *Journal of Education Science*, *142*, 29–33.
11. Pham, X. T. (2005). Testing the quality of higher education. *Journal of Education*, *115*, 1–5.
12. Scheerens, J., Luyten, H., & Ravens, J. V. (2011). Perspectives on educational quality. *Perspectives on educational quality: Illustrative outcomes on primary and secondary schooling in the Netherlands* (Vol. 1, pp. 3–33). Springer.
13. Tikly, L., Barrett, A., Angeline, M. (2013). *Education quality and social justice in the global south challenges for policy, practice, and research*. Routledge.
14. Williamson, B. (2017). *Big data in education: The digital future of learning, policy and practice*. SAGE Publications, pp. 26–47.
15. Wortham, S. C. (Ed.). (2012). *Common characteristics and unique qualities in preschool programs: Global perspectives in early childhood education* (Vol. 5). Springer Science & Business Media.

CHAPTER 10

Blended Learning and Flipped Learning: Challenges and Opportunities for the 21st-Century Students to Create a Digital Environment

RUNJUN BORAH

Department of Education, Normal School Nagaon,
State Council of Educational Research and Training, Assam, India

**Corresponding author. E-mail: runjunborah1978@gmail.com*

ABSTRACT

Over the years, the world has seen several new e-learning practices. There are millions of e-learning models that are being applied in all spheres of the educational system. Among them, blended learning and flipped learning are considered the most acceptable e-learning approaches. Blended learning combines both online and traditional ways of learning. Flipped classroom is essential for the appropriate application of blended learning. The flipped classroom is a reversed classroom where students can access the contact of the subject matter before facing the classroom intraction with their teacher. The flipped classroom is one of the finest e-learning models which has reversed classroom learning. Here, students can acquire the knowledge of new instructional content outside their classroom. A teacher plays the role of a transmitter of knowledge and facilitates pupil to acquire knowledge; here, the teacher helps students to develop skills and increase the competency level to make students become perfect. This chapter explores the blended learning and flipped classroom approaches and their implications in the practical field. These two approaches explain how value-based poems and theme-based songs are helpful in arousing interest among students through a digital environment. In this digital environment, students are getting acquainted with a virtual platform, and they are allowed to perform activities together to reach their desired destination or to achieve goals. Students use social media such as Facebook or WhatsApp as a part of the flipped classroom approach. It creates a congenial atmosphere between teachers and students when they engage themselves in discussion.

10.1 INTRODUCTION

In modern times, the pupil-centered approach is preferred by all. For making the teaching-learning process more effective, everyone should use technology. The use of technology in the teaching-learning process make this beneficial for the pupils. The growth of technology has created a revolution in the filed of education. But one should remember that the use of technology in teaching-learning process is done very carefully with the help of pedagogical knowledge.

10.1.1 BLENDED LEARNING

Blended learning is a new educational approach. It ties the knot between traditional teaching-learning process and online mode of learning. The blended learning approach enhances a perfect learning environment with computer-based activities through various technological devices such as smart phones, tablets, laptops, and desktops. Students are more involved in such an environment rather than in a traditional process. It increases the interest level of the learners. The main purpose of blended learning is to create a pupil-centered learning environment. A proper blended learning approach requires well-integrated teaching-leraning strategies. All the strategies should be apted symmetrically together. Every element of blended learning should be hooked up in such a way that they can intensify one another. The students can choose any media, technologies, strategy under the umbrella of blended leraning.

10.1.1.1 ADVANTAGE OF BLENDED LEARNING

Without taking instruction from educators, we can be able to gather expertise knowledge from other people in a blended-leraning-fostered environment. In the blended learning process, every student gets an opportunity to acquire individualized instruction. No two individuals are alike. They have their own capabilities, ways of thinking, intellectual level, interest level, etc. Blended learning helps individuals to cater their needs according to their own pace. Each and every individual can choose their own way to achieve goals. Blended learning provides skillful and knowledegous resources which are needful for the students. Blended learning reduces the monotory of the traditional lecture method.

Blended learning is always considered as a cost-effective strategy. A learner can get detailed learning information even without the presence of an educator, though blended learning helps to improve the interaction amongst the students and the teacher. Students can access the resources from online learning sources such as Facebook live, messaging on WhatsApp, and online discussion. Here, students can utilize their learning time collaboratively with other participants.

10.1.2 FLIPPED CLASSROOM

Flipped classroom is another innovative approach of the modern technology-based educational system. It is known as the inverted classroom. A student can attain their aspiration level according to their own pace. They can mould their own leraning strategies.

In a flipped classroom, the teacher provides instructions at home, and the pupils engage themselves in doing home assignment in the class. The subject matter of the flipped classroom is instructed through online, outside the classroom. And the instructor can utilize the classroom time for different thought-provoking activities which influence the students. A flipped classroom follows a pupil-centered teaching-leraning approach. Here, an instructor can use classtime perfectly, so that learners get proper space for acquiring knowledge, gathering experiences, and attaining the most appropriate skills to achieve their goals, and by this way, learning and retention are maximized. In a flipped classroom, pupils carry out their intensive studies at home. After that, they engage in the classroom with the help of the teacher who explains the gathered knowledge in an innovative way.

A flipped classroom promotes an atmostphere where a teacher and students engage themselves in an interaction process which enhances learning through application and practice. A flipped classroom focuses on the student-centered approach. Students are encouraged to learn through experimentation. Teacher provides demonstration to classes and lectures before regular class time. Students are allowed to conceptualize their leraning activities and are allotted enough time to evaluate their own learning activities after acquiring knowledge from video lectures before coming to the class.

10.1.2.1 ADVANTAGE OF THE FLIPPED CLASSROOM APPROACH

In a flipped classroom, students are considered a major determinant. The flipped classroom helps students to develop a valuable and productive approach to meet their needs and queries. Here, students are being obligated to learn through applying their erstwhile gathered information. Students can proceed on in and outside the class by communicating with teachers. The teacher encourages student to watch online videos.

Fulton (2012) explained that students' behavior can be observed by the teacher. Through observations, teacher will know various difficulties of students while in class or at home. After that, teacher can easily refurbish the curriculum.

Bergmann et al. (2013) found that flipped learning encourages students to involve in the active learning process. The need of the student is considered as a prime factor for learning. The flipped classroom approach always enhance the student–teacher interaction.

Millard (2012) revealed that the flipped classroom lays stress on interactive discussion. Teachers can act so freely, provide guidance to cater the need of the students regarding the context of the text. Team work is another essential element of flipped learning.

Schmidt and Ralph (2014) elucidated that teaching-leraning process should always be active. The teacher should try to furnish visionary and resourceful lectures to motivate the students. Theory of active learning has been given more stressed since the last century. It occurs in the flipped classroom only where teachers and students are co-operative to each other. They stressed that pupils are much egar to complete their content of learning in the class as they have gathered information and made themselves ready at home by viewing online lecture class.

Nichols (2012) explicated that the flipped classroom helps students to appraise the content of learning as they have got enough time to retrospect their gathered knowledge. Here online lecture of the instructor accelerates the total learning environment where students who cannot attend class will be able to make themselves perfect with more practice.

10.1.2.1.1 Implications of Online Video Lectures in Flipped Classroom

To collect information and to gather knowledge, people generally avail the Internet service. For billions of people in the entire world, the Internet becomes the key enabler of the newfangled and high-tech society.

Video is considered as a core element of a flipped classroom. In a flipped classroom, teachers prepare video lecture with proper content of learning and present to students through Facebook and WhatsApp. By watching online lectures at home, syudents try to acquire knowledge of the subject matter properly. By keeping themselves away from the classroom, students make themselves ready for regular classroom interactions with utmost novelty with their mentors. Here a teacher plays an important role as a conciliator to improve the entire classroom interaction with a supportive attitude and in a generous way. Teacher's cognizant and magnanimous effort through online leactures turn students enthusiastic, responsive, and co-operative to carry out their learning process in a smooth way.

10.2 METHODOLOGY

The teacher can use blended learning and flipped leraning approaches according to their own choice in the classroom. The teacher can create a co-operative learning environment where students learn and work together. Here, students share their experiences with each other. Co-operative learning helps in managing the classroom environment. Co-operative learning is very helpful for students to get their self-evaluation or assessment. It helps to develop interdependencebetween the group members.

Co-operative learning also encourages students to take their own responsibilities for their work and what they have already learned. In co-operative learning, a teacher should constantly assess the results of every group and try to find out the probable changes which are needful and to ensure efficiency. Co-operative learning also provides facilities for interatctive communication. Here, information is transmitted through digital tools used with a lesson to promote students' problem-solving capacities, development of critical thinking, and learners' elaboration of competences through co-operative learning.

With the help of the teacher's guidelines, the students at home begin their research, after viewing videos in an autonomous way. The students should proceed with a view to learn and understand whatever teachers have already designed as elements of video lessons.

TABLE 10.1 Student–Teacher Activities Continuously Increasing Through Digital Learning Methodology

Teachers Activities	Students Activities
Preparing an introduction on the subject	Acquiring introductory knowledge
Finding out the requisite qualities that are essential for the appropriate utilization of resources	Classifying the essential norms for proper application
Choosing attainable objectives that the lesson aims to achieve	Underlying the objectives and their utilities
Developing class discussion and critical thinking	Engaging wholeheartedly in classroom activities by enhancing problem solving capacities

Students collect data, analyze the gathered information, and synthesize the learned knowledge in a well-organized manner at home. After coming to the class, students build up the capacity to rationalize, to think critically.

10.3 BRIEF EXPLANATION OF EACH STUDY

The researcher used Facebook live and WhatsApp videos with the students of a higher secondary school to teach a poem and a song which have intensed learning value.

10.3.1 FIRST STUDY

On the basis of the analysis of the poem "The Definition of My Ego," written by a famous Assamese prolific writer, lyricist, and first Assamese movie maker Late Rukonwar Jyoti Prasad Aggarwala, the researcher made a video with the recitation of the poem and its explanation. The content of the poem is narrated below:

> It's all about human entity and definition of one's own self where the poet asked his ego or self about literature, song, Fine Arts, that is, what are actually these?

And lastly, he said to his self that what is self's (his) definition. In reply, the ego or self of the poet explains: 1. Whatever the self (ego) has written is literature. 2. Which one is created with Rhythm and tune by the self (ego) is song. 3. Whatever the self (ego) has drawn, written created, sung, beat, and made through which self (ego) can triumph the entire world that is Fine Art. 4. When the poet asks ego about his definition the self (ego) has replied with a smile by keeping his eyes on poet's face that the poet himself is the definition of his ego (delf). The researcher explained each and every point of the poem.

After uploading the video on WhatsApp and on Facebook, learners showed great interest on hearing the recitation and its explanation. When the learners acquired knowledge, they realized the philosophy of human entity and the relation of abstract ideologies with the concrete world and its values in human life. The learners enlightened themselves by correlating the literature song and Fine Arts that have immense values and how these three elements help human beings to inculcate great values of life. They realized the important values that are to be desired in life. Are these values rooted in reality? How can one realize them in one's experience? How does a man gain knowledge?

Literature is connected with the society and its social beings. Through literature, man can change their dream to the reality. Literature trains human society to find out truth, beauty, and goodness among each aspect of the society. Literature reflects the person's own character. That is the essence of literature.

Important songs are very helpful in expressing our feelings as well as emotions. They give relief and help human being to reduce stress. They help us to wash away all the pain, worries, and the tensions from our mind and soul. They also teach us peace and harmony which bind people together.

The Fine Arts improves self-actualization and self-concepts of human being. Fine Arts are the fundamentals of creativity through body, mind, and soul. Art is an effective way of channeling emotions, and recognizing exactly what the emotion means, we cannot deny the importance of Fine Arts in our life; it is there in our life in all its forms. Whether it is painting, music, or dance, we rely on then at some point in our life. Fine Arts soothes our soul and involves us in the culture of music. It affects our mood and mind and fill us with joy and absolute pleasure.

When the poet asked ego about his definition, the ego answered, "The Poet himself is his definition." This means that if all the human beings are well versed and well accustomed to their own culture, own history, and great social, human values concern with one's own native world, then only human beings can change the entire world with a positive way.

With the help of their explanation, learners at home began their research with the help of the teachers' guidance. In the classroom, the learners developed the ability to inculcate all the above-mentioned elements which have great values, that is, literature, song, and Fine Arts. Here, the teachers coached the learners and scaffolding in each and every sphere of the learners' program.

1. The learners were able to know the abstract ideologies.
2. Attributes of the poem which are corelated with human soceities.

Learnerers understood that the literature expands our horizons of mind, and it helps to reveal what is not in front of us:

1. They learned how to build critical thinking skills through literature.
2. They developed feelings of love and praise for other cultures and beliefs through involving themselves in literary works.
3. They formed a great concept that literature helps to revealing the truth and fact of human nature.
4. Teachers encouraged learners to use songs in the educational filed.
5. The teachers focused on the issue that for language learning, songs are very usable. Development of vocabulary, idioms, and expressions was introduced.
6. Teachers helped learners to select value-based songs according to their needs and interests.

Cultural aspects can also be introduced through songs; teachers encouraged learners in the classroom to find out the cultural differences of different races around them:

1. The teacher helped learners to use song lyrics for relating to situations of the entire world around us.

"Fine arts are great aspect in today's life." This viewpoint was already developed in learners.

The learners acquired knowledge that Fine Arts provide benefits as follows.

Sense of Beautificiation

A learner is able to know the sense of beautification that should be inculcated in mind. Thus, he/she may develop the esthetic awareness that makes a man fascinated to do anything, to solve any problem easily.

Inculcating Human Values

Fine Arts help learners to develop human values such as friendship, love, solidarity, honesty, truthfulness, goodness, empathy, sympathy, love toward inanimate objects, love and care for the nature, patriotic sense, compession, and punctuality, which make students extraordinary human beings for their all-round development in all aspects of life.

Developing Creativity

The learners acquire knowledge on how creativity can be developed in a proper way and how can that be related to the IQ level of human beings. After that, they will be able to know where, when, and how creativity should be applied according to the intelligence power of human beings.

They also acquire knowledge on how creativity is connected to interest, aptitude, and the aspiration level of human beings.

Arousing Motivation

Learners are able to understand the power of motivation, that is, how extrinsic and intrinsic motivations are helpful for arousing the interest level of the pupils in the class and also outside the classroom. They are also guided by the teacher to gain knowledge regarding intrinsic or internal motivation and how it helps students to develop their inner potentialities and enthusiasm.

Bringing Social and Emotional Integration

Learners are able to know the value of social and emotional integration, that is, how social cohesiveness is very important to make a perfect society. Social and emotional integration helps pupil to develop the feeling of brotherhood-ness, fellow feelingness, fraternity, equality, and equity in every sphere of the society.

Bringing Academic Intelligence and Achievement

Learners know how academic intelligence and achievement can also be enlightened by the perfect application of Fine Arts and how it helps to evolve the intense values and talents of the pupils.

Developing Sense of Unity in Diversity

Learners also develop the sense of unity in diversity as they are in a diversified environment where they can gather and learn more and more about diversified culture, nature, tradition, lores, social ethos, and almost all aspects of the entire society including the world.

Increasing Social Cohesion and Congruency

Learners also develop the feeling of social cohesion and congruency where they can acquire social efficiency for a better social life. Social cohesion helps people to live peacefully in the society by meeting their legitimate needs without harming their fellow people. Here every human beings are treated equally with a democratic outlook where they can fulfill their own wish and interest.

Devloping Self-pride, Self-trust, and Self-assuredness

The learners are able to know how to maintain self-pride, self-trust, and self-assuredness where they can feel complete freedom with utmost sense of loyality. Here, learners learn how to develop self-confidence among them, as self-confident pupils can achieve everything and overcome all the barriers to reach their goal or destination.

Exhibiting the Cultural Components

Learners can explore cultural components of the contemporary society, ancient societyits way of living livelihood, various developmental approaches, literature, scripts, songs, folk lores, etc. through Fine Arts.

Learners experienced that Fine Arts completely transfer the teaching-learning environment into a fantasy world with proper action where pupils can explore their inner potentialities to the maximum.

10.3.2 SECOND STUDY

The second study belongs to an educational value-based song "Namami Brahmaputra" written, composed, and sung by an Assamese singer, lyricist sri Nayanjyoti Saikia.

The researcher has added this song in a video, and after the presentation of the song, the researcher uploaded an informative lecture regarding the song and its mythical, geographical, and historical connectivity to several places, including Assam.

The content of the song has been narrated below.

The song is all about the mighty river Brahmaputra—the great river of Assam. In Assam, the river starts its journey from Sadiya and turns to southwest. When flowing through Tibet, this river is known as Tsang-Po. Its source is in the Tibetan Himalays in Chenmayung-Dugat Tamchuk Khambab Chorten. The river while flowing thorugh Himalayas is known as Siang in Arunachal Pradesh. As it enters the plains, the two main rivers Dibang and Lohit join together to form Brahmaputra.

In mythology, the River Brahmaputra was supposed to be the "Son of Param Brahma" (the creator of this universe). The Great Sage "Shantanu" and his wife "Amogha" were living in the Kailash. Lord Brahma wished to make "Amogha" as his beloved as Brahma got enchanted with Amogha's divine beauty, but she refused his proposal. Myth says Brahmaputra (watery form) was born from Amogha's nostrils after imbibing the Brahmabeej of Brahma. Santanu emplaced the watery form child in the midst of four mountains: the Kaliash, Sanvartak, Jarudhi, and Gandhamardan. The watery form drop by drop dilated into a Great Lake Brahmakunda. Sage Parasuram came to Brahmakunda by following his father Jamadagni's order to make himself clean, virtuous, and immaculate from matricide. After washing his axe in the holy Brahmakunda (the axe was attached with his hand) got unshackled. But the water of Brahmakunda became red because blood stains of the axe. Vargav Parasuram was being astonished with the miracle, and he considered the water of kundaas divine. He decied to bring it down to the plain. So he cut a channel and brought the red river (Lohitya) down to the Kamrupa or Pragjyotishpur (The City of Eastern Lights).

The Brahmaputra river flows through the Assam Valley and looks like a ribbon in the midst of the mountain ranges. The name Brahmaputra is an Aryan word. The Brahmaputra is also known as the Lohitya in Sanskrit; it means the "red river." The Ahom dynasty who rulled Assam for 600 years called this river as Nam-Dao-Phi. It means "the river of the Star God." Another source explains that the word Lohitya comes from Lauti or Tilao words of the "Kirat Tribe." The Brahmaputra has an enough number of tributaries and feeding streams which enhance the beauty of the land. The tributaries like Kapili gives tremendous water power potential to Assam and makes it "a place of Golden Crops." The great culture of intrigration also bloomed along the river Brahmaputra, where Aryans and non-Aryans mingle with utmost fraternity. Aryans and non-Aryans erected the bridge of reconciliation through assimilation.

After acquiring knowledge from the explanation of the song, the learners began enthusiastic activities at home to meet their queries about the River Brahmaputra. The classroom teacher helped students to correlate the song's content with geographical knowledge. The teacher used this value-based song to arouse the interest among students for geography by a tremendous way. The teacher used the song to explore key concepts and characteristics of each aspect of the River Brahmaputra along with its geographical entity:

1. About the River Brahmaputra and its origin.
2. About the state Assam as the River Brahmaputra runs through the valley.
3. How it becomes Assam's lifeline.
4. The main features of the Brahmaputra river system.
5. Its tributaries and islets.
6. Brahmaputra river pollution, erosion and flood which affect the environment.

With the help of teachers' guidance the learners' gathered experience about the origin of the River Brahmaputra from its myth to geography and its nomenclature according to various ethnic and linguistic group of people of the Assam Valley. The Brahmaputra runs through the valley and looks like a ribbon in the midst of the mountain ranges. It has strong under-currents. The River Brahmaputra rushes in torrents. The name Brahmaputra is an Aryan word. Later on, an eminent Assamese torchbearer of Assamese culture, Kalaguru Bishnu Prasad Rava, tried to correct the name of the river with a Bodo (an enthic group of Assam) word "Bullumbuthur," which means "river of bubbing water."

The Brahmaputra has two river islets: the one is as small as the other is big. The biggar one is Majuli in the district of Jorhat, and the other one is Umananda in the city of Guwahati. In Majuli, the Miri Tribe is living—a Tibeto-Burman Tribal population of indigenous habitation. The Majuli is a main place of pilgrimage for Vaishnavities of Assam.

The Umananda Island has a tremeadous rocky bed. During monsoons when the River Brahmaputra swells, the waters dash against it and produce a continuous roaring murmur.

The teacher encouraged students to know about the Assam—its ancient history, present geographical condition, and all about it. Students asked to find the origin of the name "Assam." In Sanskrit, "Assam" is known as "Asom." The meaning of "Asom" is "Unparalled" or "Peerless."

History reveals that in the 13th century, the Ahom dynasty entred into the Assam Valley by crossing the “Patkai” hills. They conqured and ruled Assam.

Till the 20th century, Assam was enriched with its deep woods and great swaps. With the passage of time, Assam underwent lots of changes in its texture and composition. The richest flora of Assam is found generally in the plateaus of the hills. The land of Assam is gifted with a certain timber of rare species. The climatic condition of Assam is humid in general. Assam is rich in oil and coal resources. The state of Assam is famous for tea, silk, and one-horned rhino.

The learners acquired knowledge on how the River Brahmaputra becomes the lifeline of Assam. People alongside the River Brahmaputra have been developing the civilization from the time immemorial by cultivating a variety of crops. Brahmaputra becomes the life force of people. It is the lifeline of Assam, its people, and culture. Futile lands on its bank help develop variety of crops, creating human civilization. The River Brahmaputra and its water resources potential of the country.

Teachers helped learners to know about the features of the Brahmaputra river system:

1. The Brahmaputra river emerges in Tibet, on the eastern part of the Lake Mansarovar.
2. The total length of this river is 2900 km.
3. It makes the largest delta in the world along with the River Ganges.
4. The inhabitants residing in different parts of the river valley belong to various cultures and origins.
5. The Brahmaputra leaves a lot of fertile alluvial soil deposits on its river banks.

At the same time, the River Brahmaputra can be dangerous because of its divastating flood. The teacher explained the learners about the causes of flood, its affects as the river undergoes severe floods that occur yearly which is marked as a curse for human being and its environment. Flood destroyes huge paddy fields after which farmers are unable to meet their legitimate needs for a long time. Every year during the months of June–September, flood causes severe annihilation and demise in Assam. The Brahmaputra erosion destroys nearly 4000 sq km of Assam land. According to *The Hindu*, the Brahmaputra and its tributaries have eroded away more than 4.27 lakh hectors of land since 1950, which is 7.40% of the total area of the rivers. This is one of the major problems faced by the state, according to Assam's water resources department.

The teacher helped students to know the cause of river bank erosion. Floods and erosion are interrelated. Floods are intricately linked to causes of erosion wherever erosion leads to floods in many cases. The basic reasons for erosion in the Brahmaputra is due to the fact that it is an extremely braided river.

10.4 CLASS EVALUATION

10.4.1 FIRST STUDY

Students were organized into groups and provided sheets of paper to express their views. They were asked to:

1. Differentiate among literature, song, and Fine Arts.
2. Write the importance of literature.
3. Write the impact of literature on students of adloscence age.
4. Write how literature, song, and Fine Arts are interrelated.
5. Write the essence of Fine Arts and how it influences the tender mind of the learners.
6. Write theme-based songs and value-based poems which can draw a picture of contemporary society.
7. Find out abstract ideologies from the aforesaid three elements: songs, literature, and Fine Arts.
8. Explain how songs become motivational for young students and narrate their benefit for the total development of body, mind, and soul.
9. Explore the teaching-learning implications of literature, songs, and Fine Arts.
10. Elucidate the function of literature and how it becomes the power-potential for enlightening the mind and heart of the readers.
11. Explain how literature is useful and productive to provide instruction and to teach beneficially.
12. Narrate how literature develops human's way of life, credence, and wisdom of the ages.
13. Apply the knowledge of Fine Arts to enhance the allure of concrete objects.

14. Write individualized opinion about the virtue and asset of the literature, songs, and Fine Arts.

All the group members shared what is written with each other and made a analytical report from each group. The students did synthesis of poetic ideologies and its implications for the concrete world in a summary. At last, students were asked to make a slogan, write a poem and a prose, draw some creative art, make some sculptures, write song, and write a play.

10.4.2 SECOND STUDY

Again students were asked to make groups to explore the concept of the song and its implications in the practical field. They were asked to:

1. Write a report about the devastating flood and erosion of the River Brahmaputra which affects the entire Assam almost every year and its possible remedies.
2. Write about the contribution of the Brahmaputra river and how it makes the Assam Valley fertile.
3. Write a analytical report of the Brahmaputra river pollution and its probable causes.
4. Explore how the big dam on the Brahmaputra river affects the fertilized soil of Assam Valley.
5. Write a report on the construction of a strategic tunnel under the Brahmaputra river, which has already been approved by the Government of India.
6. Find out the possibilities and opportunities of the four-lane tunnel. How the tunnel will be beneficial for military services and how it will connect Assam and Arunachal Pradesh.
7. Explain why the bridge over Brahmaputra is considered as important to connect the river island Majuli with Jorhat District of Assam or with entire Assam.
8. Prepare report on endangered water animals of the River Brahmaputra and how these water elements are badly affected by water pollution caused by human.
9. Explore about the most notable animals of the Brahmaputra Valley with special reference to Kaziranga National Park.

10. Make a study on a numerous varieties of fishes of the River Brahmaputra and how livelihood is managed through fishing by the people of river bank.
11. Narrate the climate of the Brahmaputra Valley; how it varies; and about annual rainfall.
12. Explain about hydroelectric stations and impact of hydropower potential of Assam.
13. Explore how inland navigation is more important for the River Brahmaputra.
14. Carry out research work to understand the hydrology of the River Brahmaputra with a scientific outlet.
15. Draw a Brahmaputra river map.
16. Write about the Brahmaputra River Basin.
17. Write report on correlation of value- or theme-based songs and geography along with cultural geography.
18. Do a project work on the nomenclature of the River Brahmaputra.

For the evaluation of learners' progress in learning, teachers helped learners to do project works, paper works, etc. on various topics of the program in the classroom; before that, the mentor instructed them to do research works by using web reports getting from Facebook and WhatsApp. These two innovative apps bring the revolutionary changes to the virtual world. These are two important virtual environments for any type of learning. Hence, the researcher used Facebook and WhatsApp for the learning of literature and geography in the higher secondary classes. By posting interactive videos such as lecture slides, online reference literature, teachers can create a virtual classroom environment.

10.5 FEEDBACK OF STUDIES

For teaching literature and geography in the higher secondary classes, flipped classroom was implemented. The teacher informed about the new flipped classroom methodology to the students. The teacher instructed the students to get their expectations clarified after class interactions, interviews, and discussions. All learning activities were congregated properly so that the students can be able to develop reasoning power to know how to merge both co-operative learning and the collaborative learning approach. The students showed utmost interest in such activities, and they had done all these whole heartedly. The researcher added some examples of initiatives and results.

The flipped classroom environment helped students to access the technologies and resources that were available on the web. By accessing all those resources, students were acquainted with various useful learning style and experiences.

Flipped learning helped various scholastic and nonscholastic activities. Learners acquired skills and knowledge of analyzing and synthesing literary work; choosing value- or theme-based songs which are helpful for teaching any subjects.

The students learned to make a relation between poetry and literature, song and geography, with special reference to Assam's geography and Assamese poetry which have immense contribution to the human society. The main purpose is to instruct:

1. How to interpret and explain a poetic text and their values.
2. How to interpret theme based songs, their text and how these can be explicated in the analysis of various educational subjects and its future applications with proper & perfect synthesis.

The teacher displayed all the students' activites on the Youtube channel, websites, and Facebook. Teacher encourages students to upload their project works or intensive study reports on Facebook, WhatsApp, etc. Through such activities, students updated themselves by keeping in touch with their mentors and classmates.

10.6 CONCLUSION

Learners of the modern era are more accustomed to technology. Flipping the classroom helps students in such a way that they can be able to develop self-confidence among them. Instructors need to guide their students to acquire knowledge when and why learning strategies and technological tools would assist the learners to reach their desired goals and in achieveing academic success. Flipped learning methodology enhances students' interest and does encourage students self-learning. Blended leraning and flipped classroom both are effectual and benignant for developing critical thingking, problem-solving ability, developing collaboration and communication skills, innovative ideology, global citizenship, and enhancing the ability to become lifetime learners, strengthen independent thought. The 21st-century students need to inculcate all values, skills, and develop intense knowledge of technology to cope themselves up with the virtual world.

10.7 POTENTIAL FOR FUTURE RESEARCH

In some areas, Internet facilities are not available or sometimes they are very limited. After conducting the flipped classroom approach in schools, the author provided suggestions that researchers can apply in the practical filed.

The suggestions are as follows:

1. Recorded version of lecture should be provided by the researcher where the Internet facility is limited.
2. Emphasis should be placed on the development of technologies which assist the teaching learning process.
3. Where possible smart class facilities should be implemented.
4. Stress should be laid on making regular phone calls by the teacher to encourage the students.
5. Where live interation between teacher and the students is not so satisfactory due to lack of proper Internet facilities, teacher can use WhatsApp for sending audios and text.
6. Teacher should try to evade PDF files as it requires more data for downloads.
7. Should try to provide workbooks and pamphlet to students with necessary learning contents.
8. Should prepare DVD disk which serves as an important teacing learning tool to upgrade students' knowledge.

The author's expectation is that all the suggestion put forwarded for utilizing the teachnology in the classroom can be instrumental and worth-while to other researcher in future study.

KEYWORDS

- **blended learning**
- **flipped learning**
- **social media**
- **theme-based song and value-based poem**

REFERENCES

Bergman, J., Overmyer, J., & Wilie, B. (2013, July 9). The flipped class: what it is and what it is not. *The Daily Riff.* Retrieved July 11, 2014, from http://www.thedailyriff.com/articles/the-flipped-class-conversation-689.php

Fulton, K. (2012). Upside down and inside out: Flip your classroom to improve student learning. *Learning & Leading with Technology*, *39*(8), 12–17.

Millard, E. (2012). 5 reasons flipped classrooms work. *University Business*, pp. 26–29. https://www.questia.com/read/1G1-312106284/5-reasons-flipped-classrooms-work-turning-lectures

Nichols, D. (2012, September 18). Flip classroom instruction: How to Guide Part 1. *Educational Technology Tips*.

Schmidt, S. M., & Ralph, D. L. (2014). The flipped classroom: A twist on teaching. *Contemporary Issues in Education Research*, *9*(1), 1–6.

CHAPTER 11

Digital Education: From the Discipline to Academic Opportunities and Possible Academic Innovations—*International Context and Indian Strategies*

P. K. PAUL

Department of CIS, Information Scientist, Raiganj University, West Bengal, India

**Corresponding author. E-mail: prantoshkpaul@gmail.com*

ABSTRACT

Digital affairs and engagement are increasing day by day, and various sectors and spaces are using digital tools, techniques, and technologies. As a fact, the teaching-learning process and moreover the education sector have become high-end technology-specific in the recent past. The emergence of technology and related issues lead to many innovative concepts as well as domains, namely e-learning, online education, education technology, virtual education, and blended learning. Initially, only education technology or e-learning was considered as a concept. But the gradual involvement of other areas led the concept of digital education under one roof. Later on, this was also called as digital learning. Digital education has emerged as an interesting academic innovation. It has become a field of study in the recent past. Some of the international universities have already started to offer educational programs and allied courses not only doctoral but also bachelors and master's degrees, including their subfields and allied areas (mentioned above) in the digital format. Moreover, digital education also offers courses in different subjects and in a concentration and/or a major. Due to its enormous potential, digital education is considered an important research area in many research centers and universities. It is worthy to note that the research degrees, namely PhD and even professional doctorate degrees, have started in many countries with the help of digital education. This study is concentrated on the contribution of digital education, in the recent past, in master's degrees and doctoral degrees. With adopting specified research methodologies, this chapter highlights these issues. Moreover, the chapter also describes the foundation, nature, characteristics, and the role digital education has played in the recent past. It also mentions about the future growth and stakeholders in this field, including the technologies involved and possible educational programs in the context of India and developing countries.

11.1 INTRODUCTION

Digital education is a broad area that deals with the digital tools and technologies integrated with educational aspects. Information technology and computing, namely communication technology, database technology, networking technology, multimedia technology, software technology, etc., plays a leading role in the creation of a better and sophisticated digital education system (Blikstein, 2013; Edwards, 2015; Paul et al., 2018). Digital education is also known as digital learning and is an emerging tool applicable in various areas such as higher educational institutes (HEIs), universities, and colleges internationally. However, this is important that in the developed nations, the schools system should also be integrated with digital education concepts in their wide range of areas including classes, examinations, operations, and administration (Buchanan, 2011; Gibson et al., 2015; Paul et al., 2018). The most common form of digital education may be considered with the following (but not limited to): adaptive learning-based classroom systems and technologies, electronic textbooks, learning, contents, objects, and analytics, open and free educational contents and resources, content and streaming contents, teaching and learning enabled with technology, etc.

In the present context, digital education can be treated as a field of study, and in this context, e-learning (also called as educational technology) may be considered as a branch or subfield of digital education. It is concentrated on a practicum of educational approaches and complete knowledge transfers (Dillenbourg, 2016; Dunleavy et al., 2019; Reeves, 2003) and also on new aspects of development and exchange among the learners. Recently, the use of digital education and similar technologies has also enhanced in corporate houses, companies, and organizations owing to their profit-making ability.

Digital education and technologies are used for the purpose of, but not limited to, training (with job based or on-the-job training), higher studies, and research opportunities in the companies and corporate houses. Also, the use of such technologies has increased for continuing learning and education, skills development, etc., especially regarding the blended learning. Similar to the corporate houses, the integration of digital education and similar technologies into the higher education and research centers is an important tool. Here different kind of technologies are useful in various purposes, namely education and research including training and developments in HEIs, universities, etc., and many of the programs in the field or allied fields started with bachelors, masters, doctoral degrees with different nomenclatures and titles (Elboubekri, 2017; Heller et al., 2008; Paul et al., 2018). Here it is worthy to note that apart from the digital education, similar software can be offered such as

1. ET or Education Technology or EduTech.
2. Electronic Learning or E-Learning
3. Virtual and Online Learning
4. Digital Instruction and Teaching
5. Digital Pedagogy
6. Online and Blended Learning.

Universities and HEIs can start these potential programs at different levels and in various ways such as major in the subject and allied areas, minor in digital education, honors in digital education, concentration/theme and specialization, etc. in the subject and allied areas. The details of current opportunities (based on methodology adopted), technological aspects, and educational affairs are discussed in the following sections.

11.2 OBJECTIVES

The current chapter is theoretical as well as conceptual in nature, and additionally it is the combination of management, educational science, and technological studies, and thus it is interdisciplinary in nature. The chapter thus carries the following agendas (but not limited to):

1. To learn about digital education and allied areas such as its general meaning.

2. To gather the knowledge about basic functions as well as features of digital education in the basic context, including the advantages of digital education and learning.
3. To understand the applications, integration, as well as utilization of the current theme of digital education, allied areas, various sectors, etc.
4. To gather the knowledge about the future growth of the stakeholders of digital education and its changing scenario.
5. To explore different technologies and tools regarding digital education as well as e-learning, including the future scenario.
6. To gain the knowledge regarding the methods of e-learning, including digital education and allied areas in simple sense.
7. To get about the educational and manpower-related availabilities and opportunities of the fields, namely Digital and Electronic Education, apart from the Education Technology, Online Learning, Blended Learning, etc.
8. To get the knowledge on bachelors, masters, and doctoral programs on this field and offering universities with review of basics of curriculum components.
9. To learn about the academic and professional doctorate degrees available in some countries with its emergence in digital education and allied fields.
10. To propose the degrees, nomenclatures and areas in digital education and educational technology-related areas in the Indian context.
11. To learn about the proposed concentration of the field in Information Science and Computing, including allied areas.

11.3 METHODS ADOPTED

The current chapter is theoretical and conceptual in nature. Further the chapter holds interdisciplinary features and functions, as it is to be mixed with the Educational Science, Information Science and Computing, Management Science as well. It thus carries multiple methods, namely collecting secondary sources and reviewed journals on the field and combine them into a meaningful sense and form. Further, web review is done in two phases/contexts. Firstly, gathering the visibility of the term digital education in organizations, educational institutes, etc. Secondly, gathering the educational programs offering in different universities and HEIs. Further,

Google has been used as a search engine, and the keywords searched were "Digital Education," "MSc-Digital Education," "PhD-Digital Education," ""EdD-Digital Education," and "EdD-Educational Technology." Also for each keyword, total 15 results were considered and further analyzed and reported in the form of scientific literature.

11.4 DIGITAL EDUCATION AND LEARNING: FUNDAMENTALS

Digital education is a kind of learning; it may also be treated as an emerging educational system backed by technologies—mainly computing or information technology. In the instructional practices effective technologies are being used; further enhancing the utilizations and application of practices and it is possible to carried with the integration of blended and virtual learning. Online learning, electronic learning, and digital learning are identified with the digital education, and thus these branches can be treated as a subfield of digital education (Hanna, 1998; Narciss, 2013; Robin & McNeil, 2012). Among the common technologies and strategy that may be considered as important and useful are adaptive and intelligent learning, learning analytics and delivery, learning objects and contents, open educational resources, blended learning, classroom-based technologies, e-textbooks, e-contents, technology-enhanced teaching/learning, mobile-based learning, personalized learning, including self-learning, real-time online learning (i.e., e-learning real), virtual-reality-based learning. It is worthy to note that use of these tools and strategies may require combined way and is increasing day by day in various contexts (Hiltz & Turoff, 2005; Paul et al., 2012; Robin & McNeil, 2012).

11.5 DIGITAL EDUCATION: DEFINITIONS AND MEANING

Initially, digital education was treated as a concept, but gradually it has become a tool, and moreover a field of study. Virtually, this concept is changing rapidly, and it is tough to explain digital education with a single definition. Digital education is a kind of field, practice, or technology that is dedicated to the utilization, development, and improvement of the education system with the support of IT and computing; among its common subfields common are e-learning and online education concept (Huang et al., 2019; Paul et al., 2014; Reeves, 2003). Further, digital education can be treated as a technology-supported teaching-learning system with IT and digital technologies in healthy management and administration, including the leadership of education and studies (including research). Digital education is a theory and practice of educational approaches to the teaching-learning activities. Digital education can simply be considered as educational technology that helps in communication of knowledge, including the exchange of views using IT and computing. Digital education uses various kinds of tools and systems, including information technologies, for the delivery of educational activities, namely curriculum components and complete management. Digital education can be considered as a tool for healthy educational management-based information systems dedicated to educational activities and promotion of complete educational administrations (Knox, 2016; Paul et al., 2012). Digital education or educational technologies are helping in back-office management, including administration such as namely, budget and HR management, and learning record store, official communication, network building, communication among the departments and other institutions for complete information solutions, emphasizing data storage and analysis. Digital education and similar fields are using similar technologies (e.g., IT) in the educational activities and systems. Now, it has become a strategy toward the educational development with improved, IT-enabled, and efficient teaching-learning affairs.

11.5.1 DIGITAL EDUCATION AND SYNONYMOUS AREAS

Digital education is quite a new term and has been conceptualized as a field of study in the recent past. Digital learning can also be an alternative nomenclature in this regard. However, apart from these two, following nomenclatures are also valid and important:

1. ***Education Technology:*** This is similar to digital education, is widely used, popular, and treated as a field as well.
2. ***Online Education:*** Application and utilization of computing as well as IT use to promote education online is called online education. Here, cloud-based technologies need to use for better performances.
3. ***Virtual Learning:*** Similar to the previous one, virtual learning is the ICT-based education. It may or may not be online.
4. ***E-Learning:*** It is the educational system for teaching and learning adopting IT and computing. Further, administrative activities can also be fulfilled by the e-learning.

5. ***Blended Learning:*** It is simply the combination of manual education, that is, physical (face to face), full time and part time education, distance learning, self-studies through online means.

Hence, digital education may be treated as the combination of all of the above the use of which has risen in the recent past. Moreover, it is gaining interest as a domain/field of study instead of the existing concept of tools/technologies in educational means (Hiltz & Turoff, 2005; Robin & McNeil, 2012).

11.5.2 METHODS OF DIGITAL EDUCATION

Digital education can enhance the teaching-learning activities and experience as well. Various combinational and traditional methods can be used for this (but not limited to): blended and hybrid tools, online education, e-learning, differentiated and advanced learning, individualized and personal learning, flipped learning, gamification, understanding and education by design, universal design for learning, user experienced content, content management systems, enterprise resource planning, etc. These technologies as well as tools can be suitable and useful in common pedagogies, especially in the practices of teaching which includes the technology and blended learning perfectly.

11.5.3 TOOLS AND DIGITAL EDUCATION ENHANCEMENT

As far as the promotion and development are concerned, various tools, techniques, and technological resources are used for better education systems and for bringing a newer digital learning environment, namely a smarter learning system. There are many tools and resources that might help in building digital education or smart education, namely (but not limited to), Social Readers or RSS, YouTube and similar channels, Google + Communities, including Digital Pocket, Zoom Meeting and Networking, MS Meeting, CISCO WebEx, Zotero, Skype, Cloud-based Word Processors, Cloud-based file-sharing platforms, Evernote, etc. These are the sample tools that are increasing day by day for the betterment of the creation of digital education.

11.5.4 BENEFITS OF DIGITAL EDUCATION

Digital education comes with a valuable service in building smart and healthy education systems. Moreover, it comes with various advantages and benefits as follows (Ozga, 2016; Wahabi et al., 2019).

Time: Digital education and learning is helping in reducing time as it comes with the advantages of the electronic education system. It comes with anytime and anywhere opportunity. Owing to no physical distance to cover, it helps in time management.

Place: Digital education normally does not come under the traditional classroom settings; thus, place is not an important factor here. ITs and other systems are helping in making this for healthy learning and education. Further, digital education can also be offered in traditional educational as a beneficial tool.

Path: Modern-day education is not only restricted to one pedagogy; today various advanced technologies and computing systems are important to reach a specific path of the education system. Modern and smart education is a kind of learning technology that promotes healthy and newer educational pedagogy and methods in real time. For betterment, we need better e-learning technologies.

Pace: As digital education comes with various IT components, namely, database, web technology, multimedia technology, software technology, it allows the stakeholders and users minimizing time and helps in reducing time in different activities.

11.6 DIGITAL EDUCATION AND STAKEHOLDERS

Digital education is actually not only covered by the simple laptop and/or projector, but it is the combination of technologies, computing and information/content. Instructions also play a valuable role in educational deliveries (but not limited to) (Figure 11.1).

Technology: In digital education, technologies are the most important driver to deliver online teaching and learning process, and within this, different subfields of IT are important, namely networking technology, database technology, multimedia technology, and web technology. However, apart from these, various kinds of tools and components are also important (Paul et al., 2012; Xu et al., 2019).

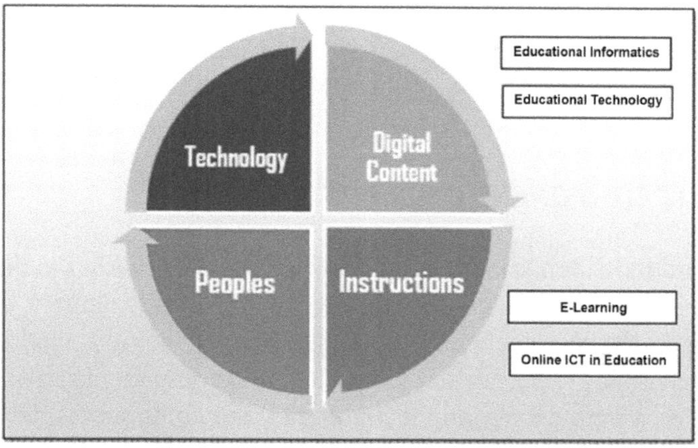

FIGURE 11.1 Stakeholders of digital education and allied nomenclatures.

Digital content: Content and information are vital for building and developing smart education systems designing and development of technologies such as Online Education, Electronic Learning, Virtual Education and Learning, and all other areas of digital education. Only computing and technologies have become valuable in digital education and systems in different sorts.

Instruction: Instruction is most valuable to offer knowledge including the education and degrees, teaching and learning, and educational delivery-based activities. Here digital education and learning plays an important role for healthy instruction and delivery.

People: People are most important stakeholders in designing and developing digital education systems, and thus this is one of the mandatory parts in this connection. However, people can be categorized into following (as far as digital education is concerned):

1. Students/Learners/Trainees
2. Teachers/Faculties/Trainees
3. Administrators/Educational Authorities
4. Mediators.

11.7 DIGITAL EDUCATION AND EMERGING TECHNOLOGIES

Different technologies are applied for the promotion of educational activities, administration, and operation, as far as digital education is concerned.

Further, within technologies, IT, computing, and management sciences are important and considerable. However, in this context, the following aspects are important to note down.

11.7.1 IT COMPONENT IN DIGITAL EDUCATION

In Information Technology main jobs are related with the information; therefore various subfields are considered important, namely, networking technology, database technology, multimedia technology, web technology, security technology, etc. Virtually the fields and components are increasing in IT fields. Among the emerging technologies and components, few important which help a lot in digital education and similar activities are following. The benefits of ICT in education including its basic and advanced components are depicted in Figure 11.2.

11.7.1.1 CLOUD COMPUTING AND VIRTUALIZATION

Cloud computing is a type of virtualization technique of IT infrastructure, including hardware, software, applications, systems, platform, contents easily available in a remote platform using the Internet and similar technologies. In a physical classroom, teaching learning, online learning, educational administration, and cloud computing can be a good tool. Admission, teaching-learning, examination and evaluation all spaces are close with cloud computing support these days (Hiltz, & Turoff, 2005; Wahabi et al., 2019).

11.7.1.2 BIG DATA AND ANALYTICS

Big Data Analytics is the combination of *Big Data Management* and *Analytics* which are required for the data management, especially the complex and large number of data and information management. Further, education means information, so that in educational affairs, particularly in digital education operation and in E Learning, it is most important and valuable. Big Data Analytics is needed in different activities in education/teaching learning. Among these, few important are: physical and virtual classroom teaching learning, educational management and administration, application and admission of the candidates, examination and evaluation, etc.

11.7.1.3 USER EXPERIENCE DESIGNING AND HCI

User experience designing is about the designing of user-friendly information systems and computation. it is commonly abbreviated as UX or UXD. It is very close to the usability engineering, human–computer interaction, human-centered computing, and man–machine interaction. In digital education and educational technology, practices mainly required in the sectors such as creation, development of contents for the students and learners, multimedia presentation for the academic events, namely seminars, conferences, workshops, delivery of user-friendly online contents, student documents, brochures, online kiosk, and information kiosk enriched with HCI principles.

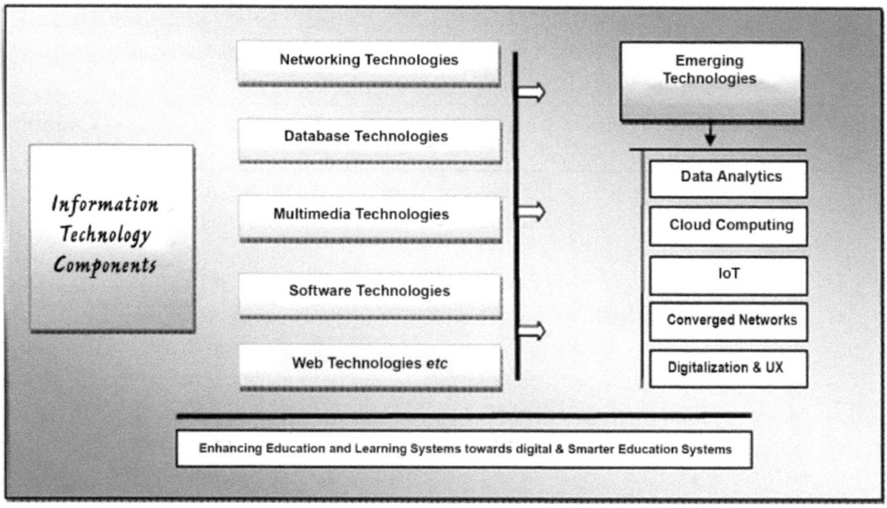

FIGURE 11.2 The components of IT and emerging areas for promoting digital education.

11.7.1.4 ADVANCED NETWORKS AND CONVERGENCE

Network is a very important component in digital education—both for physical as well as real time and/or streaming. In this case, apart from general networking affairs of education, multimedia-based contents are normally delivered by the advanced and converged network for better and healthy delivery of education and teaching learning.

11.8 INTERNATIONAL VIEW CONTEXT OF DIGITAL EDUCATION AND EDUCATIONAL PROGRAMS

Digital education as well as other similar technologies have become important in academics, institutions, and organizations. As a result, different universities and educational institutes have started to prepare skilled human resources in this field. Practically, the need of digital education and similar subjects has increased in the recent past, and thus internationally, many universities have started to offer educational programs in this field not only at the research level but also at bachelors, masters, doctoral, and post-doctoral levels. According to the concerned methodology of research regarding this work, it is noted that nomenclatures such as "Digital Education," "Education Technology," and "Digital Learning," are well explained in Table 11.1.

TABLE 11.1 Universities with Digital Education and Allied Program According to Adopted Methods: International Look

Sl. No.	Name of the Program	Institute/University	Remarks	Country
1.	Master of Science (MSc) in Digital Education	EU (University of Edinburgh)	1 year (relevant bachelor degree)	UK
2.	Master's Degree in Digital Learning (MDL)	CDU (Charles Darwin University)	2 years (The program is open to any degree students)	Australia
3.	Master of Education—MEd (Digital Education Leadership)	SPU (Seattle Pacific University)	1–2 years (any degree is acceptable)	United States
4.	Master of Education—MEd (Digital Learning and Leading)	LAMAR University	1 year (any degree is acceptable)	United States
5.	Master of Professional Practice (Digital Learning)	DU (DEAKIN University)	2 years	Australia
6.	MSc in Education (Digital Age Learning and Educational Technology)	JHU (John Hopkins University)	1–2 years (any degree is acceptable)	United States

Further, many universities even started doctoral programs on digital education or similar technologies/nomenclature. According to this research work, only professional doctorates or doctoral degrees other than PhDs were considered. Based on the result, the common nomenclatures of the subjects are education technology, instructional systems and technology, learning technologies, education technologies and leadership, etc. The list of universities in this category is increasing. It is important to note that, these programs are little different as they involve huge and rigorous coursework and taught components. These courses are suitable for the professionals already engaged in teaching learning, educational operation, administration, etc. The programs are also suitable for the fresher who wants to be a digital education or e-learning expert or professional (Table 11.2).

TABLE 11.2 Depicted Doctor of Education (EdD/DEd) Programs (Internationally)

Sl. No.	Degrees offered	Universities	Eligibility and Credit etc	Country
1.	EdD (Doctor of Education) in Educational Technology	Walden University	Master's in any field or higher degrees online mode 79 credits	USA
2.	EdD (Doctor of Education) in Educational Technology	The University of Florida	Master's in any field or higher degrees blended mode 54 credits	USA
3.	EdD (Doctor of Education) in Educational Technology	Boise State University	Master's in any field or higher degrees online mode 66 credits	USA
4.	EdD (Doctor of Education) in Instructional Design and Technology (IDT)	The University of Memphis	Master's in any field or higher degrees on campus mode 56 credits	USA
5.	EdD (Doctor of Education) in Instructional Technology • Advanced Track (Technology Leadership) • Initial Track (Basic)	Kennesaw State University	Master's in any field or higher degrees online mode 66 credits	USA
6.	EdD (Doctor of Education) in Learning Technologies	Pepperdine University	Master's in any field or higher degrees blended/hybrid mode 51 credits	USA
7.	EdD (Doctor of Education) Educational Technology Leadership	New Jersey City University	Master's in any field or higher degrees online mode 60 credits	USA

TABLE 11.2 *(Continued)*

Sl. No.	Degrees offered	Universities	Eligibility and Credit etc	Country
8.	EdD (Doctorate in Education) in IT, that is, Instructional Technology	American College of Education	Master's in any field or higher degrees online mode 64 credits	USA
9.	EdD (Doctor of Education) in Instructional Design and Technology (IDT)	Idaho State University	Master's in related field or higher degrees online campus mode 64 credits	USA
10.	EdD (Doctor of Education) in Educational Technology	The University of Phoenix	Master's in any field or higher degrees online mode 54 credits	USA
11.	EdD (Doctor of Education) Educational Leadership in Educational Technology	Wilkes University	Master's in any field or higher degrees on campus (part time) 64 credits	USA
12.	EdD (Doctor of Educational Science) in Instructional Technology and Distance Education	Nova South Western University	Master's in any field or higher degrees on campus (part time) 55 credits	USA
13.	EdD (Doctor of Education) in Educational Technology Leadership	Morehead State University	Master's in related field or higher degrees on campus (part time) 60 credits	USA
14.	EdD (Doctor of Education) in Educational Leadership and Management (*Educational Technology Track*)	Drexel University	Master's in related field or higher degrees on campus and online 60–61.5 credits	USA
15.	EdD (Doctor of Education) in Curriculum and Instructions (Educational Technology Track*)*	University of Nevada	Master's in related field or higher degrees on campus 60 credits	USA
16.	EdD (Doctor of Education) in Educational Practice and Leadership (Instructional Design and Technology Track)	University of North Dakota	Master's in any field or higher degrees online 60–90 credits	USA

TABLE 11.2 *(Continued)*

Sl. No.	Degrees offered	Universities	Eligibility and Credit etc	Country
17.	EdD (Doctor of Education) in Educational Leadership (Instructional Technology Leadership Track)	Appalachian State University	Master's in any field or higher degrees online and part time 60 credits	USA
18.	EdD (Doctor of Education) in Learning Design and Performance Technology	Florida State University	Master's in any field or higher degrees online 66 credits	USA
19.	EdD (Doctor of Education) in Digital Technologies	The University of Strathclyde, Glasgow	Master's in related field or higher degrees full time and part time 180 credits	UK
20.	EdD (Doctor of Education) in Instructional Systems Technology	Indiana University, Bloomington	Master's in any field or higher degrees online 60 credits	USA

Most of these programs are also suitable for the professional working in the fields, including education, training, and technologies. Based on this research work, it is revealed that the universities normally offer these programs for minimum three years; however, in some cases, it is three and a half year or four-year program. Further, in some countries such as UK, Singapore, and Australia, the minimum duration of the program is extended from three years, and it varies from five years to seven years in the case of the part-time program. From the masters-based study, it has been noted that digital education is available in different forms, degrees, and concentrations, as far as the masters level is concerned. Among the important common degrees are

1. MSc/MS (Master of Science)
2. MEd (Master of Education)
3. Master of Professional Practice/Studies
4. Master of Digital Learning.

So, apart from the science and education track, digital education is also available in professional studies. Social science tag also started by few universities, directly and indirectly. One-year degree program is the most common degree program, whereas few universities, namely DEAKIN University, Australia and Charles Darwin University, Australia also offer two-year degree programs. Education, technology (including IT and computing), and management sciences are most common curricula components and gradients in this regard. Technology concentration, technology policy concentration, instructions, and limited educational affairs are common in most curricula, whereas few universities have concentrated on educational aspects and policies. Few programs are also research oriented and conceptual. Sample curricula in international universities on these subjects are offered by few universities. These are listed in Table 11.3 (Paul et al., 2018; Wahabi et al., 2019).

TABLE 11.3 Digital Education and Learning-based Master's and Basic Curricula

Universities and Programs	Papers/Courses
Program: Master of Digital Learning **University:** Charles Darwin University **Territory:** Australia	Digital Technologies and Project Management in Education (Core) Research (Core) Internship (Core) Elective Papers New Media New Media Web Development Web Development and Programming System Management
Program: Master of Education—MEd **Subject/Major:** Digital Learning and Leading **University:** LAMAR University **Territory:** United States	Concepts of Educational Technology Digital Learning in Local and Global Contexts Assessing Digital Learning and Instruction Digital Citizenship Applying Educational Technology Portfolio Leading Organizational Change Disruptive Innovation in Education Creating Significant Learning Environments Resources for Digital Environments Synthesis of Digital Learning and Leadership/Capstone Instructional Design in Online Learning Special Topics - Technology
Program: Master of Professional Practice **Subject/Major:** Digital Learning **University:** DEAKIN University **Territory:** Australia	Critical Thinking Communication Digital Learning, Designing, and Assessment Research Design, Development, and Methods Digital Learning Professional Expertise Digital Literacy Innovation Problem solving Teamwork

TABLE 11.3 *(Continued)*

Universities and Programs	Papers/Courses
Program: Master of Science (MS) Education **Subject/Major:** Digital Age Learning and Educational Technology **University:** John Hopkins University **Territory:** United States	**Core:** Technology and the Science of Learning **Electives:** Culturally Responsive Teaching Evaluation and Research in Digital Age Learning Advanced Applications in Digital Age Learning Emerging Issues in Digital Age Learning Advanced Seminar in Digital Age Learning **Specialization:** (Online and Blended Learning) Gaming and Media Design for Learning Instructional Design for Online Learning Designing and Delivering Online and Blended Learning Environments (Digital Age Learning Leadership) Technology Leadership for School Improvement Integrated Media into Standards-based Curriculum Data-driven decision-making

Source: Websites of respective universities.

11.9 INFORMATION SCIENCE AND ALLIED AREAS AND DIGITAL EDUCATION/ALLIED FIELD POTENTIALITIES ADOPTABILITY

Information Science is a broad field and interdisciplinary in nature responsible for information activities (i.e., collection to dissemination). As far as the smallest part is concerned, Computer Science (CS) is important which deals about the internal affairs of computing, including the designing, development, etc. The field with deep concentration on manufacturing and application designing is treated as Computer Science & Engineering (CSE). It is further theoretical and mathematical in nature, thus also works for management of computer systems. Computer Application (CA) is an another branch which is software centric and deals with programming languages for the design and development of the applications. This branch can be considered under Information Science; it will be worthy to note that "Computing" is a field of study in many countries, internationally. IT is another sub-branch within Information Science, and here core fields are treated as Database Technology, Multimedia Technology, Web Technology, Software Technology, Network Technology, etc.

ICT or Information and Communication Technology is a study area under the field of Information Technology with special focus on Communication and Networking Technology. Information Science is treated as broader than all these fields and very similar to IT, but it is specially focused on society and sector-specific applications of IT and allied fields, namely Healthcare, Education, Business, Governance, and Administration, etc. This branch is treated as Applied Science and is interdisciplinary in nature. Apart from these, few other branches include Information Management, Information Studies, and Information Systems. It is important to note that these branches are offered as bachelors, masters, doctoral degrees in many universities. However, due to its importance, need, and the current scenario, digital education can be offered. Tables 11.4–11.7 provide the possible programs and specializations of digital education and similar technologies in respect of the subjects, namely CS (and also CSE), CA, IT (also ICT), Information Science (also Informatics, Information Science, and Technology).

TABLE 11.4 Possible UG, PG, and Research Programs of Digital Education in CS/CSE

Under Graduate	Postgraduate and Research
BSc in Computer Science	MSc in Computer Science
(Digital Education/Education Technology, etc.)	(Digital Education/Education Technology, etc.)
BS in Computer Science (Digital Learning and Education/Education Technology, etc.)	MS-CS (Digital Learning and Education/ Education Technology, etc.)
BTech/BE in Computer Science and Engineering (Digital Education/Education Technology, etc.)	MTech/ME (Digital Education/Education Technology, etc.)
BSc (Engg.) in Computer Science and Engg.	MSc (Engg) (Digital Education/Education Technology, etc.)
(Digital Education/Education Technology, etc.)	MS/MTech in CS/CSE (by Research)
	(Digital Education/Education Technology, etc.)
	PhD (Science/Engg.) in CS/CSE
	(Digital Education/Education Technology, etc.)

It is worthy to note that in the countries (such as in India and UK) where CS is available as an engineering degree, BTech/BE/MTech/ME can be introduced through digital education or similar technologies, namely Education Technology, Educational Informatics, etc. Further, Table 11.5 proposes the possible programs in CA subjects.

TABLE 11.5 Possible UG, PG, and Research Programs of Digital Education in Computer Application

Under Graduate	Postgraduate and Research
BCA (Digital Education)	MCA (Digital Education)
BCA (Education Technology)	MCA (Education Technology)
BCA (Educational Informatics)	MCA (Educational Informatics)
BCA (Digital Education and Learning Analytics)	MCA (Digital Education and Learning Analytics)
BCA (E-Learning)	MCA (E-Learning)
	MCA (by Research)
	(Digital Education/Education Technology, etc.)
	PhD CA
	(Digital Education/Education Technology, etc.)

Moreover, it is worthy to note that the subject CA is mainly offered in India and other countries. Similar subjects in the field of Computing can also be offered with these specializations (e,g,, BSc/MSc/BS/MS-Computing-Digital Education or allied branches). Table 11.6 proposes the programs on digital education with IT specializations.

TABLE 11.6 Possible UG, PG, and Research Programs of Digital Education in IT/ICT

Under Graduate	Postgraduate and Research
BSc in Information Technology	MSc in Information Technology
(Digital Education/Education Technology, etc.)	(Digital Education/Education Technology, etc.)
BS in Information Technology (Digital Education/Education Technology, etc.)	MS in Information Technology (Digital Education/Education Technology, etc.)
BTech/BE in Information Technology (Digital Education/Education Technology, etc.)	MTech/ME in Information Technology (Digital Education/Education Technology, etc.)
BSc (Engg.) in Information Technology (Digital Education/Education Technology, etc.)	MSc (Engg) in Information Technology (Digital Education/Education Technology, etc.)
	PhD (Science/Engg.)-IT/ICT
	(Digital Education/Education Technology, etc.)

Since ICT is a subfield of IT, therefore digital education or similar nomenclatures at bachelors, masters and PhDs and detailed are depicted. Table 11.7 depicts programs on digital education in the Information Science track.

TABLE 11.7 Possible UG, PG, and Research Programs of Digital Education in Information Science

Under Graduate	Postgraduate and Research
BSc in IS/ IST/Informatics	MSc IS/IST/Informatics
(Digital Education/Education Technology, etc.)	(Digital Education/Education Technology, etc.)
BS in IS/IST/Informatics (Digital Education/Education Technology, etc.)	MS in IS/IST/Informatics (Digital Education/ Education Technology, etc.)
BTech/BE in IS/IST/Informatics (Digital Education/Education Technology, etc.)	MTech/ME in IS/IST/Informatics (Digital Education/Education Technology, etc.)
BSc (Engg.) in IS/IST/Informatics (Digital Education/Education Technology, etc.)	MSc (Engg.) in IS/IST/Informatics (Digital Education/Education Technology, etc.)
	PhD (Science/Engg.) in IS/IST/Informatics (Digital Education/Education Technology, etc.)

Here, similar nomenclatures, namely Informatics/Information Science and Technology can also be offered in this context. Moreover, in Information Science or allied branches, namely CS, CA, and IT, the specializations of digital education (or allied branches, namely Education Technology) can be offered as a merged domain, namely Digital Education and Analytics, Digital Education and Cloud Systems, Digital Education and Management, Digital Education and Policies, and Digital Education and IoT.

11.10 EDUCATION DEGREES ABROAD AND INDIAN POTENTIALITIES

In the educational field, academic and professional degrees are available. Academic degrees are namely BA/MA Education, whereas in the professional context, the common degrees are Bachelor of Education (BEd) and Master of Education (MEd). Depending upon countries and norms, the style of education varies. In many developing countries, education programs are available with the digital education or education technology and e-learning concentrated. Among these specializations, based on this study, important are The University of Johannesburg, The British University in Dubai, University of Wollongong, Australia, Cape Breton University, Canada. In

India, diploma, bachelor's degree (academic and professional), master's degree (academic and professional), and research doctoral degrees are available. The nomenclatures of the programs are Diploma in Education (DEd) or Diploma in Elementary Education (D.El.Ed.), BEd, that is, Bachelor in Education, BA and MA in Education, MEd-Master's in Education, MPhil (Education), PhD (Education). The specialization/honors/major in digital education and areas with MEd in India are not common.

11.10.1 DIGITAL EDUCATION/IT: EDUCATION CONTEXT

MEd programs can be offered in different subjects and fields related to IT. Among these, few important nomenclatures are MEd (ICT), MEd (IT), MEd (Computing), etc. Here some of the programs have been illustrated university-wise based on the methods adopted. Note that universities open the criteria for both education and noneducation bachelor's degree holders; refer Table 11.8 for details.

TABLE 11.8 MEd Program in Information Science and Allied Areas in International Universities

Sl. No.	Name of the Program and University	Eligibility
1.	Master of Education—MEd (Information and Communication Technology) University of Wollongong, Australia	Bachelor's in any stream are eligible
2.	MEd (KNDI—Knowledge Networks and Digital Innovation) Charles Sturt University, Australia	Bachelor's in education preferred
3.	Master of Education—MEd (Information Technology) Cape Breton University, Canada	Bachelor's in any stream/ education preferred
4.	Master of Education—MEd (Information and Communication Technology) The British University in Dubai, UAE	Bachelor's in any stream
5.	Master of Education—MEd (Information and Communication Technology) The University of Johannesburg, South Africa	Bachelor's in Education preferred
6.	Master of Education (Digital Education) Massey University, New Zealand	1 Year (BEd/any degree)

A model digital education or similar education in education concentration should be mixed with the IT, education, and Social Science. Table 11.9 in this respect provides the list of courses/papers.

TABLE 11.9 MEd Program Structure in International Universities (Sample)

Universities	Papers/ Courses
The University of Johannesburg, South Africa	**90 Credit Coursework** Educational Reform: ICT Practice and Policy Development (A) Educational Reform: ICT Practice and Policy Development (B) Learning Ecosystems: Current ICT Issues and Future Trends (A) Learning Ecosystems: Current ICT Issues and Future Trends (B) Information and Communication Technology in Education: Learning Design (A) Information and Communication Technology in Education: Learning Design (B) Information and Communication Technology and Assessment (A) Information and Communication Technology and Assessment (B) **90 Credit Research Work (Dissertation)**
Master of Education MEd (Knowledge Networks and Digital Innovation) Charles Sturt University, Australia	**Total 64 Points** **Core Subjects** Concepts and Practices for a Digital Age [key subject] Digital Futures Colloquium [capstone subject] **Restricted Electives** *(choose six subjects)* Data Informed Practice Introduction to Educational Research Literature Across the Curriculum Digital Citizenship in Schools Digital Curation and Preservation An Introduction to Data Management: Governance, Standards and Ethics Social Networking for Information Professionals Global Knowledge Networking for Educators Literature in Digital Environments Designing Spaces for Learning Game Based Learning Extended Essay in Information Studies
The British University in Dubai, UAE	**120 Credit Coursework** Research Methods in Education Educational Policy Teaching and Learning Learning and Educational Technology Learning including the Blended Learning Managing Educational Technology

11.10.2 INDIAN POTENTIALITIES IN EDUCATION: DIGITAL EDUCATION AND ALLIED FIELDS

In India, as discussed, there are no as such specializations in the education bachelor's, master's, and doctoral degrees, both in academic and professional programs (Hanna, 1998; Wahabi et al., 2019). Hence, the digital education or educational technology programs can be started here in education and allied degrees (Tables 11.10 and 11.11).

TABLE 11.10 Proposed and Possible Digital Education Specializations in Academic Education Programs in Indian Context

Under Graduate	Postgraduate and Research
BA/BSc in Education (Digital Education)	MA/MSc in Education (Digital Education)
BA/BSc in Education (Education Technology)	MA/MSc in Education (Education Technology)
BA/BSc in Education (Educational Informatics)	MA/MSc in Education (Educational Informatics)
BA/BSc in Education (Digital Education and Learning Analytics)	MA/MSc in Education (Digital Education and Learning Analytics)
BA/BSc in Education (E-Learning)	MA/MSc in Education (E-Learning)
	MPhil in Education (Digital Education/Education Technology)
	PhD in Education (Digital Education/Education Technology)

TABLE 11.11 Proposed and Possible Digital Education Specializations in Professional Education Programs in Indian Context

Under Graduate	Postgraduate and Research
DEd/Bed in Education (Digital Education)	MEd/EdD in Education (Digital Education)
DEd/BEd in Education Education (Education Technology)	MEd/EdD in Education (Education Technology)
DEd/Bed in Education (Educational Informatics)	MEd/EdD in Education (Educational Informatics)
DEd/BEd in Education (Digital Education and Learning Analytics)	MEd/EdD in Education (Digital Education and Learning Analytics)
DEd/Bed in Education (E-Learning)	MEd/EdD in Education (E-Learning)

It is worthy to note that at the professional level in India, there is no approved nomenclature in the field of education. Only BEd and MEd degrees are available. However, there is a potentiality to offer the Doctor of Education (EdD/DEd) Degree in India due to its nature and benefits. Further, different specializations in general from education and technology (i.e., digital education or allied) could be started.

11.11 BLENDED LEARNING: THE WAY FORWARD

These programs on digital education including the subjects and nomenclature Educational Technology, Educational Informatics, E-Learning, Online Learning, both as a specializations and major, can be offered with information technology and education; and these degrees can be offered not only in regular on-campus mode but also in other mode of studies, namely online mode and private mode in some cases and context where learners may not be able to access the on-campus programs and may be at the mid age of their carrier, though the blended mode on offering these subjects would be better and specific.

11.12 SUGGESTION WITH FUTURE POSSIBILITIES

Today the teaching and learning systems are becoming smarter as well as advanced by the use of computing and latest IT. Therefore, different educational models and platforms have emerged and emerging. Educational institutes are moving with this digital-technology-enabled systems. Based on this study, following may be considered as potential suggestions and issues in this regard:

1. In the course gradients, the latest digital education and allied components should be included apart from basic IT components, namely Big Data, Data Analytics, Cloud, IoT, etc. for healthy educational systems.
2. Specialization may be started on different platform and subjects, namely Computing, CA, IT, ICT, Informatics, and other technology-allied subjects as proposed.
3. Specialization on management and similar branches can also be started, namely MBA (Educational Technology Leadership/Digital

Education/E-Learning Management), MCom (Digital Educational Venture Management/Digital Education), MBA (E-Learning and Leadership).

4. In educational allied branches, digital education or similar nomenclature is the need of hour and crucial at this very moment of smart education systems.
5. Research-oriented components, skills on communication, psychology, and philosophy also need to include as per the trend and requirement.
6. Industry and academia may reach a healthy digital education system; thus, efforts need to take care of these emerging issues.

11.13 CONCLUSION

A solid and healthy promotion of educational activities can lead the development in different ways. In this context, technologies such as IT and management systems are important stakeholders in the introduction of the new-age education systems. Digital education nomenclature has risen rapidly in the recent past not only in academia but also in industries for different reasons. Various research studies in these areas, namely Digital Education, E-Learning, Educational Informatics, Online Education, etc. are enabling modern universities as well in some contexts. In developing countries, many universities are offering IT and educational programs, but there should be a liaison between these program with IT by introducing new-age specializations. It is noted that bachelor's, master's, and doctoral degrees are offered in many countries with the specialization on digital educational aspects and fields, but full-fledged specializations on degrees may be offered in the fields such as BSc/MSc/BTech/MTech/MPhil/PhD in digital education or educational technologies. Moreover, universities may change their existing nomenclatures and fields from exiting educational technologies to digital education as well by incorporating few additional technologies, management components, etc. As digital education is interdisciplinary in nature, so allied departments within the academia should collaborate and interact with each other for better results for the promotion of the fields, and its result in the professions and institutions, accordingly.

KEYWORDS

- **educational technology**
- **EdD**
- **digital education**
- **higher education**
- **international universities**

REFERENCES

Blikstein, P. "Digital fabrication and 'making' in education: The democratization of invention." *FabLabs: of Machines, Makers and Inventors*, Vol. 4, No. 1, (2013): 1–21.

Buchanan, R. "Paradox, Promise and Public Pedagogy: Implications of the Federal Government's Digital Education Revolution." *Australian Journal of Teacher Education*, Vol. 36, No. 2, (2011): 67–78.

Dillenbourg, P. "The evolution of research on digital education." *International Journal of Artificial Intelligence in Education*, Vol. 26, No. 2, (2016): 544–560.

Dunleavy, G., Nikolaou, C. K., Nifakos, S., Atun, R., Law, G. C. Y., & Car, L. T. "Mobile digital education for health professions: Systematic review and meta-analysis by the Digital Health Education Collaboration." *Journal of Medical Internet Research*, Vol. 21, No. 2, (2019): 1–17.

Edwards, R. "Software and the hidden curriculum in digital education." *Pedagogy, Culture & Society*, Vol. 23, No. 2, (2015): 265–279.

Elboubekri, A. "The intercultural communicative competence and digital education: The case of Moroccan University students of English in Oujda." *Journal of Educational Technology Systems*, Vol. 45, No. 4, (2017): 520–545.

Gibson, D., Ostashewski, N., Flintoff, K., Grant, S., & Knight, E. "Digital badges in education." *Education and Information Technologies*, Vol. 20, No. 2, (2015): 403–410.

Hanna, D. E. "Higher education in an era of digital competition: Emerging organizational models." *Journal of Asynchronous Learning Networks*, Vol. 2, No.1, (1998): 66–95.

Heller, L., Parker, P. A., Youssef, A., & Miller, M. J. "Interactive digital education aid in breast reconstruction." *Plastic and Reconstructive Surgery*, Vol. 122, No. 3, (2008): 717–724.

Hiltz, S. R., & Turoff, M. "Education goes digital: The evolution of online learning and the revolution in higher education." *Communications of the ACM*, Vol. 48, No. 10, (2005): 59–64.

Huang, Z., et. al. "Digital health professions education on diabetes management: Systematic review by the Digital Health Education Collaboration." *Journal of Medical Internet Research*, Vol. 21, No. 2, (2019): 1–12.

Knox, J. "Posthumanism and the MOOC: Opening the subject of digital education." *Studies in Philosophy and Education*, Vol. 35, No. 3, (2016): 305–320.

Narciss, S. "Designing and evaluating tutoring feedback strategies for digital learning." *Digital Education Review*, Vol. 23, No. 2, (2013): 7–26.

Ozga, J. "Trust in numbers? Digital education governance and the inspection process." *European Educational Research Journal*, Vol. 15, No. 1, (2016): 69–81.

Paul, P. K., Sridevi, K. V., Ghosh, M., & Lama, A. "Education technology: The transparent knowledge delivery through QPN and cloud computing." *International Journal of Sustainable Development*, Vol. 12, No. 2, (2012): 455–462.

Paul, P. K., Dangwal, K. L., Garg, A. K. "Education technology and sophisticated knowledge delivery." *Techno-Learn*, Vol. 2, No. 2, (2012): 169–175.

Paul, P. K., et al. "Education technology: Emphasizing EduNxt knowledge transformation systems of Sikkim Manipal University (SMU), Gangtok, Sikkim, India." *International Journal of Embedded Systems and Computer Engineering*, Vol. 4, No. 2, (2014): 109–113.

Paul, P., Bhuimali, A., Kalishankar, T., Aithal, P. S., & Rajesh, R. "Digital education and learning: The growing trend in academic and business spaces—An international overview." *International Journal on Recent Researches in Science, Engineering & Technology*, Vol. 6, No. 5, (2018): 11–18.

Reeves, T. C. "Storms clouds on the digital education horizon." *Journal of Computing in Higher Education*, Vol. 15, No. 1, (2003): 3–12.

Robin, B. R., & McNeil, S. G. "What educators should know about teaching digital storytelling." *Digital Education Review*, Vol. 22, (2012): 37–51.

Wahabi, H. A. et. al. "Medical Doctors' offline computer-assisted digital education: Systematic review by the Digital Health Education Collaboration." *Journal of Medical Internet Research*, Vol. 21, No. 3, (2019): 1–14.

Xu, X., Posadzki, P. P., Lee, G. E., Car, J., & Smith, H. E. "Digital education for health professions in the field of dermatology: A systematic review by Digital Health Education Collaboration." *Acta dermato-venereologica*, Vol. 99, No. 1–2, (2019): 133–138.

CHAPTER 12

Smart Schedule: Training Management Approach in the Digital Education Era

BUI THI THANH HUONG*, TRAN VAN CONG, TON QUANG CUONG, and NGUYEN DUC NGUYEN

University of Education, Vietnam National University, Hanoi, Vietnam

Corresponding author. E-mail: bui.thanh.huong.vn@gmail.com

ABSTRACT

In this era of digital education, the smart schedule has proved essential in training management. On the basis of online training, the research group has proposed a new model of smart schedule serving the studying process of each student: choose a module, make up personal plans, and support students as an effective time-management method. Meanwhile, the smart schedule acts as a useful manager, which is supposed to manage teachers, students' diligence, and be able to store big data in the act of training management of the school.

12.1 INTRODUCTION

Course scheduling is an issue that all universities, as well as other institutions that involve teaching, have to deal with, and it is an essential administrative and training tasks. In the past, this was often done manually and was time-consuming and costly. Recently, with the development of computer systems, this task has been accomplished through the help of very efficient software and computing systems.

Currently, the systems of planning classrooms, courses, laboratories, and exams are critical in improving university performance and efficiency in using their resources (Mirhassani & Hooshmand, 2019). The variety of student curricula, increasing classroom, number of students, and continuous expansion of university departments have led to complicated timetables. This problem requires automated timetabling software systems (Arogundade et al., 2010). There are a lot of researchers who focus on timetabling software systems, with significant real life in schools' educational units (Figure 12.1) (Fawcett et al., 2009).

As stated by Wren (1995), "Timetabling is the allocation, subject to constraints, of given resources to objects being placed in space-time, in such a way as to satisfy as nearly as possible a set of desirable objectives." Timetabling is a list of scheduled events organized at predefined time intervals. The intervals are divided into weekly and daily, depending on the organizations or institutions. The course timetabling is a sophisticated real-time issue of any organization or institution (Mishra, 2016).

The timetabling of events, including the components such as lectures, tutorials, and seminars at universities to meet the demands of its users, is usually a challenging problem to solve effectively (Lewis et al., 2007).

12.2 SMART SCHEDULE CLASSIFICATION

To increase the efficiency and reduce the operating costs of educational systems, the need to research and develop a smart scheduling system that can meet the evolving needs of the organization is becoming increasingly urgent. For instance, we have found the trend on the IEEE explorer with searching terms "Timetabling OR Timetable." As mentioned above, regarding course timetabling in university, the objective is to allocate lectures to rooms and timeslots based on various hard and soft constraints (Figure 12.2) (Lindahl, 2017).

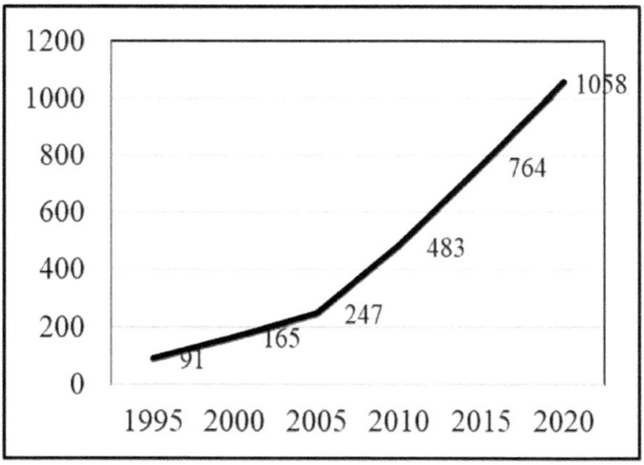

FIGURE 12.1 Number of papers on timetabling.

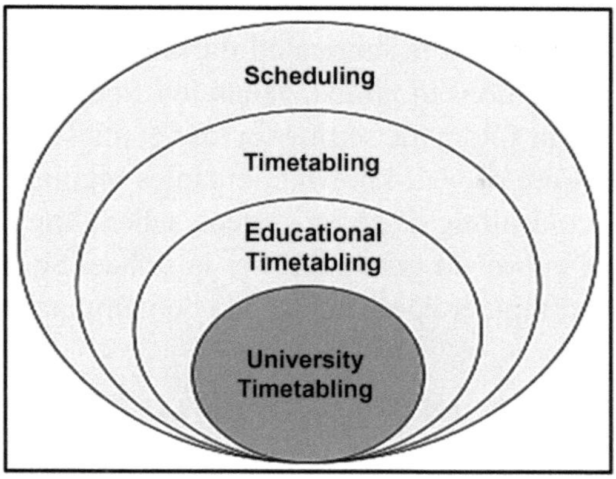

FIGURE 12.2 University timetabling and its parent categories (reprinted from Lindahl, 2017).

According to Aziz and Aizam (2018), course timetabling in university comprises a set of:

1. Event: one scheduling activity for teachers, courses, and students with weekly or daily timeslots.
2. Resource: rooms, Internet, equipment, timeslots for organizing events.
3. Constraint: restriction is in the scheduling of activities, including soft and hard limitations.
4. People: instructors and students.
5. Conflict: duplication between events, for instance, planning of more than a teacher for spontaneously a classroom.

Besides, according to Burke et al. (2002), hard constraints are rigidly enforced. Soft restrictions are required, but not critical. In real-life situations, satisfying all soft constraints is usually impossible. Both hard and soft constraints differ among institutions. Therefore, there is no general model for high education timetabling because the problem feature varies from one university to another (Mushi, 2004).

For instance, according to the survey conducted by Burke et al. (1998), the most common hard constraints include: exams must be scheduled in the precisely one timeslot and assigned to rooms of appropriate size and allocated invigilators; no student is expected to be in two different exams simultaneously. There are enough seats in each period for all scheduled exams; the individual exam is scheduled into specific timeslots or rooms; exams must take place at the same time.

Vrielink et al. (2016) confirmed that information and communication technology has become more fabricated during previous decades. Timetabling software applications modified, generated, and optimized more appropriate schedules in an automatic setting for universities. Literature in the field revealed that software of well-known companies regularly accepted the use of worksheets, a controlling database system, scheduling editors, web-based tools, and a more improved graphical user interface. So, the companies are using scheduling computer applications and contemporary techniques.

12.2.1 CLASSIFICATIONS OF EDUCATIONAL TIMETABLING

There are many ways to classify educational timetabling, depending on many different factors. For instance, according to Petrovic and Burke (2004), university timetabling issues are categorized into two main classes: course timetabling and examination timetabling. Meanwhile, Schaerf (1999) classified timetabling problems (university or school) into three main ones:

1. School timetabling (weekly plan for all classes): No teacher has two classes instantaneously.
2. Course timetabling (weekly plan for lessons): The number of covered lessons of courses which have common students is minimized.
3. Examination timetabling (plan of exams for all university courses): The number of overlapped courses which have common students are avoided, exams are spread among students are much as possible.

In another study, to solve the main dissimilarities that exist within the practice area of educational scheduling, McCollum et al. (2012) divided timetabling into three sections (Figure 12.3):

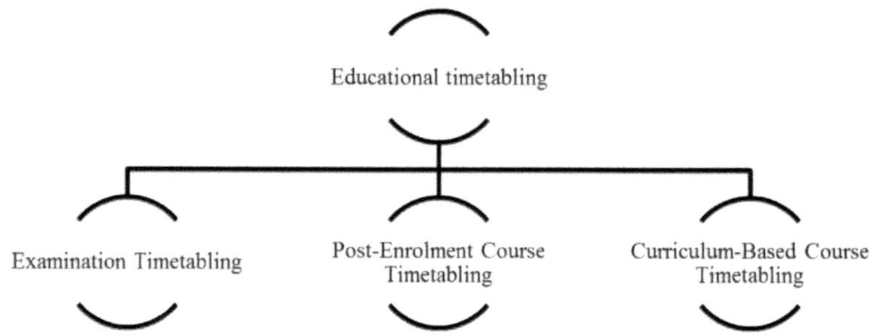

FIGURE 12.3 The three educational timetabling sections (Reprinted with permission from McCollum et al., 2012. © Springer Nature.)

Examination timetabling: Schedule exams following horizon time adapting to a set of constraints (McCollum et al., 2010). Examination timetabling contains individual examinations that are allocated in a suitable time and place, allowing all participants to take part in all the sessions that they must join. Time is typically divided into a set of nonoverlapping examination phases with equal or different lengths. An individual examination is required to be given to one certain period that is more comparable or lengthier than the one.

In the same way, examinations are required to happen in a room that is large enough. Only an exam in a room is allowed throughout the same examination period. Also, numerous other optimization criteria need to be avoided in the case of the direct troubles when lecturers and students are required to join two exams timetabled in all at once (Müller, 2016).

Postenrolment course timetabling: The goal is to put one group of events into rooms and timeslots (weekly, regularly) following the students' selections (McCollum et al., 2010).

Curriculum-based course scheduling: The goal is to put a group of events to rooms and timeslots (McCollum et al., 2010). The main difference is that all the constraints and objectives in the curriculum-based course scheduling are related to the conception of the program, which is the complete workload that is made of a cluster course for a group of students. On the contrary, the concept of curriculum is absent in postenrolment course timetabling, and objectives and constraints are built on the basis of the enrolments to the courses (Ceschia et al., 2012).

12.3 RESEARCH APPROACHES ON SMART SCHEDULE

In this section, we classify the approaches according to Carter and Laporte (1995, 1998) and extend the classification presented by Petrovic and Burke (2004) and Norberciak (2006).

12.3.1 MATHEMATICAL PROGRAMMING APPROACH

Researchers have approached the university timetable problem from the field of operational research and computer science. Direct heuristic was the first approach used to solve the timetable problem, followed by an algorithm based on integer programming and the network flow method. Researchers from the field of artificial intelligence (AI) were involved in solving this kind of problem using search techniques (Schaerf, 1999).

For instance, Bucco et al. (2017) proposed a mathematical model to resolve the university course timetabling issue and optimize classroom utilization. An algorithm divided the problem was developed, addressing it with mixed-integer linear software. Computational experiments applied to a real database of a Brazilian public university confirmed the excellent performance of the proposed approach, which significantly reduces the number of assigned classrooms.

Regarding the advantages and disadvantages of this approach, mathematical models provide a quicker, more powerfully process to a large range and in a shorter period. It can help prevent human mistakes because of the complication and timing of the procedure. It can provide a suitable tool for the staff to implement the organization's rules and control resources. Additionally, it uses Internet platform facilities for the interaction between teachers, learners, and educational institutions (Mirhassani & Hooshmand, 2019). On the other hand, the high computational demand of these methods makes them less attractive to large timetabling problems (Petrovic & Burke, 2004).

12.3.2 SEQUENTIAL METHODS

According to Norberciak (2006), the research community in AI is reasonably proactive in the scheduling and timetabling areas. They have developed a range of methods for solving scheduling issues. Timetabling was divided into four categories: cluster methods, sequential methods, constraint-based procedures, and metaheuristic methods.

In sequential approaches, events are ordered using domain heuristics and then they are successively allocated into valid periods that are called timeslots. Hence, no events in the period are in conflict with each other. Events are typically put in such a way that the events that are most challenging for scheduling will be assigned to timeslots first (Burke et al., 1998).

Sequential methods order activities based on how difficult it is to schedule them and then assign the activities into valid resources in sequence. Graph coloring is the most common sequential method to resolve the timetabling problem (Lee et al., 2005).

Graph-coloring techniques can be used to model and solve the university timetabling problem. Old-style graph coloring for timetabling includes graphs that a vertex symbolizes a course to be arranged, an edge that represents a pair of courses that conflict (i.e., cannot be in the same period), and the vertex that symbolizes color (Redl, 2007). This technique is increasingly improved to overcome the disadvantages of former applications. For example, Burke et al. (1995) conducted a university timetabling system using constraint manipulation and graph coloring. Burke et al. (2004) offered the application of graph-coloring methods for timetabling. They highlighted the role of graph-coloring methods in the timetabling literature over the last 40 years or so. They pointed out that graph-coloring methods are incorporated into several modern hybrid metaheuristic techniques. Redl (2007) presented another graph-coloring method for institution scheduling that integrates room assignment during the conflict graph creation and coloring procedure. This method includes the creation of another type of course-conflict graph, accompanied by a somewhat dissimilar method of coloring it. Hussin et al. (2011) used the graph-coloring approach to ordering which exams are selected. They also incorporate graph coloring, cluster heuristic, and sequential heuristic. Akbulut and Yilmaz (2013) developed a university examination scheduling system using the graph-coloring algorithm and radio-frequency identification technology.

In mathematical and computer illustrations, it is typical to use the first few positive or nonnegative numerals as the "colors." Anyone can use any limited set as the "color set." The number of colors determines the coloring problem but not on what they are. Graph coloring applies many practical applications as well as theoretical challenges. Limits could be set on the graph, or on the way a color is assigned, or even on the color itself (Cauvery, 2011), for instance, the software of the graph-coloring algorithm is limited to a highly constrained real-world occurrence of the college course timetabling problem. By using this software, we get an adequate preparation of classes and events where complex combinations of resources must be assigned efficiently to timetables. It tests all the constraints applied and also checks all the clashes and eliminates it (Gajbhiye et al., 2018).

12.3.3 CLUSTER METHODS

In cluster methods, two different events in a specific group do not conflict with each other. These approaches have some limitations like the groups of events are shaped and fixed at the beginning of the algorithm. These limitations may result in a reduced quality timetable (Norberciak, 2006).

For instance, Thomas and Tajudin (2006) use the clustering method as a part of exploratory data analysis. Clustering operates by analyzing and grouping dataset into clusters so that the dataset within a cluster is similar to each other. The purpose of clustering is to ease sequencing in each process. Every component in the same cluster has some similarities, and this forms the basis of grouping the components. The examination and the timeslot represent each cluster. The points will be given to each sequence of a cluster. Aljasmi (2016) used a color-coded pattern clustering model combined with an automated system for the courses scheduled. A color-coded pattern cluster distribution model that applies a clustering method has been used to allocate the labs of course clusters together as a cluster to form pair clusters. The lab blocks can be one or more depending on the availability of labs to create one or more lab clusters. This method has generated a pattern for lectures and lab clusters that helped in minimizing lecture–lab or lab–lab clashes. The same color code is used to specify the lecture cluster and its accompanying lab cluster. Most lecture clusters contain courses required in different semesters of the program plan.

12.3.4 CONSTRAINT-BASED APPROACHES

In approaches based on the constraint, timetabling is a group of variables like events that have a restricted domain to which values (i.e., resources like time) must be allocated to fit with many restrictions. Several rules are defined for allocating resources to events. When no rule applies to the present partial solution, backtracking is accomplished until a solution is found that fits all restrictions. As the fulfillment of all obstacles might not be possible, algorithms are permitted to break some restrictions in controlled ways to produce a complete schedule (Do & Kambhampati, 2001; Norberciak, 2006; Yakhno & Tekin, 2002).

12.3.5 METAHEURISTIC APPROACHES

In the metaheuristic approach such as "tabu" search, simulated annealing, evolutionary algorithms, and hybrid approaches have been investigated for scheduling. Metaheuristic practices started with one or additional original solutions and use exploration strategies to find the optimal solution, trying to avoid local optima in the procedure (Norberciak, 2006).

12.3.6 *MULTICRITERIA APPROACHES*

In multicriteria approaches, each criterion assesses the violation of corresponding restrictions. The rules differ importance levels in different circumstances and organizations (Petrovic & Burke, 2004). For example, Burke et al. (2000) addressed university examination timetabling issues as multicriteria decision issues. The approach was divided into two phases. In the first phase, the goal was to find high-quality schedules concerning each standard distinctly. In the second phase, trade-offs between criteria values are carried out to find a compromised solution regarding all the requirements instantaneously. This approach considered an ideal point in the criteria space, which optimizes all criteria at once.

12.4 SMART SCHEDULE SOFTWARE

Many scheduling software packages have been developed in recent years (Table 12.1). According to Fahmy et al. (2014), timetabling applications can be categorized into three types:

1. Commercial software is *designed and developed for sale to the general public* (e.g., Wise, Asc TimeTables, and Keith Johnson's TimeTabler).
2. Freeware software is *a piece of software that can be downloaded, passed around, and distributed without any initial payment.*
3. Open-source software is *grant access to the source code* (e.g., UniTime, Free Educational Timetable (FET), Timetable Plus).

TABLE 12.1 List Some of the Common Scheduling Software in the University

Software Name/ Type	Description
FET: Free Timetabling Software/ Freeware, Open-Source	The FET project was initiated by Liviu Lalescu on October 31, 2002, using a genetic algorithm to develop a scheduling tool. At the beginning, the author used genetic algorithms to find suitable solutions for optimal combinatorial problems. However, this algorithm is not really effective because of its slow calculation ability and it can only fit into simple timetables. URL: https://lalescu.ro/liviu/fet/

TABLE 12.1 *(Continued)*

Software Name/ Type	Description
Mimosa Scheduling Software/Sample Commercial	With the function of preventing users from making mistakes, Mimosa's interface helps users to receive valuable hints (answers to "What-If" questions). It is easily for users to configure these distinctive needs through the user theme and apply the templates and files. The capacity of Mimosa covers up to 255 meetings scheduled (30 timeslots per day in a 7-day week), 300.000 timetables (courses, exams, meetings, etc.) and resources (teachers, classrooms, equipment, etc.) in a single small file. URL: https://mimosa-scheduling-software-trial.soft112.com/
Lantiv Scheduling Studio 7/ Commercial	It resolves the problem of calendar of people, resources, and activities. It controls struggles and permits scheduling using drop and drag and shows available timetables. The workspace of the application is very convenient. URL: http://schedulingstudio.com/
Tandem/ Commercial	Tandem provides the user's entire scheduling team the ability to enter everything into one master calendar. It will automatically alert users when an event will conflict with another before the event gets scheduled on their calendar. URL: https://www.tandemcal.com/
CalendarWiz/ Commercial	CalendarWiz tailors user's calendars based on the industry purchasing their software, such as nonprofits, small and large businesses, clubs, and in this case, primary/higher education. This software gives schools the option to either integrate their calendars on their websites, leave their calendar up on the CalendarWiz website, or both, depending on the needs of students, parents, teachers, and administrators. The one difference between this calendar software and others on this list is the emphasis on events and announcements rather than student assignments. URL: https://www.calendarwiz.com/index.php
Google Calendar/ Freeware	Google Calendar is the most famous application with easy functions to use, all informations saved on the cloud computing system, exists on web and available on IOS and Android so that user does not need to download any special software to access the calendar URL: https://sites.google.com/site/cliotech/google-calendar
Localendar/ Commercial	It was shaped to help users organize their events and share them online. Unmatched customization options and interactive theme editor make getting the right look easy for both beginners and web designers. URL: https://localendar.com/elsie
Wise Timetable/ Commercial	The wise timetable is easy to use and powerful timetable software that helps users to generate and maintain timetables. Users can easily input data about professors, lecture rooms, study program. URL: http://www.wise-t.com/product-wise-timetable.php?lng=en

TABLE 12.1 *(Continued)*

Software Name/ Type	Description
Edval/ Commercial	Edval is designed to make scheduling easy. From storing information, timetable construction, auto-staffing, and auto-rooming, managing lesson spreads, fixing clashes, balancing classes, etc. URL: https://www.edval.education/
Keith Johnson's TimeTabler/ Commercial	Keith Johnson's TimeTabler is a fast and friendly computer program. Many hundreds of schools and colleges, in over 80 countries, now use TimeTabler to do the actual scheduling of their timetables on a computer. URL: https://www.timetabler.com/timetabler/
Timetable Plus/ Freeware (Timetable Plus Spring Lite)	Timetable Plus is a demand-driven schedule software application for all levels of schedule managers for resolving all kinds of schedule scenarios. Timetable Plus Spring Lite is free. At the same time, Timetable Plus Spring Pro is the advanced form of the Lite version, which is cheap and affordable for schools. URL: https://www.timetableplus.com/#
iMagic Timetable Master/Freeware	iMagic Timetable Master is easy to use. It has been in use in schools and colleges across the globe for over three years. It will shape and connect all the relevant information. URL: http://www.imagictimetablesoftware.com/

Within the scope of our literature review, we found that the number of studies evaluating the effectiveness of scheduling applications in universities is still limited. This is a necessary research path to help administrators, teachers, students, and parents choose to use the app to meet their needs. It is also a suggestion for builders to review, improve, and upgrade the features of the application.

12.5 EXAMPLE OF APPLICATIONS OF SMART SCHEDULES ON TRAINING MANAGEMENT

12.5.1 FET-5.44.0

FET has automatically scheduled the timetable of a school. The timetabling algorithm it uses is fast and efficient. On March 19, 2006, Volker Dirr participated in the project. He found out that FET could not solve his difficult data file, but other software could. Therefore, he emphasized on the continuation of a better algorithm for FET. As of June 24, 2007, he applied the heuristic algorithm for FET-5.0.0. Based on the basic principles: brute force principle (suitable for large search spaces), greedy principle (select optimal criteria on a global scale), order principle (make an optimal selection based on a reasonable structure of constraints space). FET is licensed under the GPL version 3 or later. FET can solve a complicated timetable in 5–20 min maximum. To process a simpler timetable, FET may take less time, which is normally under 5 min. In some cases, it only takes FET several seconds. For more complicated schedules, it may take FET more time to process, sometimes up to hours.

Heuristic algorithms are techniques that use heuristic functions to find and discover constraints in order to offer an optimal solution. With unrealistic survey research, heuristic algorithms are used to speed up the search process with appropriate solutions through reduced thinking to reduce problem awareness when making quick decisions. They can solve difficult schedules. This algorithm has been used to simulate the timetable searching procedure, starting with FET version 5.0.0. The latest version has been updated on March 27, 2020. FET is an application that uses the Qt and C++ toolkits to develop the cross-platform application with open source. FET can be run on any Qt-supported platforms, such as Linux, Windows, Mac OS X, and other platforms.

Using FET to timetabling for a training institution is based on basic data: list of subjects, list of teachers with teaching subjects, list of students of different grades and class, number of classrooms, number of seats in each classroom, number of lecture hours per day, and number of sessions per week with associated constraints.

We illustrate the scheduling using FET-5.44.0 to build the smart schedule for training management in VNU University of Education, Vietnam National University (UEd-VNU) as shown in Figure 12.4.

In order for FET-5.44.0 to automatically schedule, the following issues must be listed:

The basic information includes the number of schooldays in a week, number of lessons in a day, list of subjects, list of lecturers, list of grades, list of classes of different grades, list of lectures, and list of classrooms.

The constraint information is classified into two categories: compulsory constraints and optional constraints.

- ✓ Compulsory constraints: The maximum number of teaching days per week for all teachers, the maximum number of gaps per day for all teachers, the maximum number of gaps per week for all teachers, maximum teaching hours per day for all teachers, and minimum teaching hours per day for all teachers.

✓ Optional constraints: Subject constraints (number of periods for a subject), practice subjects, teacher constraints (do not teach continuously 5–6 periods/day, number of day-off per week, teachers are raising children), student constraints, constraints on facilities.

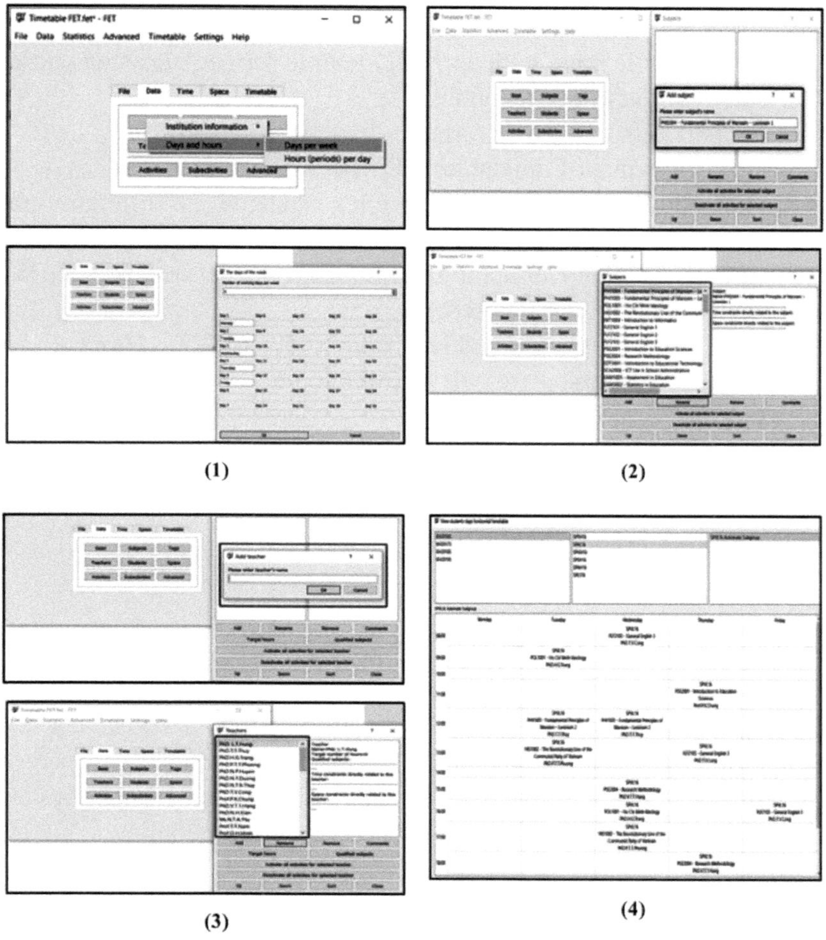

FIGURE 12.4 Application of FET on building smart schedule for UEd-VNU.

Through using FET-5.44.0, the authors recommend some advantages of FET such as

✓ As an open-source and cross-platform application, FET gives users the freedom to use, modify, improve, and release.
✓ FET is free, suitable for units with limited financial resources.

- ✓ It is localized in 36 different languages such as Arabic, Catalan, Czech, Danish, German, Greek, British English, American English, Spanish, Basque, Persian, French, Galician, Jewish, Hungarian, Italian, Indonesian, Japanese, Lithuanian, Macedonian.
- ✓ The creation algorithm is fully automatic, allowing semi-automatic or manual allocation.
- ✓ Flexible file input with an XML module format, allowing editing by the XML editor or manually.
- ✓ Import data in the .csv format.
- ✓ Export format of timetables: HTML, XML, CSV.
- ✓ The student structure is flexible, arranged in different groups: courses, groups, subgroups. FET allows overlapping courses and groups and nonoverlapping subgroups, even defined as specific as to individual students (as separate sets).
- ✓ Each constraint is weighted, from 0% to 100%. However, some special constraints are only allowed to have weighted 100%.

However, using FET, limitations such as heuristic algorithm, number of constraints, complex import data, long processing time, and complex export timetable can also be faced.

12.5.2 GOOGLE CALENDAR

Google Calendar (GC) belongs to the ecosystem, such as Google Docs, Sheets, Drive, and Gmail. These Google apps are all available in a school-based format for free under Google Apps for Education. It is easy for a personal user to use a free online calendar by workplace (timetable classes, conference list, scheduling for meetings and events, scheduling work, events list, etc.).

In recent years, UEd-VNU has been using GC for listing activities/events for managers and staff (working week calendar for staff and schedule for students and lecturers). Specifically, GC is suitable for setting up small plans for many students and lecturers join, such as internship plans, field trip plans, and communication plans. Each participant can be involved in a joint plan and regularly reminded every day, and every week following the duties planned. GC has been in use as a tool to manage the lecturer's plans, which also integrate with the plans of UEd-VNU. The university lecturers have exploited GC for teaching with the following specific features:

1. Create a personal schedule

- ✓ Create each lecturer's favorite or preferred calendars using the "add a coworker's calendar" feature and then select "Browse the calendar of interest" in the GC events setting mode.
- ✓ It is very convenient for lecturers to customize the calendar view so that lecturers are not distracted into other purposes that often take a lot of time to find the events. Lecturers can choose the "custom view," which is suitable for a specific purpose.
- ✓ Using GC to make personal plans of lecturers (even long-term plans for the entire life of the lecturers) is popular in UEd-VNU. The plans are categorized into many different classes as follows: undergraduate teaching, postgraduate teaching, scientific research, research group, high school teachers' improvement, internship guidance for students, training, advisory activities, other administrative meetings, etc. (Figure 12.5).

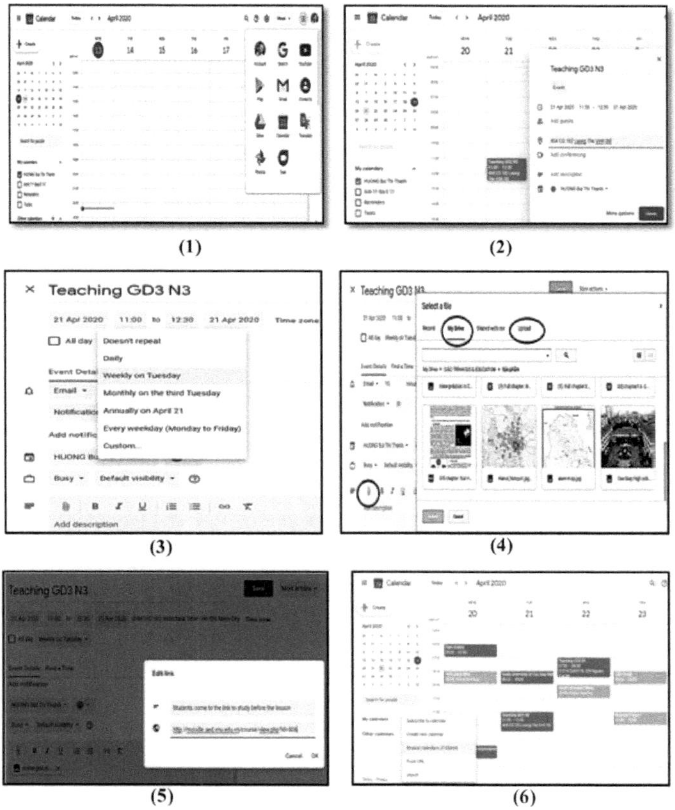

FIGURE 12.5 Building private working plan of lectures by GC.

✓ Lecturers can save time by using keyboard shortcuts to manage their plans at this address (Figure 12.6):

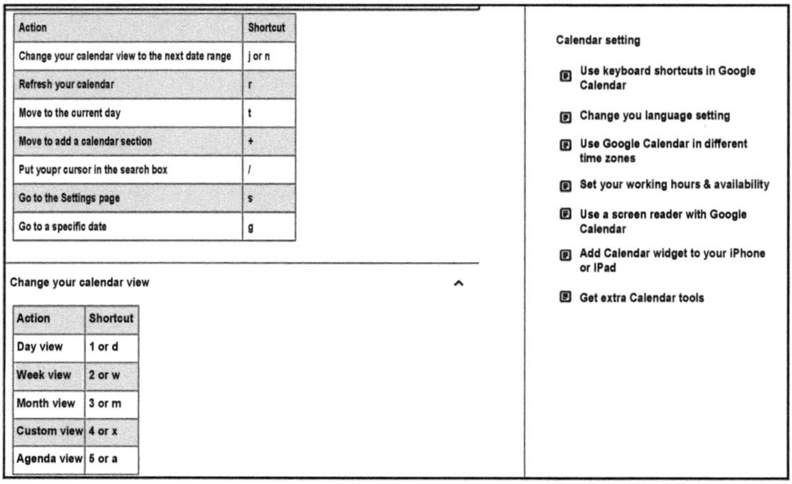

Action	Shortcut
Change your calendar view to the next date range	j or n
Refresh your calendar	r
Move to the current day	t
Move to add a calendar section	+
Put youpr cursor in the search box	/
Go to the Settings page	s
Go to a specific date	g

Change your calendar view

Action	Shortcut
Day view	1 or d
Week view	2 or w
Month view	3 or m
Custom view	4 or x
Agenda view	5 or a

Calendar setting

- Use keyboard shortcuts in Google Calendar
- Change you language setting
- Use Google Calendar in different time zones
- Set your working hours & availability
- Use a screen reader with Google Calendar
- Add Calendar widget to your iPhone or IPad
- Get extra Calendar tools

FIGURE 12.6 Using keyboard shortcuts to organize working plan with GC.[1]

✓ Synchronize GC with application software that lecturers are using in their teaching and research work such as LMS-Moodle online teaching platform, online teaching room Zoom.us or other teaching applications (Figure 12.7).

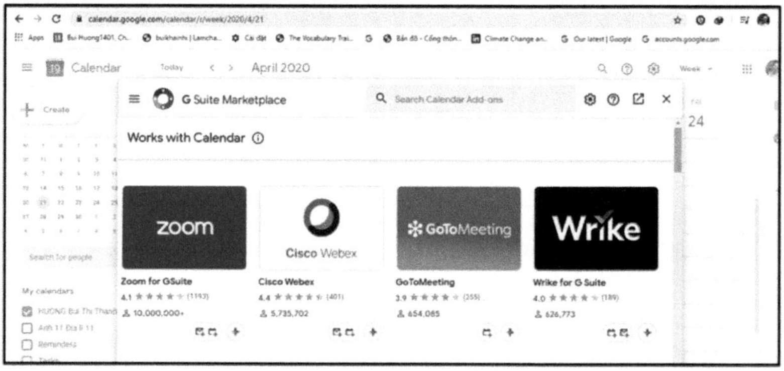

FIGURE 12.7 Many applications of teaching add-on GC.

[1]https://support.google.com/calendar/answer/37034?__hstc=20629287.cc67e775a6a0f8ab368277939a6f5b0f%20.1572842339642.1572844816274.1572853895909.3%20&%20__%20hssc%20=%2020629287.1.1572853895909%20&%20__%20hsfp%20=%20621399751

2. Organize research group events in a smart way

✓ Make a group work plan for lecturers: Lecturers are often busy, so finding a suitable time to organize group work is often difficult. Therefore, lecturers can exploit the "Find a time" feature by making public the free time of the team members to suggest a reasonable time to organize the team meeting (Figure 12.8A).

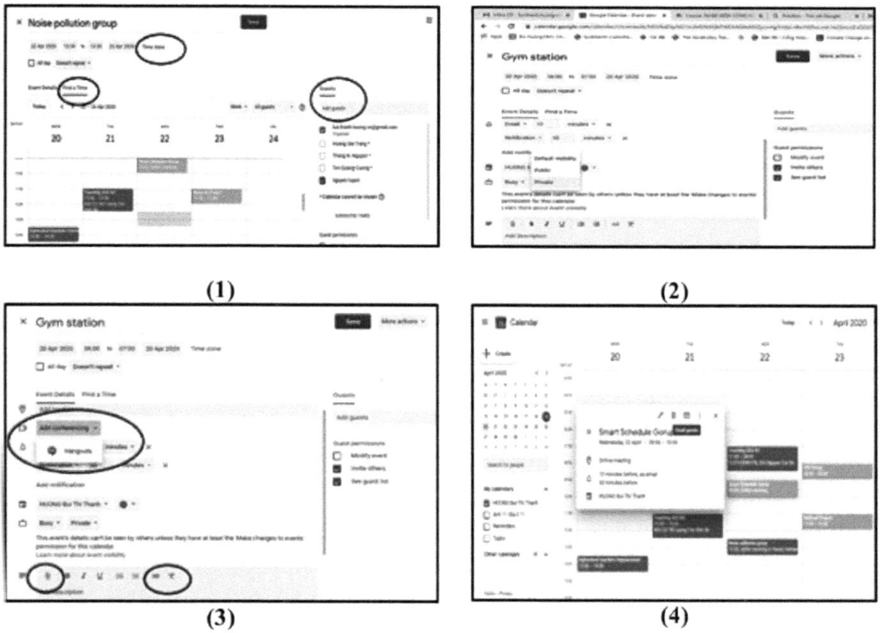

FIGURE 12.8 Smart organization for a co-working plan of lecturers.

✓ Besides, the lecturers also feel secure with the "hide events" feature of GC. To keep the privacy and "sensitivity of the event," the personal calendars of lecturers can be set at "private" and labeled as "busy" so that colleagues do not schedule research group meetings during their busy personal time. These are the smart features of GC (Figure 12.8B).

✓ Set up remote meetings using the Google Hangout feature by using the "add video call" feature to each lecturer's events in the schedule. Add documents as an attachment to the event. This will help lecturers save time and effort from searching in Mail, Google Drive, or any other shared place. It is of great convenience that lecturers only need a smart schedule, which includes full documentation/video call links for the meeting in the event deployment process (Figure 12.8C).

- ✓ Features to adjust the time, location as well as the time to remind events for participants. This is an automatically installed feature according to the settings of the event organizer.
- ✓ Set the time of the event according to Greenwich Mean Time by the option "show world clock" in the calendar view of each participant. Thus, the team, which consists of members from different countries, can still easily participate in the event.
- ✓ Send an email reminder of the calendar to each event participant. This feature is Event Reminder, an outstanding feature of GC, synchronized in all Gmail lecturer's devices. For the event organizer, it is easy to just click the envelope icon in the event, and the reminder information will automatically be sent to all members.
- ✓ The Event Reminder feature is on the desktop of each member participating in the event. Although this feature is not a good feature because it violates personal freedom, event organizers in some urgent cases can use it by turning on the "Desktop Notification" feature in Event Setup (Figure 12.8D).

Through using GC, the authors recommend some issues as follow:

By using GC, teachers and staffs are easier for making plan when it is integrated with Contacts, Drive, Hangout, Sites, and especially Gmail. Users can edit and view schedules from their phones, laptops, or tablets.

Understanding the schedule built up by GC as its simple theme is easy to understand. Users can create multiple calendars, for example, for businesses calendar, individual calendar, teaching schedule, studying schedule in the one calendar by assigning different colors for each duty. Even though multiple calendars can be overlayed on top of one another, we still can toggle a calendar off to hidden.

It is smart scheduling for meetings with events quickly checking when colleagues are free and showing their calendars in a single view. Users are able to share their calendar so that everyone can see the full event details or just see if they are free. GC is easy to send invitations to other people or share with anyone we choose by grants or allows editing the contents. It helps users pick the right content and gives control to them to share the calendar. In addition, user can transfer schedule from a different scheduling application of smartphone, laptop, iPad, or Google Account to GC and to any similar program. Users are also able to use GC to book the nearest conference room. These benefits help users save time so that they can improve productivity on works.

Utilities of GC can sync with Apple iCal, Microsoft Outlook, and similar programs and can be used on computers, smartphones, or smart tablets. Smartphones can be synced with GC to perform all the functions that computers exclude so that even without a computer, everything still stays organized.

Besides accessing to the calendar, users can schedule meetings and integrate meeting applications, so that they easily make event links to calendar invites and set reminders for any scheduled meeting.

However, GC still might have minor glitches such as reminders for events that are not properly saved or get notifications about the meetings at weird times, but that is still the thing in control. Although GC can be easily accessible by phone with the browser, users need to sign in the Google account and must have access to the calendar.

12.5.3 ZALO-TO-DO

Since 2019, UEd-VNU has implemented the time management tool initiative by the Zalo Group cloud platform. Zalo is the biggest Vietnamese app publisher with more than 45 million users. This company owns more than 16 apps that have together more than 243.98 million downloads + times (https://bit.ly/2Rm4RNq). Besides free message and call application on mobile devices, the mobile app ZaloTo-Do provides a useful tool for timetable management offered to lecturers, administrators, and students in three aspects: learning schedule, time management, and project management individually, in a group, and interaction between learning participants (to define activities, delivery and sequence activities, estimate activity durations, schedule development, project monitoring, and reminder). This app is not only supportive for timetable management for students but can also be used in mobile learning. Firstly, Zalo-to-Do sets a group learning path to locate students' tasks, time, contact person in the learning process accurately. Secondly, it can display students' task learning status, learning reports, and deadline reminders. Definitely, Zalo views for all participants the monitoring process, chart of analysis of their learning task.

12.5.4 TRELLO

For efficient project management, Trello has been offered (in both mobile and desktop) by some academic faculties and research group at the UEd-VNU. By using the Trello app, educators, teachers, and students will keep on top of teaching and research tracks, learning tasks, and streamline various entire project in instant collaboration. The core functions of Trello are as follows:

1. Card (including Title, Description, Task, Feature, Question, Note. In particular, each card can be broken into a checklist of detailed activities.
2. List (including card in progress) may be categorized as List Doing and List Done.
3. Boards are composed of a series of lists and may be offered as projects. Board contains lists that are categories, phases, or tasks and time management within the projects.
4. The organization contains all Boards and Team members.

12.6 CONCLUSION

In this era of digital educational transformation, the smart schedule has proved essential in training management. There are many ways to classify educational timetabling, depending on many different factors. Smart schedule organized into some main kinds: examination timetabling, school timetabling, and course timetabling. Regarding course scheduling in a university, the goal is to place lectures into rooms and timeslots given various options.

This chapter provided an overview of research approaches to smart schedule, including approaches based on mathematical programming, sequential methods, cluster methods, constraint-based approaches, metaheuristic methods, multicriteria approaches, and case-based reasoning approaches. This chapter also introduced some smart schedule software. Many scheduling software packages have been developed in recent years such as commercial software, freeware software, open-source software.

By using smart schedule software on training management in UEd-VNU, Hanoi, this chapter summarized advantages and limitations of each software. FET has automatically scheduled the timetable of a school, high school, or a university. It uses a fast and efficient timetabling algorithm. However, FET has also some limitations such as the heuristic algorithm, number of constraints, complex import data, long processing time, and complex export timetable. Using GC, teacher and training staff can save time for planning and integrate everybody with Gmail and sites such as Hangouts and Drive. User can view and revise schedule from their laptops, smartphones, or tablets. Although GC can be easily accessed by smartphones. However, with the browser, users need to sign in the Google account and must have access to the calendar. Since 2019, UEd-VNU has implemented the time management tool initiative by the Zalo Group cloud platform. As an efficient project management, Trello has been offered (in both mobile and desktop) by some academic faculties and research group in UEd-VNU. By using the Trello app, educators, teachers, and students can be on top of teaching and research tracks, learning tasks, and streamline various projects in instant collaboration.

KEYWORDS

- **smart schedule**
- **training management**
- **big data**
- **time management**

REFERENCES

Akbulut, A., & Yilmaz, G. (2013). University exam scheduling system using graphcoloring algorithm and RFID technology. *International Journal of Innovation, Management and Technology*, *4*(1), 66–72.

Aljasmi, L. (2016). Using color coded pattern clustering model combined with automated system for courses scheduling. *International Journal of Computing and Network Technology*, *4*(03).

Arogundade, O. T., Akinwale, A. T., & Aweda, O. M. (2010). A genetic algorithm approach for a real-world university examination timetabling problem. *International Journal of Computer Applications*, *12*(5), 975–8887.

Aziz, N. L. A., & Aizam, N. A. H. (2018, September). A brief review on the features of university course timetabling problem. In AIP Conference Proceedings (Vol. 2016, No. 1, p. 020001). Melville, NY: AIP Publishing LLC.

Bucco, G. B., Bornia-Poulsen, C. J., & Bandeira, D. L. (2017). Development of a linear programming model for the university course timetabling problem. *Gestão & Produção*, *24*(1), 40–49.

Burke, E., Bykov, Y., & Petrovic, S. (2000). *A multicriteria approach to examination timetabling*. Heidelberg, Berlin: Springer.

Burke, E., Elliman, D., Ford, P., & Weare, R. (1995, August). *Examination timetabling in British universities: A survey*. In International Conference on the Practice and Theory of Automated Timetabling (pp. 76–90). Heidelberg, Berlin: Springer.

Burke, E. K., Kingston, J., & De Werra, D. (2004). Applications to timetabling. *Handbook of graph theory* (p. 445). London: Routledge.

Burke, E. K., Newall, J. P., & Weare, R. F. (1998, August). *A simple heuristically guided search for the timetable problem*. In Proceedings of the International ICSC Symposium on Engineering of Intelligent Systems. Nottingham: ICSC Academic Press.

Carter, M. W., & Laporte, G. (1995, August). *Recent developments in practical examination timetabling*. In International Conference on the Practice and Theory of Automated Timetabling (pp. 1–21). Heidelberg, Berlin: Springer.

Carter, M. W., & Laporte, G. (1998). Recent developments in practical course timetabling. E. Burke, M. Carter, eds. *Practice and Theory of Automated Timetabling. Springer Lecture Notes in Computer Science Series*, 1408, pp. 3–19. Heidelberg, Berlin: Springer.

Cauvery, N. K. (2011). Timetable scheduling using graph coloring. *International Journal of P2P Network Trends and Technology*, *1*(2), 57–62.

Ceschia, S., Di Gaspero, L., & Schaerf, A. (2012). Design, engineering, and experimental analysis of a simulated annealing approach to the post-enrolment course timetabling problem. *Computers & Operations Research*, *39*(7), 1615–1624.

Do, M. B., & Kambhampati, S. (2001). Planning as constraint satisfaction: Solving the planning graph by compiling it into CSP. *Artificial Intelligence*, *132*(2), 151–182.

Fahmy, S., Ruhaida, I., Akhyari, N., & Teknologi, F.S. (2014). Open-Source Academic Timetabling System: FET Implementation at TATI University College.

Fawcett, C., Hoos, H. H., & Chiarandini, M. (2009). *An automatically configured modular algorithm for post enrollment course timetabling*. Vancouver: University of British Columbia, Department of Computer Science.

Gajbhiye, S., Shende, N., Sahu, U., & Behar, Y. (2018). Automated timetable generator for educational institutions using graph colouring technology. *International Journal of Applied Research*, *4*(4), 42–45.

Hussin, B., Basari, A. S. H., Shibghatullah, A. S., Asmai, S. A., & Othman, N. S. (2011, September). Exam timetabling using graph colouring approach. In 2011 IEEE Conference on Open Systems (pp. 133–138). New York, NY: IEEE.

Lee, J., Ma, S. P., Lai, L. F., Hsueh, N. L., & Fanjiang, Y. Y. (2005). University timetabling through conceptual modeling. *International Journal of Intelligent Systems*, *20*(11), 1137–1160.

Lewis, R., Paechter, B., & McCollum, B. (2007). Post enrolment based course timetabling: A description of the problem model used for track two of the second international timetabling competition. [Working Paper]. Cardiff Working Papers in Accounting and Finance. Cardiff: Cardiff University.

Lindahl, M. (2017). Strategic, tactical and operational university timetabling. DTU Management. Retreived from: https://orbit.dtu.dk/en/publications/strategic-tactical-and-operational-university-timetabling.

McCollum, B., McMullan, P., Parkes, A. J., Burke, E. K., & Qu, R. (2012). A new model for automated examination timetabling. *Annals of Operations Research*, *194*(1), 291–315.

McCollum, B., Schaerf, A., Paechter, B., McMullan, P., Lewis, R., Parkes, A. J., ..., & Burke, E. K. (2010). Setting the research agenda in automated timetabling: The second international timetabling competition. *INFORMS Journal on Computing*, *22*(1), 120–130.

Mirhassani, S. H., & Hooshmand, F. (2019). *Methods and models in mathematical programming*. London: Springer Nature.

Müller, T. (2016). Real-life examination timetabling. *Journal of Scheduling, 19*(3), 257–270.

Mushi, A. R. (2004). Mathematical programming formulations for the examinations timetable problem: The case of the University of Dar es Salaam. *Asian Journal of Science and Technology, 5*(2), 34–40.

Norberciak, M. (2006). Universal method for timetable construction based on evolutionary approach. *International Journal of Applied Mathematics and Computer Sciences, 3*(3).

Petrovic, S., & Burke, E. K. (2004). University timetabling. In J. Y.-T. Leung (Ed.), *Handbook of scheduling: Algorithms, models, and performance analysis* (Chapter 45), pp. 451–460, London: Chapman & Hall.

Redl, T. A. (2007). University timetabling via graph coloring: An alternative approach. *Congressus Numerantium*, 187, 174.

Schaerf, A. (1999). A survey of automated timetabling. *Artificial Intelligence Review, 13*(2), 87–127.

Thomas, J. J., & Tajudin, A. (2006, June). *Visualizing the examination timetabling data using clustering method and TreeMaps*. In Proceedings of the 2nd IMT-GT Regional Conference on Mathematics, Statistics and Application University Sains Malaysia (pp. 454–458).

Vasugi Mudaliar, R. K., & Mishra, M. K. (2016). An overview of the heuristic approaches for university course timetabling system. *International Journal of Engineering Research & Technology, 4*(21), 1–6.

Vrielink, R. A. O., Schepers, D., Jansen, E. A., & Hans, E. W. (2016, August). *Practices in timetabling in higher education institutions*. In PATAT 2016: Proceedings of the 11th International Conference of the Practice and Theory of Automated Timetabling.

Wren, A. (1995, August). *Scheduling, timetabling and rostering—a special relationship?*. In International conference on the practice and theory of automated timetabling (pp. 46–75). Berlin, Heidelberg: Springer.

Yakhno, T., & Tekin, E. (2002, October). *Application of constraint hierarchy to timetabling problems*. In Eurasian Conference on Information and Communication Technology (pp. 11–18). Heidelberg, Berlin: Springer.

CHAPTER 13

Information Technology and Student Success

DHARMPAL SINGH*, IRA NATH, and SONALI BHATTACHARYYA

*Department of Computer Science & Engineering,
JIS College of Engineering, West Bengal, India*

Corresponding author. E-mail: dharmpal1982@gmail.com

ABSTRACT

Students' success and information technology are the two sides of the same coin. These can be used in an efficient manner to obtain the good results for the betterment of any society or nation. The use of technology plays a vital role and transformative support to the success and outcomes of people that make them more responsible and thoughtful. It has been observed that nearly one-third of undergraduates (UGs) leave the institutes after their first year, and many students require six years to complete their degree. Students who complete the degree are not able to provide the basic idea of some read subjects and also forget the things after a while. These problems are mainly due to the old teacher-centric method that is unable to create great zeal and zest among the students due to the lack of information technology infra-structure. Therefore, the goal of this chapter is to provide the new teaching-learning method and the digital learning platform for teacher to present the topics in well manner to create the interest among the students. This chapter is also focused on new techniques that change the teacher-centric learning to student-centric learning to enhance the students' success ratio.

13.1 INTRODUCTION

Information technology (IT) is a business sector that deals with computing, including hardware, software, telecommunications, and generally anything involved in the transmittal of information or the systems that facilitate communication. It plays a major role in many areas of life. The use of IT in education makes the life of a student good and also helps them learn things in a faster way.

13.2 INFORMATION TECHNOLOGY

IT involves many things, and people with varied jobs and diverse tasks use it. These diverse tasks series from the possession of networks and information protected to the possession of system's up to date. IT professionals are the persons who are responsible to give information as input. They control files and records and also are good programmers. The chief information officers resolve the issues related with the IT department and also decide how the IT department will execute any plan and what are the necessary equipment required.

IT also organizes information which is available in various forms. It can include belongings associated with the Internet. IT flourishes with a novel idea; Internet is its kingdom. IT includes the transmission of information through Internet, making Internet inevitable. IT has now become an integral part of our daily life and is growing to flourish into a new kingdom. IT also plays a vital role in education. With the help of IT students can get any details from anytime and anywhere in an easier way. They also learn the things in the form of audio and video and get the real effect of learning. They create and design the new topics from the different aids of IT.

13.2.1 USAGE OF IT IN OUR DAILY LIFE

It is a familiar truth that the purpose of IT in everyday living has been altered noticeably over the preceding years. IT is necessary in each corner of our daily life. Modern technologies are upgraded by modern business policies related with IT. This increases the possibility to work more efficiently through various applications of IT.

The way communications have changed affects us significantly. Long ago, writing was the only way of communication. But today, modern IT technologies such as landline, cell phones, and high-speed Internet are essential means of communications. The more advance applications of IT in this field are text messages, e-mails, MSN messenger, Facebook, and Twitter. IT has created massive changes in communications all over the world.

Over the past few years, the area of computing has changed drastically due to IT. Nowadays, most of the people are using iPod, computers, etc. The cloud computing and virtual servers are the modern inventions of IT through which society has changed drastically toward a massive success. Long years ago, a personal server was a basic requirement to run or processing a program. Online resources are now available to do it at a lower cost. Now, it is possible to earn billions of dollars residing in any part of the world.

IT has carried out a novel execution in the area of teaching learning. We can receive any kind of data at our own house when we require. The renowned books are easily accessible simply via Internet. Each day we receive novel and current data with the assistance of IT. Nowadays, IT is also utilized in e-banking systems. We can maintain our finances with the help of the Internet. IT has also produced wonderful results in the area of healthcare.

IT in the present world has been developed so much that you can relish it in all the spaces. Now, cars are manufactured with built-in satellite navigation, climate control, and digital displays. We cannot imagine our vehicles without these features.

IT has also evolved the way using which we purchase necessary goods. Long ago, an ATM was required to draw money and buy needed things with money. Nowadays, utilizing a PDQ machine/chip and pin machine or a credit card, anyone can buy anything. It has certainly made our living so much simpler. IT has presented a modern age in the area of e-commerce. E-commerce is an approach of purchasing and buying products online. With tremendously eventful lifestyles and a lack of time, e-commerce has transformed the method of purchasing our goods. It has transformed so much that businesses are continually financing in online policies from online shops.

As we can realize, the applications of IT in our regular life have certainly altered the way we lead our daily lives. Long ago, we uniquely interconnected person to person and required to capitalize that for our commercial success. IT applications have certainly altered the techniques of transmission, businesses, teaching and the methods that we have adopted for living our daily lives. This involvement of IT in education has also significantly improved the leaning rate of the learners and keeps a major impact of their employability.

13.3 CHALLENGES IN THE UTILIZATION OF IT IN EDUCATION

The foremost trial is the requirement for continuous regeneration in terms of content, which indicates that enlarging the National Achievement Test-based curricular needs to lodge prevalent requests that have developed during the last few years. As technology evolves, the environment becomes more user-friendly, which indicates that there is a least requirement of thorough acquaintance of working procedures. In straight percentage, however, there is an even superior claim for practical knowledge in the utilization of software, the establishment of transmission of data through network, and abilities in making credentials. The function of software related with graphics for the development of multimedia applications has enlarged accompanied by the implication of controlling the audio, visual, and music redrafting programs. Though the NAT construction is "tough," that is, it does not propose tangible tools and procedures, because of enhanced technical progress, local prospectuses must be changed in an uninterrupted manner so that pupils can become more well acquainted with the up-to-date practical knowledge.

13.3.1 CHALLENGES IN SCHOOL/COLLEGE TEACHERS' ADAPTATION

It has been observed that schools and colleges are facing the problem to adopt the new teaching base challenges. These are furnished as follows:

Improvement in the professional field: There is the absence of adequate, constant professional improvement for educators who have to integrate novel technologies into their classrooms but are incapable due to the deficiency of perceiving the significance of these modern technologies.

Oppose to alter: There is a "comfort with the status quo" which suggests that familiarizing with modern technologies is the external part of the job description of faculties and school rulers.

MOOCs and other modern models for schooling: Massive Open Online Courses (MOOCs) are the modern techniques in the field of teaching learning. The technique has inspired many higher education institutions to smoothly conduct the teaching-learning process and achieve grant success. K-12 schools should find out various methods to impart this knowledge to make education more accessible to ordinary pupils also.

Providing informal knowledge: Informal learning is found more frequently in nontraditional classrooms like flipped classrooms, which supplies a combination of formal and informal learning. It has much more nontraditional tactic that hugs multimedia into its own standard syllabus.

Disasters of personalized learning: In a real situation, there is a gap between the knowledge of differentiated personalized learning in the form of instruction and the skills obtainable to make it occur. Although K-12 faculties observe the requirement for personalized learning, they do not make it happen in reality due to less availability of tools, time, and infrastructure.

Disaster to utilization of modern technology to provide effective decisive evaluations: Nowadays, the examination system is used for evaluating a student. Teachers also engage themselves for teaching the students only for the test. According to the requirement of our society, the curriculum and technical skill developments have been adapted. So, in turn, the present examination system has become an obsolete evaluation technique. Modern technologies and our realizing of various learning outlines require to use novel techniques through which students will be able to earn points for the understanding of and gaining knowledge about a particular topic.

Although the implementation of technologies for K-12 also has grown enormously, there seems to be a problem with widespread applications. Older faculties have lack of realizing of how modern technology executes. This lack of realizing of new technologies is infuriated when an older faculty attempts to teach a pupil who raised up utilizing that modern technology.

13.3.2 CHALLENGES IN COORDINATION OF VARIOUS LEARNING METHODS OF SCHOOL/COLLEGE TEACHERS

Many challenges have been observed in the school/college teacher coordination, and few have been furnished below.

13.3.2.1 STUDY THE KNOWLEDGE BASED ON TEACHING AND LEARNING

We have depth knowledge about the subjects of our fields. As we start our new career in teaching, we have to be well versed with the modern methods of teaching and learning. It comprises evolving active instructional policies to attaining students and teaching with the help of technology. Books, articles in educational magazines, newsletters, conferences, and online materials deliver plenty of help. We can take the benefit from our organizations for teaching and learning or other specialized and skilled improvement resources.

13.3.2.2 ACCEPT ALL WHO ARRIVE AT THE CLASSROOM ENTRANCE

All the students in your class are not of the same category. Some students are excellent, some are average, and others are poor. The educator in the classroom must think about the requirements of all categories of students and set the strategies accordingly. An ungraded pretest can be executed for the learners to know about their knowledge of the topic that you want to deliver next. All the learners in the class require help of the instructor to know further about the topic. The instructor must rethink about the appropriate strategies. After all, pupils should not be criticized if they do not clarify anything that were already skilled in high school. Admit pupils with their present knowledge and assist them to proceed further. They all require a skill and education from college which will enhance their career.

13.3.2.3 POLICY FOR INSTRUCTIONAL MANAGEMENT

For a long time, college administrations did not think about the management of classrooms properly and scientifically. Sometimes they planned to execute this on paper and pen, but never implemented it. But this is the time to change the motive as pupils are required to know what is going on. A visual outline of what will be executed during the class can be sketched, which will assist pupils to follow the lessons and solve the more complicated assignments related to the topic. Different aspects of teaching, such as dispensing papers, taking attendance, and making a timeslot for students to ask questions and give appropriate answer to them are required to be part of course scheduling. You apply strategies how to increase the attendance percentage of students in a particular class, how can you reduce the disturbances inside a classroom, etc. and discuss those strategies with the pupils to get review.

13.3.2.4 TEACH WITH A DIVERSITY IN POLICIES

Various policies can be adopted such as education based on learner-centric, directed investigation, active learning by the students, lecture, group discussion, and discussion through online. Among the above-mentioned learning techniques, which one is producing best result in terms of the delivery of contents and student's learning requirements. The greatest assistance is to be pictorial, followed by keeping students thinking, writing, comparing, and applying novel skills dynamically. Pupils learn more easily when they have been given the justification for what they are learning, and they recognize then why the allocated educator has selected this particular instructional procedure and learning techniques for teaching them.

13.3.2.5 USAGE EVALUATION TO ENLIGHTEN PUPILS OF THEIR ACHIEVEMENT

Today's pupils are checking their grades online so that they recognize their current position at any time in the semester. Grading strategies require to be vibrant and grading scales should be easy to use. You can share your grading strategies in a written form on the curriculum and then display accurately how it executes after the test, paper, or assignment. You must remind pupils that evaluation is more than the assigning a grade (Guskey, 2013). Assessment helps us to know about pupils' attainment and helps teachers to know their necessities.

13.3.2.6 PRESERVE THE DESIRE

It is very informal to become discouraged by student criticisms, absence of administrative help, economical reductions, and profession uncertainty. However, what is it that sketched you to your discipline formerly? For most of us, it was an accurate desire for the subject, a hunger to learn all about it, and a further eagerness to share that knowledge. In higher education, we have chances to learn, research, teach, and outline the upcoming disciplines and effect the greater world through our disciplines. Successful college faculties distinguish that many of today's college pupils have the requirements for learning. Adopting activities like these provides assistance to them to accept those contests effectively.

13.3.3 CHALLENGES IN SCHOOL/COLLEGE STUDENTS' ACCEPTANCE

Although time spent at college is a recollection of glad memories, the student life is not smooth always. Some problems are always present in their college life which they have to face at school at least once (Wormeli, 2014).

In the following sections challenges with the solutions have been furnished.

13.3.3.1 MANAGEMENT OF TIME

Problem: College is a place where students facing challenges in education. There are many subjects that need more attention and energy than that required in schools for academic purpose. It has been observed that some course contents of two years in colleges are merged to deliver it in one year. So many pupils like a complete semester with 15 credit, while others attempt to earn up to 18 or even 21 credits. At times, it appears impossible to stay on top of it all.

Solution: The first point that each student must have to do is that he/she has to know his/her capacity. If it is not possible for you to tackle 18 credits in one semester, then you obviously yield only 15 credits. The main aim of college academics is to gather knowledge as much as you can during this period. It does not imply that students will be engaged in their study always. It is also essential to fix a time period every day for your entertainment and recreations. This pauses also help you to retain your mental health fresh and healthy. So, we can take all these techniques to remove our anxiety related with academics.

13.3.3.2 SPREADING YOURSELF TOO THIN

Problem: Nowadays, as most of the courses at colleges are costly due to high college tuition fees, many students eager to do jobs while studying. Juggling a job, 15 to 18 credits, affairs, and extracurricular activities is tremendously problematic. Many students attempt to do all of these activities into one day and do not get sufficient sleep for proper rest. Without adequate rest, students are susceptible to the health problems related with physical and mental.

Solution: Every student must take proper decisions to know what is vital. They have to make a proper timetable for studies, games, meetings, and social events according to their event priority. You must be conscious of your choices when receiving an appointment letter of any job. Universities frequently offer jobs which is suitable for a student's schedule.

13.3.3.3 HOMESICKNESS

Problem: Especially, in the first year of college, students who are staying far away from their homes for a long time can suffer from homesickness.

Solution: If you are staying at a hostel or a mess for your education within the journey of three or four hours from home, plan to go to your home once or twice in every month. Also, you ask your friends and family members to email, make phone calls, and send care parcels. These steps should significantly help in lessening sufferings of homesickness. The campus environment can also help students to remove these sorts of feelings. In the campus, students make their own friends, talk with them, and share their feelings, sufferings, and experiences. This can help them to reduce their homesickness. In this way, they can help each other and grow their career.

13.3.3.4 DEPRESSION

Problem: Each problem mentioned in this section can increase the anxiety level of a student. This can lead to depression in the long run.

Solution: Proper counseling is the most important treatment at this situation. One needs to find out professional support. Many campuses in India are present who are eager to give free counseling programs for students. Counselors are proficient enough to pay attention to the sufferers and assist students to recover to return to their normal life.

13.3.3.5 CONDITION OF HEALTH

Problem: The main causes of health problems are anxiety, negligence of self-care and deficiency of sufficient sleep. Staying in very adjacent residences may also create various infectious illnesses.

Solution: Every student should eat balanced and healthy foods regularly. They must give importance to self-care and take complete rest at night with sufficient sleep for minimum 6–7 hours every day. They are advised to wash their hands frequently. They must contact the nearest doctor in the campus without any delay if any one feels sick.

13.3.3.6 SOCIAL PROBLEMS

Problem: A student can make many new friends. Making good friends are very important in each student's life for spending time with classmates and roommates, creating a good social network, and discussing various problems with them to get solutions. However, spending a long time together can spoil a student. Then social relations can be converted into a disturbance.

Solution: The students can take a break from their normal schedule to get away from their campus to visit a nearby coffee shop, mall, park, or restaurant for their refreshment. Students must give highest priority to their education and self-care about their mental and physical health. If any severe problem arises which they cannot solve by discussing with their friends, then they should contact their local administrations or inform their local guardians for further steps.

13.3.3.7 PARTYING

Problem: Enjoyment in itself is not a problem at all. Enjoyment can be a solution for students to remove stress. However, sometimes enjoyment can create enormous problems. Students can be addicted to drugs and alcohol which have deadly impact on their behavior, health, and career. Having sex without sober consent and without taking the essential protections can be shocking, unsafe, and even unlawful.

Solution: While enjoyments are vital, parties should be attended in an accountable manner and authorized way so that you do not make glitches for yourself or others. You must know your capability. You request for a drive to reach your home safely if you are drunk. Also, you have to keep track of your friends and also make indisputable that they are harmless. You have to add enormous water while drinking alcohol. You must carry a condom if it is essential.

13.3.4 CHALLENGES IN INFRASTRUCTURE OF SCHOOL/COLLEGE CLASSROOM

The national IT and telecommunications can improve the infrastructure of the educational technology of one country. The following will be considered before any information and communication technology (ICT)-based program is launched.

There are many countries with a huge number of ancient school buildings and widespread retrofitting where suitable electrical wiring, heating/cooling, proper ventilation, and security would be required. Additional elementary need is the accessibility of electrical energy and telephony. In developing countries, huge areas are still deprived of a reliable supply of electricity, and the adjacent telephone connections are far away from people's homes. This reality can be seen in few countries of Africa. A wireless technology (such as Very Small Aperture Terminal) can be a solution here which is costly enough. This choice should be analyzed for other emerging countries with very poor telecommunication infrastructure.

ICT is inexpensive and simpler to familiarize a new procedure of technology into academics, and keeps it working, where teaching is depending on the large-scale growth by governments or the private sector. TV can also be used for teaching purpose; computers in schools can be preserved once profitable and private utilization has been extended to the fact where there is a recognized service industry.

The advancement of faculty experts should have the following five efforts:

1. Deep knowledge with specific applications;
2. Integration into present syllabi;
3. Syllabus changes connected to the usage of IT and industrial needs;
4. Changes in faculty role; and
5. Supporting educational concepts.

There are many countries where preservice teacher training in ICT has been introduced for improvement in teaching quality. ICTs are rapidly developing skills; however, most ICT-fluent faculties require to enhance their knowledges uninterruptedly and keep themselves up to date with the modern technologies and good practices.

Investigations on the usage of ICTs in various settings of academics over the years have always categorized as a block to achieve the incapability of educators to understand why they should use ICTs and how precisely they can use ICTs to assist them to teach the students in healthier and improved ways. Unfortunately, most educators with ICT expertise are emphasizing on "teaching the tools" and on "using the tools to teach."

English is the leading linguistic of the Internet. Approximately 80% of digital documents are available in English. In the world market, a huge percentage of software related to education is in English. Students are affected most to achieve the benefits of WWW in the urban side in Asia-Pacific, where peoples are not habituated with English at all.

Technical standardization of content is now a critical issue in on-line learning. Standardization permits various applications to exchange materials and on-line teaching-learning systems. Provisions in materials, construction, and exam setups are suggested so that interoperability may happen among various administration systems, outcoming in few cost-effectiveness. Standards must be common enough to support all types of on-line learning systems and materials.

13.4 TEACHING APPROACH

13.4.1 TEACHER-CENTERED APPROACH TO LEARNING

It has been observed that in teacher-centered education, teachers talk and students listen to them. This approach will discourage the collaborative learning among the students. In this type of learning, teachers act as a center of knowledge and students are passive learners and receive information during the class. The main role of this type of learning from a student's point of view is that the teachers provide information to students and evaluate the copy/assignment of students. Students are considered as the "empty vessels" which will be filled up by the knowledge imparted by the teachers. Here, teachers and professors are considered as a sole supplier of knowledge, instruction, and systematic, scripted lesson plans to students, and students have to follow them. In this type of learning, information provided to students is limited to teacher's knowledge of textbooks, workbooks, and PPTs, and notes. Assessments in this type of learning mainly carried out as summative and not formative evaluation. Henceforth, this type of learning rarely addresses qualitative issues of the learner's progress.

13.4.1.1 HABITS OF HIGHLY EFFECTIVE TEACHERS

It has been observed that the teachers who have problem to connect with students emotionally can give many excuses like students' lack of motivation or maturity during class and not try to learn the things. In this way, they create the gap between students and grow disconnected from them which further creates an adversarial relationship with them. They used the world like I must have hold on them to bring into the shape, It is enough for me, either them or me, these are hopeless students, why should I bother for them they are very bad students?; their parents have not provided the good culture and that is the reason they have created the bad environment.

Therefore, these types of behavior from teachers are not good for students. They have to develop constructive responses in their own affective needs as teachers and prepare their students to do the same. These types of behaviors make good relations between teachers and students, and students will remember it even after their degree. The author Stephen Covey (1989) has furnished the seven habits of highly effective teachers. This will be discussed in the subsequent sections.

13.4.1.1.1 Find Joy in Others' Success

It has been observed that many people climb the mountain and resting at the top without worrying about the others who are struggling to reach there. Except the guardians/parents, a teacher is the only person who becomes glad on their students' success; therefore, teachers must have the quality to feel and participate in the joy of students. So, teachers must motivate the students and tell them to do the work if stunts fail in the work so without angry, they have to make them understand.

13.4.1.1.2 Cultivate Perspective and Reframe

Teachers must have the perspective and reframe quality to deal with the students. They must know the difference between empowering optimism and defeatist isolation. It may be possible that students have shown the disrespectful behaviors to teachers, so instead of taking to him/her personally, we have to realize that he/she is only 14 years old, so he/she required to control his/her impulse. Teachers must focus on the positive behaviors of young adults, and if they misbehaved with anyone, then let them know how his/her words have actions and consequences. Therefore, guiding him/her to amend and restoring their mistake and built their trust and tell them to start afresh mind set up. Perspective behavior in teacher provides hope to do more for students when they have less chance to do anything. This little help to students creates the long haul with teachers.

13.4.1.1.3 Ditch the Easy Caricature

From the beginning of mankind, humans have been known for categorizing the things and people. Based on the behaviors of the students, teachers create a picture about the students and they think that they are the same as they are thinking. As a result, the gap between teachers and students are created and teachers may reject the ideas of students who have some errors/mistakes in the project.

Therefore, teachers need to visit students' homes to know more about them and try to really understand their behaviors beyond the classroom. This type of practice may be helpful to know the students better. Whenever, teachers see the students to play a game or cultural program, then they can see their extended effort and intellectual fortitude.

13.4.1.1.4 Explore the Ethics of Teaching

Teachers must show ethics about their teaching. Teaching is known as the novel profession, and teachers must keep this one in the mind.

There are no bad students but bad teachers, so the teacher must keep in mind that they can motivate the students, they have the capability to make the students good, and they also have the power to make the students to work with society because teachers have universal respect for human life and also be guided by principles of caring. In fact, teachers have novel duty to their students in the best ways. Teachers must follow the ethics of teaching and have good relation with students (Wormeli, 2013).

13.4.1.1.5 Embrace Humility

It has been observed that teachers do not accept the ideas of the students because there is a gap between thinking of students and teachers. So teachers must accept a new idea and try to know that what they want to say. It has been seen that if students criticize teaching, then many teachers take the things in wrong ways. They feel that students humiliate them. Therefore, teachers must grow comfortable with the idea of students and why they are saying so. Try to cope with the problem and discuss with the students that we may be wrong. Teaching is the one type of profession where every time we learn new things in life. It may be possible that we can learn from seniors/ juniors or from students.

13.4.1.1.6 Value Intellect

It has been observed that few teachers have been teaching some subjects since last five years and following the same process of teaching. They feel very ease to teach the same subject because they have good knowledge about it and not much problem or effort required to create the lesson plan, assignment, and many more. This mainly happens because they need room for their own lives and interests because they give much time to students for different activities.

Schools and policymakers must recognize the good teachers and give awards for that. They also motivate them to go for new technology. School/ college management also motivates them to recognize outside pursuits and gives chance to translate into better teaching style.

13.4.1.1.7 Maintain Passion and Playfulness

Teachers must have passion when they teach students because it may happen that some students ask or act in such a way that teachers may get angry. So, this will create the bad impact of teachers in the minds of students. Therefore, teachers must add fun while teaching their subjects to students. They should give assignments and work in such a way that students find themselves engaged in the assignments and learn in funny ways. Teachers should even invest in their learning and will see that it will elevate their spirits. In today's world, a lot of stress and problems are there, so teachers must have true passion and playfulness nature to tackle the students and bring back much-needed humanity.

13.4.1.2 PROBLEM OF THIS APPROACH

This approach has many advantages and disadvantages.

Pros: In the case of a teacher-centered classroom, students remain orderly, quiet, and teachers have full control of the classroom and its activities. In these ways, students learn based on teachers' instructions and make their own decisions. In direct classroom activities, teachers are not worried about students missing an important topic. Here students learn in groups, and they learn the work in a collaborative manner and try to solve the problems in their own ways. Thus, they become independent.

Cons: It has been seen that when you work alone, you will get success alone. So, in this world, we have to work in collaborative ways. If you are alone, then you are not able to learn with other students, and your communication skills may suffer. The teacher-centered approach may bore the students, and they will wander and miss important facts. The teacher-centered approach is confined to students express themselves, ask questions, and learn on their own way.

13.4.2 STUDENT-CENTERED APPROACH TO LEARNING

In the student-centric learning approach, students know what they are learning and why. It is a blend of individual, team, and a large group work, which is literally designed, both in structure and in instruction, to be a blend of individual and team work. Here they will among the groups or one-on-one instruction with their teachers.

13.4.2.1 STUDENTS USE PERSONALIZED TECHNOLOGY TO PRODUCE AS WELL AS CONSUME

Student-centered learning demands students to spend time on personalized learning tasks or projects. It has been observed that due to personalized learning tasks, students have designed many projects which are good for the college and have the impact on society.

This type of personalized learning ensures that the learner who works in a group is not just individualized but also that it is meaningful and connected to the real world.

13.4.2.1.1 Students Have Some Opportunity to Work at Their Own Pace and Explore Their Own Interests

The student-centered environment is beneficial for finding the students' interests about tasks, addressing competencies, or skills, so that they can do so at their own pace. In this type of learning, students flow in and out in their group, work on individual projects for designed on the group's interests. They also work for mastering their competencies in different fields of their interests and earn while they learn. They have opportunity to focus on their learning about communities and causes that they are most passionate about. Here, students may often work on a similar set of projects, but their ways of learning are different.

13.4.2.1.2 Problem of the Student-centered Approach to Learning

The problem of this type of approach is that students miss their own instructions and become unconscious about the subject of the course. They think that they are the "blank slate," and teachers will guide them to write knowledge upon. Most of the students start complaining about "busy work" if they do not understand the point of the student-centered activity, or if that activity does not challenge what they already know. All the learners have the same ideas about the subject of the course, and gaps also exist in their communications. So, this is the onus of instructor which sees it and creates a bridge between the gaps.

But due to the avoiding nature of some teachers, many students are not able to get good knowledge about the subject which they deserve. Some teachers consider themselves as the highest authorities to teach, direct, instruct, and control students in their class, and they believe that students will do the tasks based on their instructions. These types of teachers may fear to do the things differently or feel threat to their status and profession. Most of the teachers have the perception that they have a good approach to teaching, and there is no change required to. Students have to mold themselves to adjust to teaching style since these teachers feel that they work best and are not open to new ideas or other possibilities. On the other hand, some teachers try to work according to the minds of the students and try to do the task according to their ways. These teachers are too eager to take changes and inspire students to do the same changes. These types of teachers are not worried about the culture and realities of their classrooms. It has been seen that some students reject the student-centric approach because they think that teachers are more knowledgeable than students and they will provide good ideas. Such students feel that knowledge must be transferred from one person to another and are happy with get spoon-fed.

13.5 STUDENT LEARNING METHOD

Students learn in classroom studies in schools where their willingness to learn increases because they learn there in a team structure. Students prefer group activity and whenever they are given any assignment, many times it has been seen that they feel confident to accomplish the job when they do it within a group.

13.5.1 COOPERATIVE LEARNING

Co-operation means helping each other. In co-operative learning, the total class is divided into several divisions. In such a setting, there might be one or two students who are good enough to understand the delivered materials easily and complete their assignments early enough to help other slow learners in his or her group, to make them capable of doing their assignments also. In another case, the whole job may be divided into several parts and assigned to different divisions or groups. Each group is assigned a specific subtask, so total processing of the class is dependent on all subtasks of all groups, and every group will process their own subtask as well as help other groups for the completion of the whole task for a group benefit.

13.5.2 PERSONALIZED LEARNING

Personalized learning depends upon the self-interests and the speed of learning of the learners. It varies from learner to learner, depending upon their willingness, speed, and tenacity to learn. They may opt the subjects offered, the timing, mode, etc. as they wish. This learning is mostly supported by technology. The modern-age students are highly tech-savvy. Maximum students nowadays prefer this type of learning rather than classroom learning. Plenty of resource materials under various courses are available through web, and many of those are offered with the certification after successful completion of the courses.

13.5.3 LEARNING BASED ON PROJECTS

In project-based learning, students generally learn by doing the projects. Students are generally engaged here in critical thinking with hard problems and acquire knowledge by practically doing the projects. The projects should be authentic as well as the students' thinking and answer also have to be genuine and authentic.

The steps of handling the projects would be finite. Students would present the project by defining the problem statement, then the abstract and the methodology they have used. But the problem might be open-ended, and the actual answer may be obtained by the experiment and research only.

Nowadays there are many organizations which motivate and ask the project idea from the students on any real-life problem and students taking the challenge, study and learn, simultaneously acquire much knowledge on it. In many cases, organizations grant funds to students to implement their ideas. Depending upon various criteria, project groups are awarded also such as innovativeness, environment friendliness, commercializability, and potentiality of the social impact.

13.5.4 FLIPPED LEARNING

These days, our classroom has been changed into flipped classroom, and it is very much appreciated by academicians throughout the globe. Students also have opted this method very much. There are three activities for flipped class: preclass, in-class, and postclass activities.

13.5.5 *FLIPPED CLASSROOM PROTOCOLS*

1. Establish spaces and time frames that permit students to interact and reflect on their learning as needed.
2. Educators continually observe and monitor students to make adjustments in the pace, content, and collaborative learning as appropriate.
3. Educators are flexible in their expectations of student timelines for learning and more importantly in their assessments of student learning.
4. Give students opportunities to engage in predetermined meaningful activities without YOU being central.
5. Remember that flipped learning is a major powershift from teaching to helping students learn. Hence, in-class time must be dedicated to exploring the topics in greater depths and creating rich learning opportunities for the students by engaging them in group learning activities such as group discussions. As a result, students are actively involved in knowledge construction as they participate in and evaluate their learning in a manner that is personally meaningful.
6. Prioritize concepts used in direct instructions for learners to access on their own.
7. Educators use intentional content to maximize classroom time in order to adopt methods of student-centered, active learning strategies, depending on the grade level and subject matter.
8. Educators make themselves available for individuals, small groups, and class feedback in real time as needed.
9. Educators continually observe their students, providing them with the feedback relevant to that moment in time and assessing their work.
10. Conduct ongoing formative assessments during class time through keen observations.

Students are evaluated with the following 10 parameters with maximum grading point 10 for each under each flip learning classes:

Par. 1: Digital content – critical thinking

Par. 2: Flow, quality, and balance of digital content—creativity

Par. 3: Understanding of the digital content including complex problem-solving

Par. 4: Collaborative learning—group discussion and negotiation

Par. 5: Cognitive flexibility—attitude and aptitude

Par. 6: Leadership qualities including judgment and decision-making

Par. 7: Audience engagement

Par. 8: Pitching—to succeed in choosing/achieving

Par. 9: Body language

Par. 10: Voice modulation

This method helps to identify slow learners, moderate learners, and fast learners.

13.5.6 MEASURES TAKEN FOR IMPROVEMENT

13.5.6.1 SLOW LEARNERS

1. Additional class note with solved problems provided.
2. 3-min video assignment submission.
3. Organizing extra classes.
4. Remedial and tutorial classes are held to prepare them for remedial exams.
5. Assistance from classmates and senior students are arranged.
6. Providing tutorial assignments, lectures uploaded on web, and extra reading material to improve the basic understanding of the subject
7. Encouraging them to study courses on developing soft skills to master the understanding of language.
8. Encouraging them to participate in various activities to develop social skills.
9. The institute uses varied evaluations to test both quality and values.
10. It has a continuous evaluation system with different types of assessments spread.
11. Throughout the semester. The assessment of values is also a continuous process with focus on discipline, conscience, dignity of labor, respect for religion and culture, and community participation.
12. It implements well-defined models of evaluation with varied types of assessments to effectively evaluate outcomes, both at the course and the program levels.

13.5.7 SCHEMES FOR ADVANCED LEARNERS

The institute offers several opportunities for advanced learners to augment their talent and meet their learning needs:

1. Awarding students: To encourage bright UG students awarded for attendance, score in semester, best performance in co-curricular, and extracurricular activities.
2. International internship: Students with outstanding performances in semester-end examination (SEE) are selected for attending international internship at Asian Institute of Technology, Bangkok; Moscow Institute of Physics and Technology, Russia; etc.
3. Students are encouraged to make research contributions through paper publication at national and international conferences under mentors' guidance.
4. The institute organizes technical symposia and conferences where students take part as participants or organizing members.
5. Financial support is provided to students for participation in national and international conferences.
6. Earn while you learn provision has been made for UG, PG, and research students for part-time jobs in the large number of projects running in the institute. This helps them learn state-of-the-art techniques and also get remuneration.
7. Students are encouraged to help slow learners in their class and in junior classes.
8. The institute has a unique system of providing pathways for vertical and horizontal progression by clearing requisite tests and meeting eligibility criteria.
9. Alumni support is also available to students. Further, the institute aims at fostering global competencies among students having diverse needs and abilities by encouraging them to participate in seminars, group discussions, and active learning to develop public participation and social skills.
10. The institute encourages innovation to solve local problems and entrepreneurship with the opportunities to earn while they learn.

13.5.8 SCHEMES FOR SLOW LEARNERS

1. A unique value system of peer group teaching, learning, and assistance from classmates and senior students is arranged as one of the special measures to support relatively slow learners.
2. Other significant measures include organizing extra classes, special tutorial sessions, and remedial teaching.
3. Extra reading material is provided to improve the basic understanding of the subject, and efforts are also being made to upload video lectures on the web for students to review to understand difficult concepts.
4. Soft skills courses are offered to improve reading and writing skills in English.

13.5.9 STUDENT-CENTRIC EXPERIMENTAL, PARTICIPATIVE, AND PROBLEM

Students involved in the following activities mentioned with directions under FLIP class conduction.

1. Directions of GD, PPT, and MCQ for flipped teachers and the suggestions of the active learning procedure in flip classrooms.

A. GD:

- Topics should be assigned to every group. End time should be assigned for the discussion.
- Each group members should start wring on the topic and balance content among them.
- Members in each group are allowed to collect information from Internet, books, etc.
- At the end, one member from each group should summarize about their information collected.

B. MCQ:

- Declare the start time, end time and full marks for MCQ.
- Provide MCQ preferably in the Google form.
- Display marks and send the soft copies of result through mail.

C. PPT Presentation:

- Students group should develop PPT after group discussion performed in a class and share it to teacher's code on Blendspace.
- PPT should contain the first slide with the topic name, students name, section, and roll number.

- PPT should contain text in bullet points, images with clear labels, and video clips of not more than 3 min.

2. Recoding of GD and PPT presentation of students inside the classroom and sharing it to the students on a class-to-class basis (a copy is attached with this email).
3. Exploring of new learning management systems (LMSs) like Canvas and Open Learning.

13.5.10 LEARNING IS BETTER WITH FLIPPED LEARNING: STUDIES

In many universities throughout the world, the number of studies have been done on the acceptability of different methodologies of learning. But on top of everything, blended learning has been the most preferred. As online learning and traditional teaching in the class are both unidirectional, where only the teacher speaks, but in blended teaching and flipped learning, communication occurs in both ways: (a) teacher–student and (b) student–student.

13.5.11 FLIPPED LEARNING PROMOTES OWNERSHIP OF THE STUDENTS TOWARD LEARNING

Blended learning is the mixture of classroom teaching learning (where students learn from teachers face to face) and online learning, whereas in flipped learning, in the preclass concept, students study the materials, including teacher's video which teachers had already made for those students on a particular topic at home.

In the in-class concept, students interact in the class, they ask teachers wherever they have doubts, and teachers help them to understand the concept and initiate some group activities such as group discussion, quiz, and small seminar. Here, we can see co-operative learning between the students. Also, slow learners can keep up pace with the class by repeating the same video at his or her pace.

In the postclass concept, teachers give the assignments, quizzes, etc. to the students for practice at home.

The traditional teaching-learning process needs teacher's intervention, but the blended learning and flipped learning mostly need the ownership of the students.

13.5.12 STUDENTS' ADAPTATION FOR TECHNOLOGY-ORIENTED LEARNING

With the help of technology, the materials of blended learning and flipped learning are made. Today's world is so tech-savvy that students' communities are ready to prepare themselves for this technology-oriented learning.

13.5.13 COST-CUTTING OF FLIPPED LEARNING

As flipped learning and blended learning are much more dependent on online classrooms or virtual classrooms, there is no need of building as well as the maintenance of the building, electric consumption, etc., except for the in-class activity time. With the help of video conferencing also students can interact with teachers through web as many colleges and schools are doing during the current COVID-19 situation.

It reduces the building cost as well as the electrical expenditure. The class can be better managed online from the teacher's perspective, so a higher student–teacher ratio can be maintained.

13.5.14 COLLABORATION EXPANDS BY FLIPPED LEARNING

In flipped learning, during the teacher–student interaction, there are good possibilities of collaboration between students and teachers. Online teaching learning promotes students to have and interact with teachers from anywhere in the world. So, there is also a huge possibility of collaboration between students and teachers.

13.5.15 FLIPPED LEARNING CAN ENGAGE STUDENTS AND EXTEND THEIR INTEREST

Many of the students find the online mode more interesting than traditional classroom teaching. If the students can be engaged through critical thinking and given some application-oriented job, they are found to be more prone to engage in learning. This can improve their interests and motivation toward learning also.

13.5.16 HOW FLIPPED LEARNING IS ADVANTAGEOUS FOR STUDENTS

First and foremost advantage to the students is that they have not to move out and travel to reach the school. They can have their lessons sitting at their homes at ease. So, there will be energy saving for the students, and they can utilize that energy to other creative work.

In the case of traditional teaching, classroom was dependent on the teacher. Whatever teachers would speak, students had to listen to that only. Even if any student would know well enough about a topic, he or she could not switch on to the next topic. But in the case of flipped learning as the class is dependent on students' activity, they can enjoy, co-operate, and collaborate with the teacher as well.

As the topic is already circulated with the teaching materials in the preclass session, students have adequate time to discuss between themselves and ask the instructor if any kind of confusion arises.

13.5.17 HOW FLIPPED LEARNING IS ADVANTAGEOUS FOR TEACHERS

When students come to class with full preparation, there is little to no need for teachers to address content-related questions. Instead, they can support students in better understanding the concepts through practical applications.

Teachers can identify the individual learning needs of students, making sure that they all use the class time engaged with the learning process. And this is harder than the traditional teaching-learning practice.

13.5.18 RECORD THE ATTAINMENT OF COURSE OUTCOMES OF ALL COURSES WITH RESPECT TO SET ATTAINMENT LEVELS IN FLIP LEARNING

To assess the attainment of course outcome (CO), a variety of tools are used throughout the semester such as mid-term tests, assignments, mini projects, reports, and presentations. Respective evaluations are contributed to the internal marks sheet for a student which is finally reflected in the end semester results. The following list provides the set attainment levels:

DIRECT ASSESSMENT Rubrics for the Overall Assessment of a Course

Criteria (% of total students)	Set Attainment level
70% students must score 60% and above	Target level 3
60% students must score 60% and above	Target level 2
50% students must score 60% and above	Target level 1

1. Threshold value set for CO attainment using the tools identified under direct assessment are 60% score for class test, assignments, and semester examination each (for students having at least 75% attendance).
2. Rubrics demonstrating the set attainment levels are shown in the above table.
3. Calculation steps of the attainment of CO for a section of students
   a. Minimum attainment of a question under evaluation is securing 60% marks.
   b. Attainment percentage of a question under the evaluation process is the ratio of the sum of the students securing 60% marks to the total number of students of the same section attempted that particular question. Participation may be the total number of students appearing for the exam when questions are compulsory; else it is the number of students attempted.
   c. Each CO is related to multiple questions, the percentage of students attained a particular CO is derived from the average attainment percentage of mapped questions in Cumulative Internal Examinations (CIEs) and assignments.
   d. Attainment of each CO is calculated separately, considering the above conditions and percentage of students attaining different attainment levels are assessed in each case.
   e. Accordingly, decision is made on CO attainment, that is, Yes (with level 1/2/3) or No.
   f. Final attainment of each CO for a course is the fractional average of attainment of each CO for SEE and CIE derived.

Different assessment tools for evaluating the CO were discussed earlier. Firstly, direct assessments based on CIE, assignments. and SEE have been shown for the calculation of CO attainments. A sample attainment calculation is shown for the course.

13.6 CASE STUDY

JIS College of Engineering (JISCE)—The Gamechangers—has implemented the flip learning in India. The institute, based in Kalyani, West Bengal, is the flagship institute of JIS Group which is Eastern India's largest educational conglomerate. This institute is the first institute in India that has disrupted the higher education by introducing digital transformation through flipped learning since 2016.

They have been endorsed as the leaders for digital education in India by the Ministry of Human Resource Development (MHRD); National Project Implementation Unit (NPIU); All India Council for Technical Education; the World Bank Group; UNESCO; Microsoft, MEF University, Istanbul; and Maulana Abul Kalam Azad University of Technology, Kolkata.

Last April, they successfully organized an *International Conference on Digital Pedagogies* at New Delhi. We had both speakers and participants from across the world.

As a key takeaway of the conference, MHRD announced funding of 100 select institutions from Technical Education Quality Improvement Programme 2 and 3 for digital transformation based on the JISCE initiative of flipped learning. Last November, a select team of academicians from across India was sponsored by NPIU for training in digital transformations here at JISCE. Topics of the three days' workshop included digital pedagogical approaches, smart board usage, digital LMSs, myths about digital pedagogies, future learning spaces for digital pedagogies, and role of built environment, including biophilic classrooms in digital transformations.

The sequel in 2020:

Now that almost a year has passed and remembering the glowing faces of participants raring to initiate the digital transformations as they left the last venue of our last conference, they have decided to go in for its sequel called *International Conference on Digital Transformations.*

Conference aims:

The conference aims at three individual expert tribes of the future, each based on strategic core areas of digital transformations:

1. Leverage Industry 4.0
2. Drive digital business and
3. Enable future tech and AI.

The verticals:

The verticals have been designed taking into view the challenges and competencies that we will require as we step into the age of Industry 4.0 around 2025. As per the competencies charted out by the World Economic Forum for each year beginning 2020 to until 2025, we need to visit each set of competencies which is largely based on automation and digital data exchange, changing the way we live, work, think, manufacture, sell, communicate, perceive, exist, and even dream!

1. Digital transformations and impact on organizational transformations
2. Digital transformations and impact on teaching and learning
3. Digital transformations and impact on built environment
4. Digital transformations and impact on employability with reference to Industry 4.0
5. Digital transformations and impact on e-commerce

13.7 CONCLUSION

This chapter has taken an effort to provide the new teaching-learning method and the different IT tool/techniques used by teachers to create the interest among students. This chapter has also focused on new techniques and models to inculcate long-life learning knowledge in students for their employability. In this chapter, we have tried to furnish the pros and cons of the techniques and also tried to provide the resolutions for them. How the students get the benefits of different learning methods also stated in the book chapter.

KEYWORDS

- **information technology**
- **flip learning**
- **student-centered approach**
- **faculty-centered approach**
- **ICT**

REFERENCES

Covey, S. *The seven habits of highly effective people*. New York, NY: Free Press,1989.

Guskey, T. H. The case against percentage grades. *Educational Leadership*, 71(1), 68–72, 2013.

Wormeli, R. Take time for yourself—and for learning. *Educational Leadership*, 70(9), 14–19, 2013.

Wormeli, R. Teacher coach: Teachers' intellectual lives. *AMLE Magazine*, 1(9), 39–40, 2014.

CHAPTER 14

Perception of Information and Communication Technology (ICT) by Postgraduate Learners of B-Schools in Sikkim

APPALLA NAG SHANKAR

Department of Commerce, Sikkim University, Gangtok, Sikkim, India

Corresponding author. E-mail: anshankar@cus.ac.in

ABSTRACT

The advent of information and communication technology (ICT) has influence over every walk of life. Over the decades the processes, transactions, and the behavior of organizations have transformed by responding to the demand. World over the educational institutions has attempted to meet the ever-changing demand for blended services. Distance learning has been successfully using ICT as a platform. With a sample of B-Schools operating in Sikkim, this study attempts to know the implications and impact of ICT on students at remotely located institutions established in recent past. The perception of students captured by way of Unified Theory of Acceptance and Use of Technology model and notions of students are observed to offer an insight into pedagogy, attitude, and the use of ICT by students. An inquiry into the association between students' perception and behavior with their demographic attributes is presented. As the growth and development of students is a function of their interaction and exposure to the adjoining environment, it is revealed in this study. The home, educational institution, the community, and the location have implications on some or other aspects of their adaptation to the ICT. However, the results suggest that the students by and large are vigilant about the materials and talks delivered blended with ICT.

14.1 INTRODUCTION

Educational technologies have advanced to include the innovative gadgets that assist the teaching–learning process. Among many information and communication technology (ICT) gadgets have transformed the process of teaching and learning by eliminating the information-sharing component and shifting emphasis on the knowledge, innovation, and skill development. The paradigm shift in education system is observed pervasively across the globe. Since the phenomena have taken the toll of print media, and the television, the generations present in the educational institutions on either of the learners' chair or the teacher are finding it difficult to cope up with the gap, of understanding the knowledge, among the teachers' and students. Merely knowing the sources of information and the citations may not suffice the test of knowledge, understanding and skill, matters for true education.

Though the phenomena of ICT explosion is four decades old its diffusion in teaching and learning process in the developing countries is at nascent stage. The developing countries with remote locations, characterized by a rural population with poor literacy rate are yet to comprehend and reap the benefits of the technology. The educational institutions in these developing regions have recently (a decade ago) witnessed spur of the technology with the competition of the private participation stepping in the economy. When referred to India the governments at the local level have distributed gadgets at free of cost to the students of 10th and 12th grade. At the school level, training and development of teachers to handle the technology and its success rate are yet to be enquired across India. Even at institutions of higher learning in graduate colleges the distributions of laptops have been witnessed in states. However, if the gadgets are put to correct use remains unaddressed. When related to the teaching and learning in institutions of higher education it is observed that the contents of curriculum do demand variety of (Lindblom-Ylanne, et al., 2007; Lamera et al., 2012) pedagogy for imparting knowledge and skills essential for the concerned discipline. As observed the pedagogy for Natural Sciences and Physical Sciences, it is the exposure of learners to the experimental state that enhances their comprehension and understanding of the state of being or matter enquired into.

While the overview of contents for social sciences it is a mosaic of pedagogic skills that are demanded from teachers to impart knowledge gain attention and ensure comprehension and impart skills.

Teachers in Management and Commerce education at institutions of higher learning have witnessed tremendous variations in the need for appropriate pedagogy for the dissemination of knowledge and sharing essential soft skills among the learners. Technological advancement has a significant bearing on the shift of methods of teaching and their application in suitable pedagogy. To comprehend the shift one needs to classify the various types of modules that the Management and Commerce teachers need to disseminate among the learners. The later-described classification will be helpful to justify the application of relevant pedagogy.

For Management, Commerce, and Business learning the following (see Table 14.1) is arbitrarily evolved on the basis of nature of the subject, its broad objective for being part of the curriculum, the pedagogy demanded and the intensity of use of ICT beyond sharing information over the networks.

TABLE 14.1 Nature of General Course and Pedagogy Demanded

Course	Nature	Objective	Pedagogy	Intensity of ICT
Quantitative Techniques	Numerical, analytical, dynamic	Aptitude, skills, for managerial decision-making	Problem solving, cases	High level
Finance	Theories numerical, analytical, dynamic	Knowledge of skills for financing decisions	Cases, problem-solving, critical thinking Role play	
Accounting	Regulations, theories Relatively less dynamic	Skills to systematically report the organizations' performance	Sharing models and formats to check the recollection and application skills	
Operations Research	Numerical logical theories Dynamic			
Human Resources Management and Organizational Behavior	Theoretical and applied, dynamic	Leadership skills, interpersonal relations, and decision-making	Role play, games, cases, Group activities in interactive classes	Medium level
Marketing Management				

TABLE 14.1 *(Continued)*

Course	Nature	Objective	Pedagogy	Intensity of ICT
Taxation	Regulations, theories Relatively less dynamic	Skills to systematically report and draw inferences for corporate decision-making	Assignments, formats, trend analysis Current events	High level
Business Law	Regulations, applied, Theories Dynamic			
Business Economics				
Business Environment				
Financial Institutions and Markets				

Source: Author (2020).

From Table 14.1, it can be understood that the use of ICT in classroom teaching is demanded more for the numerical subjects as the cases are lengthy and time taken to elaborate on each of the dimensions of the numerical models could be saved. Time of experts can be better utilized in the (Renate and Katharina, 2004) face-to-face classes to cater to learner-centric difficulties in comprehending the problem related to the cases. Methods of role-play and participative games would consume time to cover the expanse of the knowledge to be imparted over a limited time span for these subjects.

It can be noted that the application of ICT is essential to discuss on the theoretical and game-based cases in subjects like Human Resources Management, Marketing Management, Business Law, Business Economics, Business Environment, and Financial Institutions and Markets. Instead, there are participative methods that could be further supplemented to impart skills and comprehension to specialize in these areas. Role-playing, management games, and quizzes are preferred over monotonous sharing the information over the animated power point slides and interactive videos.

A description of module wise need for pedagogy is as follows:

1. *Abstract knowledge*: Business Statistics, Accounting, Financial Management, Cost and Management Accounting. This knowledge is essential for improving on the philosophy and vertically contributing

to the body of knowledge based on inductive and deductive research approaches. The origin of theory and various approaches are disseminated in details to encourage the learners' ability to identify the gaps and research to innovatively fill in the gap existing in the literature.

a. Pedagogy needed in these areas demands a lot of well-articulated oration skills and relatively minimum of interactive technological devices. These subjects need a timeline of the evolution of the constructs and their interrelationship with the current thoughts and practices. Development and evolutionary stages could be covered with the help of a simple Power-Point presentation, consisting of origin of the concepts, constructs and approaches with their interdependence matrices and examples on the current practices.
b. However, fundamentals of research and the philosophy remain intensively skill-oriented teaching and learning. This needs an explanation of logic and its leading trails for philosophy, which demands creating interest and interactive tools for disseminating knowledge psychomotor domain and cognitive domain of the learners are to be well sensitized thus timing, duration, the choice of background colors, the text, and the stories are to be well managed by the teachers. Absence of any of the aspects may cause disorientation of the learners' mind and lack of interest.

2. *Skill-based knowledge*: Research methods, Financial Accounting, Financial Management, Investment Management, Cost and Management Accounting, Operations Research, Marketing, Human Resources Management, and others. It is essential to evolve performers and decision makers. The numerical exercises based on cases do speed up the decision-making and analytical traits of the learners. However, the interpersonal skills are specific to the human resources managers, group activities, role plays and cases are essential for inculcating these skills.

i. *Advanced knowledge with skills:* In subjects like Taxation, International Trade, Project Planning and Control, Business Law, Business Environment, Managerial Economics/Business Economics, and Financial Systems, it is essential for enabling learners scanning of propositions for business success. As the expanse of the literature and the regulating institutions are enormous the need for "WHY, HOW, and WHY" needs to be further supplemented by cases dominated by the implications of regulatory authorities and success or failure of business propositions.

3. *Skills of negotiation and leadership aptitude*: For subjects like Human Resources Management Organizational Behavior, and Marketing demand more of the interpersonal qualities for retaining the good employees and motivating the aspiring ones in practice. Marketing Management and Human Resources Management professional needs to understand human psychology, practical laboratories may be set up to specialize in the subjects and computer-aided data may be used to help the learners understand the patterns of behavior and then the matching proposition for consumers and employees may be drafted to promote growth. This needs a specialized application of computer-based knowledge, more intense than modeling for Accounting, Costing, and Financial decisions.

Perhaps the common (nonspecialized) classes need simple slides and specialized areas need intensive application of the ICT for the dissemination of skills. This is attributed to the learners' choice, aptitude, and attitude toward the (Mishra and Koehler, 2006) subject.

This study advances to enquire into the learners' perception about the ICT implementation in teaching, user behavior, social influence, infrastructure availability, and how do they perceive it for their future. This chapter is designed to capture and present the Business School students' of higher education grades perception regarding the implications of use and behavior for the ICT. The following section on the reviewed literature guides toward the gaps.

14.2 REVIEWED LITERATURE

Infusion of ICT in educational institutions is must evolve innovative mechanisms that synchronize with emerging contents (Lambert and McCombs, 1998), to enhance learners' analytical aptitude. In order to further self-actualization is something the learners expect while comprehension of concepts, it is possible if the experimental approach is followed while using ICT as a mechanism. During the early period of introduction of ICT (1997–2010) to the global community, the emphasis was on the exposure of the ICT and allied gadgets to the learners. Literature was mostly drawn from the nations that had affordability and were keen to improve access of the technology to the next-generation learners. To further the argument over the transformation of educational technology, Heider et al. (2009) have emphasized on learner-centric approach via a constructivist approach. The questions hovering on teaching pedagogy, using ICT, is attributed to the perception of ICT, by the teachers and the learners. Much of the arguments posed during the inception of ICT in teaching was whether the technology has to undergo a transformation or mere shift of medium of sharing knowledge is digitized? The literature in the later period (2010–2016) investigates to integrate the educational technology with ICT such that more appealing methods of sharing knowledge and acquiring skills can be evolved. As observed by Blewett (2016), educational technology is dynamic and the need for transformation in the use of "ICT as a medium" to "ICT as a measure to inspire learners" is of paramount significance. It is clearly eminent from the (Redpath, 2012) earlier works that there is a need to harness the fullest potential of the ICT in the field of imparting education.

Regarding the application of ICT (Shulman, 1987; Irani, 1998) argued that unless the perception and attitude of the learners and the teachers toward ICT are comprehended, the level of technology intervention in pedagogy will remain ambiguous. The need to (Dooley, 1999) diffuse technology and transform the pedagogy by bringing in the much-needed incorporation of ICT advent is strongly argued. In a study by Heider et al. (2009), the argument is strongly supported by way of the survey on the significance of digital teaching material at institutes of higher learning.

The plethora of literature reviewed on ICT adoption in educational technology and innovation of the pedagogy in tune with ICT intervention have stressed on the need to change. Some emphasized on the teachers' attitude and perception, some on the subject-based need to incorporate the change. The study on the need for teachers' involvement with ICT has been well proved across the literature both institutional initiatives and teachers' attitude are important.

Roberts (2003) revealed that the use of the Internet for gathering the material for instructional purposes have a favorable impact on the learners and the abundance of information with the multiplicity of perspectives adds a new dimension to improve on the pedagogy. For the students of Taiwanese Higher Secondary school, it was observed (Lee and Tsai, 2010) that students prefer more of ICT contents in pedagogy. The intensity of technology adoption demands convenience and infrastructure. Young persons do get adapted well to the developments whether the results prove the same for the elder students? It is addressed in a work on the pedagogy of medical teachers and ICT (Zayim, 2006) reveals that with the help of ICT the students get a better perspective of the contents that are to be imbibed for practice. Medical sciences and engineering are ought to follow the developments in technology so it is immaterial to opine on their deferred views on the aspect.

Similarly, an approach to involve students with innate motivation is equally argued over a period of two decades by the researchers and they pointed out some common aspects as the drivers for students to adapt to the ICT environment of teaching and learning. Drivers for students can be enumerated as infrastructure availability, secured career with the knowledge of ICT, assistance at the learning level, and convenience. These could be well transmitted to the behavior of students (Winke and Goertler, 2008) in due course of time and intention to apply the technology in career building is positively associated with perception about ICT in the learning environment. Students' perception regarding ICT intervention as studied by Neo and Neo (2009) in Malaysian institute reveals that the more autonomy the students' get to operate in the ICT environment, better is the learning outcome. This study was followed by gadget specific perception of learners and the factors that (Liang et al., 2012) have bearing on favorable outcomes of the application of gadgets in pedagogy. The researchers in the United States investigated into the future of interactive whiteboards and found its most rewarding for the institution to opt for. The learners and the teachers both have an appetite to use them in the best possible ways to suit their growth and development in the learning process. It can be inferred that the time gap and the culture have a negligible impact on the preference for ICT aided educational technology. Off late rich task using ICT has been an interest of investigation in the United States, which reveals that the alignment of students' perception and application of educational technology by teachers is welcome. Pedagogy for each of the teaching module across the variety of subjects would be different; ICT could act as a catalyst to enrich the delivery of instructions (Lindblom-Ylanne et al., 2007). The findings of the researches across the time are in conformity that the transformation of educational technology is much in demand.

In addition to the use of ICT as a medium and perception of teachers and learners, the institutional commitment from the governments has successfully attempted to ensure that the flow of education is pervasive and mitigating the impediments of spatial, economic, age, or gender. To further the approach, private sector partners and players have skillfully placed themselves on web front by catering to individual needs of the target community. Online learning has been popularly in place since the 1990s. Though the technology was commonly used in the western world, as regard to developing countries, the confidence has emerged recently.

14.3 STUDY ON INFORMATION AND COMMUNICATION TECHNOLOGY: AN ANALYSIS

14.3.1 SIGNIFICANCE OF THE STUDY

In order to bring about the desired behavior among the learners, strategies are applied by the teachers. In the institutions of higher learning, it becomes difficult for upcoming teachers to comprehend the use of appropriate educational technology to impart knowledge and skills. Educational technology has been evolving over a period of millennia; however, the implication of information and technology is comparatively of the recent origin. How do the practitioners manifest in the teaching environment is a function of the learners' aptitude and attitude toward the methods adopted. In this context, it is felt essential to know the students' perception about ICT and their notions as to the use of ICT exposure in teaching–learning environment. To further the findings a cross-sectional analysis across the demographic attributes of students is felt essential such that need based on one to one nurturing can be taken care.

Perception about ICT intervention is studied for developed nations and established its need. Most of the earlier works were on the students and teachers with relatively better exposure to teaching and learning environment. Characteristics of feeder institutions have been ignored. The parent's and ward's exposure to the ICT environment has not been well acknowledged in the literature. The findings were thus converged toward the technology adoption in pedagogy. The stage of advancement of culture and practices depends on the maturity the civilization achieves with consistent exposure to changes. The villages and hamlets have different set-up than the well-networked cities. There are pros and cons of the dichotomy of connectivity and poor connectivity. Remote places in India like Sikkim have relatively recent exposure to the advanced higher educational institutions. Training of teachers in the feeder schools within the state of Sikkim for the B-Schools operating there in is a recent phenomenon. The problem persists with the students in the higher educational institutions characterized by limited exposure to the application of technology in career building.

In order to investigate the students' perception about ICT in teaching–learning process, the following broad objectives are worked out.

14.3.2. OBJECTIVE OF THE STUDY

1. To identify the causes of variation in the perception of ICT across the demographic fabric of students of Business Schools in Sikkim.
2. To know the notions of the B-School students operating in Sikkim.

14.3.3 HYPOTHESES

Following hypotheses are evolved out of the objective to offer empirics:

H_{01}: Mean level of perception of ICT does not significantly vary across the demographic fabric captured in terms of sex, parent's/guardian's occupation, urban/rural, and community.

H_{02}: Mean level of perception of ICT among students does not significantly vary across the Program.

H_{03}: Students notion measured in terms of percentage, regarding application of ICT in teaching and learning process is independent of their demographic attributes.

14.3.4 METHOD OF STUDY

To limit this study, the observations and description in this chapter are confined to study of ICT perception, user behavior, infrastructure, and social impetus for students in higher institutions of Business Learning in Sikkim that offer programs like Master of Business Administration (MBA) and Master of Commerce (MCom). The technical higher educational institutions draw students from a variety of academic background and places as the admissions are open for all. The nontechnical programs do draw a relatively less number of students from other states and are usually of vertical nature to top-up the existing knowledge in the related field opted during the graduation program. Convenience sampling is opted but it comprises of 70% of total students enrolled in the state for the MBA and MCom Program. Choice of the researcher is minimal as the number of institutions is five in all and four of them allowed to collect information on schedules. Two are from Central Government run Department of Commerce and Department of Management, two are from self-financed institutions. Confidentiality of data sensitive to demography needs to be maintained, hence names of the institutions are not disclosed.

The schedule was designed for unstructured interview. Further 120 students responded for the interview and schedule. An analysis is done in two stages; first of all using mean and coefficient of variation for identifying the level of perception for 120 students (see Table 14.2). The respondents dispersed gender wise as 64 females and 56 males.

TABLE 14.2 Sample Break Up Across the Demographic Characteristics

Rural 68							**Urban 52**						
Demographic Characteristics			**Male**	**Female**	**M. Com**	**MBA**	**Demographic Characteristics**			**Male**	**Female**	**M. Com**	**MBA**
Tribal 19	Business 2	Sr	0	2	0	2	Tribal 19	Business 1	Sr	0	1	0	1
	Service 12	Sr	2	7	4	5		Service 11	Sr	2	6	0	8
		Jr	2	1	3	0			Jr	1	2	3	0
	Farmer 5	Sr	0	2	0	2		Farmer 7	Sr	6	0	1	5
		Jr	0	3	3	0			Jr	0	1	0	1
General 18	Business 5	Sr	2	1	0	3	General 15	Business 8	Sr	2	2	0	4
		Jr	0	2	1	1			Jr	3	1	3	1
	Service 6	Sr	4	1	3	2		Service 5	Sr	3	1	3	1
		Jr	0	1	1	0			Jr	0	1	1	0
	Farmer 7	Sr	3	1	2	2		Farmer 2	Sr	0	1	1	0
		Jr	0	3	2	1			Jr	0	1	0	1
Others 31	Business 12	Sr	3	3	3	3	Others 18	Business 5	Sr	2	1	1	2
		Jr	3	3	5	1			Jr	1	1	1	1
	Service 14	Sr	3	5	2	6		Service 9	Sr	2	2	1	3
		Jr	5	1	6	0			Jr	2	3	4	1
	Farmer 5	Sr	2	2	3	1		Farmer 4	Sr	0	1	0	1
		Jr	1	0	1	0			Jr	2	1	3	0
Total			**30**	**38**	**39**	**29**	**Total**			**26**	**26**	**22**	**30**

Across the category considered by the Government of India: tribals, generals, and other communities have been adopted for the convenience of applying suitable statistical techniques. In this study, there was only one scheduled caste candidate in all hence it is found feasible to merge with other communities. The government in India provides for constitutional reservation for the tribals, other communities, and castes that were deprived since times immemorial. This may have an impact on the perception of the technology.

Location of a home determines the feeder for the students' orientation toward technology and the environment that varies between urban and rural settlements. In India, especially North Eastern states have a significant difference due to poor connectivity and casual infrastructure for ICT.

Numbers indicated against the category name are the total of the category split over respective subcategories. In all 61 students of MCom and 59 students of MBA administration have responded to the schedule and unstructured interview. Education begins at home and hence the parents, siblings, and guardians are of significance to encourage children toward technology. In India usually the service class people are aware of ICT. Businesspersons and farmers have an option to be aware or not and they may rely on the third party to help them out with the fundamental needs. Thus the parents'/ guardians occupation is taken as a parameter to capture the impact of home on the students. Occupation is broadly categorized into business, service, or farming (see Table 14.2).

The participants comprise of 60% of total students enrolled in Business Schools operational in Sikkim. Occupation of parents or guardians is one of the factors for knowing if students are picking up technology from home. Vocational program and the theoretical programs have predefined differences in objectives. The MCom program is nonvocational thus students therein may or may not show a keen interest in technology and teachers' expectation are to the extent of foundations. However, the vocational program like the MBA has an applied significance and the need of the day is to be well versed with the ICT for application in near future. The orientation of students in these two programs is different hence a classification is felt essential. There were 75 respondent students of senior (Sr) batch and 45 students of junior (Jr) batch in all (see Table 14.1), the fresh students and the old students may have perception variation, thus this classification is felt essential. Data collected on the interview schedule was analyzed using the Chi-square test.

14.3.5 *ANALYSIS*

The data collected was with the objective of applying factor analysis; however, the Keiser–Meyer–Olkin test failed for the sample adequacy. Thus on the information collected on the Unified Theory of Acceptance and Use of Technology (Venkatesh et al., 2003) model to know the level of technology adoption of students across their demographic characteristic (attributes) mean of the response on the 7-point Likert scale has been taken and *F* test is conducted to identify the implications of demographic variables on the perception of ICT by respondents.

TABLE 14.3 Ranks of Combined Mean of Scores Across the Broad Information and Communication Technology Adoption Statements and Parameters

Parameter		Rank	Combined Mean of Scores
Perceived importance	Regular use	1	5.436
	Variety of courses	4	5.208
	Time saving	3	5.273
	Output enhancement	5	4.726
	Career significance	2	5.273
Convenience	Understanding	1	5.203
	Skill	3	5.119
	Simple application	2	5.196
	Easy to learn	4	4.993
Social influence	Family	4	4.703
	Friends	1	5.226
	Teachers	2	5.223
	Other important persons	3	4.904
Resources to access	Have the resources	3	4.667
	Institution provides	2	5.019
	People to assist	1	5.155
	Special assistance to learn	4	4.245
Behavioral intention	Will use after 6 months	3	5.525
	For academic activities	1	5.759
	Extracurricular activities	2	5.635
	Plan to use after 6 months	4	5.465

TABLE 14.3 *(Continued)*

Parameter		Rank	Combined Mean of Scores
Use behavior	Material for academics	1	5.452
	To match with classroom teaching	4	4.792
	Personal purpose	3	4.931
	Interactive purposes	2	4.961

Source: Author analysis.

14.3.6 INTERPRETATION

The results are surprising and some of them are contradicting the Null hypotheses set forth in the method of study. To begin with the Table 14.3 on the arithmetic mean of the scores obtained on the 7-point Likert Scale the following exceptions are observed that are briefed in Table 14.4:

1. As regard to "Perceived importance" of the output of ICT knowledge is rated relatively poor by the students placing it in the fifth position among others in the category. Students assign relatively more importance to regularity and followed by the career quotient of ICT. Moderate ranks of third and fourth are assigned to time saving and the variety of subjects it is used for. These scores assigned are independent of their demographic attributes. This reveals generation-wide indifference about the role of ICT in output enhancement. The range of mean score lies between 4.56 and 4.88. There is an element of indifference with technology as a medium to bring in enhanced output. As they are technology savvy the students are not perceiving its innate significance. *F*-test justifies the indifference across the demographic attributes other than for career 10% levels of significance (LoS). Perception of "career orientation" of ICT varies across the parents' occupation, wards of businessmen accord high scores followed by farmers, and least being assigned by the wards of people in service (see Tables 14.4 and 14.5).
2. When observed the perception about "convenience" (see Table 14.4), the scores assigned to "understanding" the ICT-based fundamentals are convenient but the "easy to use" quotient remains lacking. This reveals that even though the comprehension is fine but the application is relatively difficult. This holds true as the practice is needed to

TABLE 14.4 Mean Scores Across the Information and Communication Technology Adoption Parameters and Students' Demographic Attributes

Demographic Attributes		Program		Seniority		Locality		Gender		Community			Parents' Occupation		
Parameters		MCom	MBA	Senior	Junior	Urban	Rural	Male	Female	Tribal	General	Others	Business	Service	Farmers
Perceived importance	Regular use	5.46	5.44	5.52	5.33	5.37	5.51	5.46	5.44	5.42	5.42	5.49	5.42	5.58	5.23
	Variety of courses	5.10	5.34	5.27	5.13	5.19	5.24	5.13	5.30	5.18	5.06	5.35	5.27	5.23	5.13
	Time saving	5.28	5.25	5.21	5.36	5.29	5.25	5.20	5.33	5.24	5.48	5.14	5.09	5.30	5.40
	Output enhancement	4.74	4.69	4.71	4.73	4.69	4.74	4.63	4.80	4.82	4.70	4.65	4.88	4.56	4.83
	Career significance	5.20	5.32	5.13	5.47	5.23	5.28	5.32	5.20	5.16	5.30	5.31	***5.64***	***5.07***	***5.20***
Convenience	Understanding	***5.02***	***5.39***	5.27	5.09	5.19	5.21	5.13	5.27	5.13	5.27	5.20	5.36	5.09	5.23
	Skill	5.15	5.08	5.07	5.20	5.10	5.13	5.13	5.11	***5.00***	***4.82***	***5.41***	5.21	4.96	5.30
	Simple application	5.18	5.22	5.29	5.04	5.12	5.26	5.11	5.28	5.18	5.00	5.35	***5.52***	***5.05***	***5.13***
	Easy to learn	4.90	5.08	***5.11***	***4.80***	5.02	4.97	4.91	5.06	4.82	5.03	5.10	***5.36***	***4.81***	***4.93***
Social influence	Family	***4.87***	***4.51***	***4.52***	***4.98***	4.87	4.56	4.64	4.73	4.47	4.67	4.88	4.82	4.60	4.73
	Friends	5.36	5.07	***5.05***	***5.49***	5.31	5.15	***4.96***	***5.44***	5.03	5.12	5.43	5.36	5.07	5.33
	Teachers	5.13	5.32	5.31	5.09	5.19	5.25	5.18	5.27	5.29	4.94	5.37	5.45	5.07	5.27
	Other important persons	4.97	4.85	4.88	4.96	4.94	4.88	4.95	4.88	***4.53***	***4.85***	***5.24***	5.24	4.86	4.63
Resources to access	Have the resources	4.59	4.73	4.69	4.60	***4.90***	***4.47***	4.66	4.66	***4.32***	***4.82***	***4.82***	4.88	4.51	4.70
	Institution provides	***5.26***	***4.75***	5.05	4.93	4.77	5.19	4.98	5.03	***4.55***	***5.48***	***5.04***	5.24	4.77	5.20
	People to assist	5.21	5.08	5.11	5.22	5.12	5.18	5.02	5.27	5.32	5.24	4.96	5.09	5.16	5.20
	Special assistance to learn	4.20	4.32	4.35	4.11	4.13	4.35	***3.98***	***4.50***	4.26	4.15	4.33	4.42	4.28	4.03

TABLE 14.4 *(Continued)*

Demographic Attributes		Program		Seniority		Locality		Gender		Community			Parents' Occupation		
Parameters		MCom	MBA	Senior	Junior	Urban	Rural	Male	Female	Tribal	General	Others	Business	Service	Farmers
Behavioral intention	Will use after 6 months	5.38	5.68	5.53	5.51	5.48	5.56	5.52	5.53	5.50	5.67	5.45	5.45	5.56	5.53
	For academic activities	***5.52***	***6.00***	5.88	5.56	5.87	5.68	5.73	5.78	5.50	5.97	5.82	***6.09***	***5.67***	***5.57***
	Extracurricular activities	***5.46***	***5.83***	***5.65***	5.62	5.60	5.68	5.52	5.75	***5.37***	***5.70***	***5.82***	***5.97***	***5.63***	***5.30***
	Plan to use after 6 months	5.43	5.53	5.49	5.44	5.38	5.54	5.39	5.55	5.42	5.39	5.57	5.42	5.54	5.40
Use behavior	Material for academics	5.54	5.39	5.56	5.31	5.37	5.54	5.38	5.55	5.39	5.45	5.53	5.33	5.58	5.40
	To match with classroom teaching	4.64	4.93	4.88	4.62	4.85	4.74	4.71	4.84	4.79	5.00	4.63	4.91	4.65	4.90
	Personal purpose	4.77	5.07	4.96	4.84	***5.15***	***4.74***	5.09	4.77	***4.74***	***5.36***	***4.76***	4.91	4.88	5.00
	Interactive purposes	4.92	5.02	4.92	5.04	5.12	4.85	***4.77***	***5.14***	4.68	5.00	5.16	5.18	5.02	4.63

TABLE 14.5 Results of F-Test of Scores Across the Demographic Attributes and Information and Communication Technology Adoption Parameters

	Demographic Attributes Parameters	MBA, MCom	Junior, Senior	Urban, Rural	Gender	Community	Parents' Occupation
		df 1/118	df 1/118	df 1/118	df 1/118	d/f 2/117	d/f 2/117
	Regular use	0.012	1.171	0.783	0.025	0.077	1.429
		(0.913)	(0.281)	(0.378)	(0.874)	(0.926)	(0.244)
	Variety of courses	1.343	0.383	0.042	0.678	0.642	0.121
		(0.249)	(0.537)	(0.839)	(0.412)	(0.528)	(0.886)
	Time saving	0.023	0.739	0.056	0.672	1.547	1.048
		(0.879)	(0.392)	(0.813)	(0.414)	(0.217)	(0.354)
Perceived importance	Output enhancement	0.044	0.016	0.044	0.716	0.235	1.079
		(0.834)	(0.899)	(0.834)	(0.399)	(0.791)	(0.343)
	Career significance	0.360	2.428	0.053	0.319	0.213	***2.699***
		(0.550)	(0.122)	(0.818)	(0.573)	(0.809)	***(0.071)***
	Understanding	***4.291***	0.887	0.005	0.588	0.174	0.814
		(0.040)	(0.348)	(0.942)	(0.445)	(0.841)	(0.446)
	Skill	0.134	0.568	0.044	0.008	***4.608***	1.507
		(0.715)	(0.453)	(0.835)	(0.928)	***(0.012)***	(0.226)
Convenience	Simple application	0.060	2.200	0.820	1.133	1.505	***3.003***
		(0.808)	(0.141)	(0.367)	(0.289)	(0.226)	***(0.053)***
	Easy to learn	1.041	***2.778***	0.072	0.710	0.942	***3.569***
		(0.310)	***(0.098)***	(0.790)	(0.401)	(0.393)	***(0.031)***
	Family	***2.990***	***4.585***	2.111	0.188	1.336	0.410
		(0.086)	***(0.034)***	(0.149)	(0.666)	(0.267)	(0.665)
	Friends	1.707	***3.597***	0.500	***4.544***	1.285	0.770
		(0.194)	***(0.060)***	(0.481)	***(0.035)***	(0.280)	(0.465)
Social Influence	Teachers	1.153	1.410	0.103	0.237	2.061	1.683
		(0.285)	(0.237)	(0.749)	(0.627)	(0.132)	(0.190)
	Other important persons	0.292	0.109	0.072	0.103	***4.030***	2.121
		(0.590)	(0.742)	(0.789)	(0.748)	***(0.020)***	(0.125)
	Have the resources	0.424	0.180	***4.198***	0.000	***2.471***	1.085
		(0.516)	(0.672)	***(0.043)***	(0.983)	***(0.089)***	(0.341)
	Institution provides	***3.537***	0.174	2.295	0.031	***3.479***	1.326
		(0.062)	(0.677)	(0.132)	(0.861)	***(0.034)***	(0.269)

TABLE 14.5 *(Continued)*

Resources to access	People to assist	0.312 (0.577)	0.237 (0.627)	0.069 (0.793)	1.167 (0.282)	0.988 (0.375)	0.061 (0.941)
	Special assistance to learn	0.264 (0.608)	0.879 (0.350)	0.791 (0.376)	***4.656 (0.033)***	0.168 (0.845)	0.689 (0.504)
Behavioral intention	Will use after 6 months	2.207 (0.140)	0.011 (0.916)	0.143 (0.706)	0.004 (0.948)	0.386 (0.681)	0.096 (0.909)
	For academic activities	***6.288 (0.014)***	2.666 (0.105)	0.933 (0.336)	0.063 (0.802)	1.881 (0.157)	***2.377 (0.097)***
	Extracurricular activities	***6.147 (0.015)***	0.038 (0.845)	0.269 (0.605)	2.317 (0.131)	***3.276 (0.041)***	***5.396 (0.006)***
	Plan to use after 6 months	0.239 (0.626)	0.054 (0.816)	0.609 (0.437)	0.576 (0.450)	0.316 (0.730)	0.211 (0.810)
Use behavior	Material for academics	0.557 (0.457)	1.426 (0.235)	0.766 (0.383)	0.718 (0.398)	0.162 (0.851)	0.583 (0.560)
	To match with classroom teaching	1.447 (0.231)	1.048 (0.308)	0.202 (0.654)	0.279 (0.598)	0.743 (0.478)	0.544 (0.582)
	Personal purpose	1.772 (0.186)	0.248 (0.620)	***3.501 (0.064)***	2.097 (0.150)	***3.130 (0.047)***	0.098 (0.907)
	Interactive purposes	0.199 (0.656)	0.296 (0.587)	1.394 (0.240)	***2.886 (0.092)***	1.719 (0.184)	1.735 (0.181)

Note: Figures in the brackets are corresponding levels of significance (LoS). Table values of F at 1/118 degrees of freedom (df) are 3.92 for 5% LoS and 2.75 at 10% LoS. Values of F at 2/117 df for 5% LoS 4.79 and 2.35 for 10% LoS.

master the skills as simply comprehension of the application of ICT does not hold the key to success.

(a) Further, the skill enhancement component of convenience is ranked third, this significantly varies across the community to which the respondent students belong. This contradicts the Null hypothesis of no difference in perception of convenience at a 5% level of significance (see Table 14.5). Students from other backward communities accord high score to the "skill enhancement" quotient of ICT perception followed by respondents belonging to tribal communities and least scores are accorded by the students belonging to communities not privileged by the constitution (general).

(b) The test also reveals that scores assigned for "convenient in understanding" significantly varies across the programs contradicting the Null hypothesis of no difference at 5% LoS. MCom students rate less to the understanding and the MBA students rate high for the quotient. Of course, the nature of subject matters. MBA being of applied nature needs more of ICT in practice than the MCom.

(c) The parameter scores differently across the parents' occupation on "simplicity in application." It scores similar to that of "career orientation" where wards of business persons accord high rates and the service class accord lowest, moderate scores are given by the wards of farmers.

(d) The statement on "easy to learn" scores differently across the "seniority" of students at 10% LoS and the "parents' occupation" at 5% LoS, respectively and thus contradicts the Null hypothesis of no difference on perception about ICT. The senior students accord high score as they are already exposed to the mechanism of teaching, and juniors accord it low. The occupation of parents is again reflected as causing a difference in assigning scores to this parameter, the wards of business persons accord high score and that of service class accord lowest.

3. The "social influence" on ICT adoption is family assistance. It is found that friends are of foremost scorers, followed by teachers, other influential people, and last is the family members. The age group usually matters in scoring on the social front. It is natural and revealed.

(a) The scores on family members as the influencing agent for ICT adoption scores differently "program" wise at 10% LoS and "seniority" wise at 5% LoS. This contradicts the Null hypothesis of no difference in ICT perception across the demographic attributes. Program wise MCom students accorded higher score as compared to MBA, and seniority wise the juniors have accorded high score. It can be inferred that the family have a bearing on the ICT adoption across the program and the seniority of students, though it scores less among all in general.

(b) Scoring on “friends” as influencing factor has bearing over the “seniority” and the “gender” of students. The students have ranked friends as the best motivator for ICT adoption but with a variation across the gender at 5% LoS and seniority at 10% LoS. The juniors have assigned a higher score to friends causing the variation. The girls have accorded higher score than boys for friends as the influencing factor for ICT adoption. Both the gender and seniority attributes of respondents’ demography reject the Null hypothesis of no difference on this quotient.

(c) “Other influential people” though scores third and varies across the “community” of the respondent students at 5% LoS. The respondents belonging to tribals assign the lowest score for “other influential persons” as influencing factor and the students from “other community” accord highest. The students belonging to general community accord moderate score to influential people. This causes failure in acceptance of Null hypothesis of no difference in scoring on ICT perception by students across the communities.

4. Resource accessibility as a parameter of ICT adoption is one of the important parameters. Within this is the respondents assigned a highest mean score to “people to assist” followed by the “availability of resources” and “institutional provision,” least being the “specialized assistance."

   a. Within the accessibility parameter the quotient of “institutional provision” varies significantly across the “program” at 10% LoS. The MCom students accord high score followed by the MBA. “Parents’ occupation” has proved to matter at 5% LoS the scored assigned by students’ belonging to parents of business persons accord high score and the wards of the service persons assign low score. The Null hypothesis of no difference across the demographic attributes stands no accepted for this quotient of ICT adoption within the availability of the resources.
   b. The statement on “special assistance” as part of resources availability varies across the gender at 5% LoS, where the female students accord high score than their male counterparts. The Null hypothesis with regard to this quotient of resource availability stands not accepted at 5%. Its noteworthy that either the

male students do not take interest or the female students are more cared for?

5. Students "behavioral intentions" are captured in terms of the possibility of use of the learned skills in next semester or plan to use. Whether intended for academic purpose and for "extracurricular activities." Scoring on this parameter reads that academic activities are more preferred followed by extracurricular activities. The "plan to use" is ranked last as most of them will be applying knowledge in next semester itself.

   (a) The intention to apply knowledge on ICT for academic activities significantly varies across the demographic attributes of "program" at 5% LoS, and "parents' occupation" at 10% LoS. This contradicts the null hypothesis of no difference of demographic attributes on the perception of ICT adoption. MBA students accord high score for "academic activities, and wards of business person assigned high scores followed by the wards of service persons and it was least rated by the wards of farmers.
   (b) The intention to apply knowledge on ICT for "extracurricular activities" significantly varies across the demographic attributes of "program," respondents' "community" and "parents' occupation" at 5% LoS. This contradicts the null hypothesis of no difference and the sample statistics failed to accept the Null hypothesis at 5% LoS. When observed program wise, the students of MBA accorded high score, community wise the wards of tribal accorded low score followed by the general, and highest score was assigned by the other constitutionally privileged communities. On observing the mean scores across their parents' occupation it is found that wards of business persons assign the highest score and the farmers assigned lowest score. This reveals the lack of overall perception of ICT among the students based on their parents' occupation and community to which they belong.

6. The "use behavior" component of the ICT adoption is measured in terms of scores collected on the following quotients, "study material collection," "match with classroom teaching," "personal purposes" and "interactive purposes." It is observed that the mean score is

highest for academic purpose followed by "interactive purpose," "personal purpose," and finally for "matching classroom teaching."

(a) The application knowledge for personal purpose varies significantly across the location to which the student belongs (urban/rural). The students from urban places assign higher scores than the students from rural places. Further the community-wise scores significantly at 5% LoS the students belonging to general category assign more score followed by the other backward communities and the students belonging to tribal communities. This is noteworthy and its inferred that opinion of students use of "ICT for personal purpose" does not negate the Null hypothesis thus the sample information fails to accept the Null hypothesis of no difference.
(b) The opinion of students and their scoring on the use of ICT for interactive purposes varies significantly across the gender at 10% LoS. Female student accord high score to the component thus rejecting the Null hypothesis of no difference of demographic attributes on the ICT perception and its use.

To conclude the birds' eye view of the mean scores across the ICT adoption parameters and its perception by the student respondents across the program reveals that students accord the lowest score to the availability of special assistance. This reveals in addition to general foundation certain special assistance is needed at the institutional level and personal front. Their expectations are higher than the generations preceding. Other low scoring was on the availability of resources. It seems the resources at their disposal are not adequately explored or adequate knowledge as to access to the online resources is not disseminated. Students scored poorly on the family as the social influence factor. It shows that the age group prefers more of the peers than the home solutions or the socialization at the family level is weak. The expected result of ICT as output enhancement is not well comprehended as the generation of students enquired into were mostly born during the peak of ICT introduction and they grew up in the environment where Internet is everywhere.

When noted across the Hypotheses the following brief cannot be made:

$\boldsymbol{H_{01}}$**:** Mean level of perception of ICT does not significantly vary across the demographic fabric captured in terms of sex, parent's/guardian's occupation, locality of students' brought up (urban/rural), and community.

There are several sub-Null hypotheses to this broad Null hypothesis that can be broken down as follows:

H_{01a}: Mean level of perception of ICT with regard to "regular use" does not significantly vary across the demographic fabric captured in terms of sex, parent's/guardian's occupation, urban/rural, and community, and so forth. In brief, Table 14.5 would guide the readers through the failed to accept the Null hypotheses of no difference with the respective levels of significance in brackets.

1. The nature of program MCom or MBA have a significant impact on awarding scores to " ease of understanding," "family" as influential actor, "institutional provision for resources," intention toward " academic activities," "extracurricular activities."
2. The seniority of students in a program has implications on " convenience in learning" "family."
3. As social influence factor, and "friends" as social influence factor.
4. Location to which student belongs has implications on "availability of resources" and "personal purpose" use behavior.
5. Gender has implications on "friends" as a social influencer, "special assistance" accorded for learning, and "interactive purpose" of use behavior.
6. The community has a bearing on "convenience in skill enhancement," other important persons" as social influencers, "having resources," "resource provision by institution," "extracurricular activities," and "personal purpose" of use.

Finally, parent's occupation has bearing on the "career perception" with ICT knowledge," simplicity in the application of technology," "easy to learn" perception of convenience, intention for " academic activities," and " extracurricular activities."

14.3.7 DISCUSSION

The survey being preliminary in nature first of its kind in the context of Business Schools of Higher Learning operating in Sikkim. Study reveals that despite of abundant resources from the government, parents, and institutions there is a difference in perceptions regarding technology. The pedagogy of teachers has a significant role in the dissemination of knowledge. Students' perception for use of technology mostly for academic purposes as observed needs to be curbed and put forth in a manner pervasively beneficial for them.

The ICT resource allocation perception is found to be even across the gender and the exactness and equality surprise the findings with *F*-value =0 at 98% LoS. So gender has limited bearing on the possession of ICT resources. The society is advancing in terms of gender equality at B-Schools.

Students in B-Schools are well included as far as ICT adoption is concerned. The scores on the 7-point Likert scale have been by and large toward agreement. The perceived importance of ICT is a welcome sign as it does not vary across any of the demographic attributes. However critical, analytical and creative thinking needs to be addressed as there is much to draw from the ICT than the academics. The out of the box response to the questions is missing. This is where one needs to step in for grooming the teachers to transform students' orientation toward technology and its application and use beyond the academics.

Out of 150 possible Null hypotheses of no difference on the perception of ICT adoption across the demographic profile of the B-School students in Sikkim only 23 proved to contradict at 10% LoS and 127 remains negated. In brief, one can say the perception of students is to some extent a function of the community, program they are studying, location from which they hail, their parents' occupation, gender, and the duration of their exposure to the institution for higher learning in B-Schools of Sikkim. In brief we find the following:

1. Community attribute of students' of B-School have implications on six components as follows: 2 from "resources allocation" perception, 1 each from perception of "convenience," "social influence," "behavioral intention," and "use behavior,"
2. Parents' occupation and Program to which students belong affects five aspects (1 Importance, 2 Convenience, 2 behavioral Intention),
3. Program to which students belong affects five aspects (1 Convenience, 1 Social Influence, 1 Resources, and 2 Behavioral intention),
4. Seniority in educational institution affects three aspects (1 convenience and 2 social Influence),
5. Gender affects three aspects (1 Social Influence, 1Resources, and 1 Use behavior) followed by
6. Location of students affects two aspects (Resource Allocation and Use Behavior).

This study remains unprecedented as the statistical tools proposed to be adopted, have lost feasibility on grounds of sample adequacy. Alternatively arithmetic mean, consequentially the *F*-test, and ranks were resorted to describe the pattern of perception regarding ICT adoption in students of higher learning at B-Schools operating in Sikkim.

To further the enquiry, an interview was conducted with the students and the discussion lead to the following:

On the question related to the preference for the subjects and their goals it was revealed that only 58 of the respondents opted for specialized services. Of the 58 only 29 have opted for services like that of Accounts, Finance Officer, Bank, Stock Broking, and Income Tax. Thus interpretation as to the numerical content may portray gloomy picture. However, the rest of the students opted for specialized services were for Marketing and most dominating was the Management of Human Resources. Approximately, 50% of the 62 respondents opted for Business and remaining for Academics and Research. It can be seen that the orientation of students across the B-Schools for higher learning is scattered.

In order to capture their preference of ICT and their opinion on pedagogy of ICT, a questionnaire was administered with dichotomous responses on Yes or No. This has been analyzed using percentages and the Chi-Square Test of the responses was conducted across their demographic attributes.

The objective of this enquiry was to know whether the students actually know, apply, and understand the ICT applied in B-Schools for their learning?

Broad Null hypothesis of no difference is set forth as

H_{03}: There is no significant difference in B-School students on application of ICT across their demographic attributes described in terms of; “program,” “seniority,” “community,” “parents’ occupation,” and “locality.”

The results of percentage analysis across the broad questions are presented in Table 14.7. These are interpreted in the order of their serial number in Table 14.7.

1. In all 99% of students like use of ICT as pedagogy intervention when split across the demographic attributes its observed that all the students of MBA like it. Only 2% of MCom students decline, indicating need-based ICT intervention. Across other demographic attributes, it seems that the variation is 2% only; among the wards of business class, urban dwellers, and other backward community differ by 3%, 2%, and 2%, respectively. Table 14.6 reveals the insignificance of the variation. Thus need for teachers’ in the field is

significantly high to orient their dissemination of knowledge via the media wherever suitable. The results of χ^2 test are in hypothesized direction and for the given sample of 120 students at 10% LoS we fail to reject the Null hypothesis of no difference of students' notion on the liking of ICT across their demographic attributes. This warrants teachers' concern for the application of ICT to make the learning playfully meaningful and attractive.

2. In aggregate 86.7% students replied yes to the application of ICT for theoretical content delivery. The 15% of students feel the need to limit the application of ICT in delivering theoretical content and 85% are of the MBA students opine to extend. A total of 88% of the students of MCom opine to extend the theoretical content over ICT and 12% declined. The teacher needs to balance the application of ICT, and the preference by junior students for ICT is less than the preference by the senior students by 7%, though statistically insignificant, there may be an impact of maturity level. 90% of wards of farmers prefer ICT application in theoretical content delivery and least (78%) preferred is by the wards of business community, 89% of the wards of service persons prefer ICT for theoretical contents. The differences do exist in preference of ICT though statistical tests of Hypotheses do not show the significance in difference. A small learners' community spread across the institutions do demand change in pedagogy probably content-wise organization is essential.
3. Animations applied in the content do attract of students' attention in the class as responded by the 96.7% students. This implies that the students do prefer the animated presentations. The expectations may be met by the teachers for relevant modules so as to motivate students and orient them toward classroom teaching. The preference is even across the demographic attributes of the students, revealing students' response being independent of their demographic attributes. So the Null hypothesis of no difference set forth;
   "Students' Preference of animation for gaining attention in the class is independent of their demographic attributes" could not be rejected at 10% LoS (see Table 14.7).
4. A total of 70.8% of students agree to the notion that the same method of ICT intervention in teaching for all the contents. When observed its split across the demographic attributes of students it is found preferred unevenly where 34% of MBA students need different methods for different modules, 24% of the MCom students prefer

TABLE 14.6 Results of Null Hypotheses-Test of Scores on Information and Communication Technology Adoption Parameters Across the Students' Demographic Attributes

Demographic Attributes Parameters		MBA, MCom	Junior, Senior	Urban, Rural	Gender	Community	Parents' Occupation
Perceived importance	Regular use						
	Variety of courses						
	Time saving						
	Output enhancement						
	Career significance						*REJECT (0.071)*
Convenience	Understanding	*REJECT (0.040)*					
	Skill					*REJECT (0.012)*	
	Simple application						*REJECT (0.053)*
	Easy to learn		*REJECT (0.098)*				*REJECT (0.031)*
Social influence	Family	*REJECT (0.086)*	*REJECT (0.034)*				
	Friends		*REJECT (0.060)*		*REJECT (0.035)*		
	Teachers						
	Other important persons					*REJECT (0.020)*	

TABLE 14.6 *(Continued)*

Demographic Attributes Parameters		MBA, MCom	Junior, Senior	Urban, Rural	Gender	Community	Parents' Occupation
Resources to access	Have the resources			*REJECT (0.043)*		*REJECT (0.089)*	
	Institution provides	*REJECT (0.062)*				*REJECT (0.034)*	
	People to assist						
	Special assistance to learn				*REJECT (0.033)*		
Behavioral intention	Will use after 6 months						
	For academic activities	*REJECT (0.014)*					*REJECT (0.097)*
	Extracurricular activities	*REJECT (0.015)*				*REJECT (0.041)*	*REJECT (0.006)*
	Plan to use after 6 months						
Use behavior	Material for academics						
	To match with classroom teaching						
	Personal purpose			*REJECT (0.064)*		*REJECT (0.047)*	
	Interactive purposes				*REJECT (0.092)*		

Source: Author analysis.

different methods, 29% of juniors and 30% of seniors refuse to prefer similar method. 35% of urban and 25% of rural students need different methods for different contents. 38% of tribals, 20% of other privileged communities, and 33% of General community students prefer different methods. The Null hypothesis set forth regarding the preference of the same method; across the demographic attributes there is no significant difference in notion toward the preference of the same method of ICT applied for delivering all the modules. This stands not rejected as the sample students responded though differently it's not statistically significant. On an average there is not much deviation in the opinion of students across demographic attributes. However, since the aggregate percentage of agreement is only 70.8, teachers need to meet the needs of the one-third of alert students.

5. To the tune of 75% of students across their respective demographic attribute prefer ICT aided numerical content delivery and there is a scope of non ICT methods of teaching as preferred by 25% of students. When observed across the community 81% of wards of tribal community opine that highest, wards of the general community prefer it only to the tune of 70%, wards of farmers to the tune of 79% prefer ICT must be used to deliver numerical content. This reveals a preference for blackboard teachings still exists among the 25% of students. This emphasizes that in addition to the ICT blackboard method or one to one explanation of the numerical is preferred by 25% of the students. The Null hypothesis set forth; Students preference for ICT as method for teaching Numerical subjects do not significantly vary across their demographic attributes. This stands not rejected as the sample did not provide adequate evidence though the differences exist may be treated as by chance.
6. 88.3 % of the students across the demographic attributes prefer delivery of case study via ICT as a medium.

   Within their categories of demographic attributes though there is minor difference its negligible and by chance. However, the 12% of students in all need other methods of delivery of case studies. Null hypothesis of no difference set forth: Students prefer "ICT as teaching aid for deliberating on case studies" is independent of their demographic attributes. This stands not rejected at 10% LoS with a given sample studied.
7. Opinion as to ICT as a medium of study material sharing is preferred by 84.2% of students. When revealed from the χ^2 test of hypothesis

on "students' notion on ICT medium for sharing study material does not vary across their demographic attributes" is contradicted significantly by the students across their "parents' occupation;" more number of wards of (97%) business persons accord preference, lowest (81%)being opined by the wards of service persons. Null hypothesis stands rejected at 10% LoS (see Table 14.7).

Though statistically insignificant, less 82% of juniors prefer ICT as a medium when compared to seniors its 4% less, the difference between MBA students and of MCom students is 6%, it is more preferred by MCom students. Students from rural and tribal background prefer less than their counterparts.

8. The question on the justified quantity of e-material on the web as relevant to their courses, 63.3% of students opined for it, this is relatively less when compared to their interest in ICT as a method of teaching and animations. It reveals that the students are vigilant over the classroom delivery and independent browsing on the web for study materials is less preferred . This low preference significantly varies across the "program" and "seniority" at 5% LoS.

   This reveals that either materials suggested by the instructors are inadequate or the students remain puzzled with the mosaic of possibilities that they encounter over the websites. More number of MCom and seniors opine on justifiability of materials on the web and less number of MBA and the juniors agree to it. So the Null hypothesis:

   "students' opinion about 'quantum of material on web justify the course needs' does not vary across the MBA or MCom," stands rejected at 5% LoS and alternatively one can state that "the program to which student belong have a significant bearing on their opinion about the 'quantum of study material available on justifies the course needs'." Surprisingly the vocational students dislike the browsing through the net for the materials, 25% of MCom students exceed the preference when related to MBA students. This indicates a significant course-wise difference in exposure to web contents.

   Likewise "students' opinion about 'quantum of material on web justify the course needs' does not vary across their seniority in the institution" stands rejected at 5% LoS and alternatively one can state that "the seniority level to which student belong have a significant bearing on their opinion about the 'quantum of study material available on justifies the course needs'." Seniors exceed by 19% in preference when compared to juniors. Juniors are from a variety of backgrounds of feeder need to tune up by the passage of time.

This question has roughly 36.3% of students who do not opine for the notion rather they decline and it becomes the cause of concern for teachers to take due care while suggesting the websites for gathering the study materials. Even students need to be analytical in selecting the contents and matching it with the course.

9. The query on the "interactive participative methods make learning interesting" scores 88.3% liking the playful pedagogy. Preference by MBA students is 9% more than that of MCom, further split across their seniority more juniors outnumber seniors by 7%. Wards of service class outnumber preference by 4% than business person wards, urban students exceed preference by that of rural students by 9%. Across the communities the difference is negligible. These differences are statistically insignificant at 10% LoS.

   So the Null hypothesis:

   "Students' opinion about 'interactive participative methods make learning interesting' does not vary across the demographic attributes." This stands not rejected at 10% LoS with a given sample studied. There is a sense of unanimity across the demography of students regarding "interactive participative methods make learning interesting" ICT .

10. Response over the question on "skill enhancement is emphasized through ICT" is similar to that of "interactive participative methods make learning interesting" 88.3%, there exists disagreement by 11.3% of the students spread across the demographic attributes. The cause of indifference or contradicting to the statement may be attributed to the overall environment that is experiencing the ICT. So the Null hypothesis:
   "Students' opinion about 'skill enhancement is emphasized through ICT' does not vary across the demographic attributes." This stands not rejected at 10% LoS with a given sample studied. There is a sense of unanimity across the demography of students regarding the skill component associated with ICT.

11. Around 92.5% of students are with a positive notion on the statement "attention is gained due to attractive presentation." Observation across the demography unveils the significant variation across the program and seniority level at 10% LoS. The Null hypothesis on the percentage of students is in favor of the statement: "Attention

is gained due to attractive presentation" do not vary across the "program" and the "seniority level," stands rejected at 10% LoS. Alternatively one can say the MBA students prefer more of the power point presentations than the MCom students and the junior students accord more interest than the senior students. This draws the attention of the teacher to prepare as per the program and the level the students are exposed to the program. Power points may not be equally attractive for all the students.

Percentage of wards of business persons preferring the power point presentations is greater and lowest preference is by the wards of service persons, percentage of students with urban background is higher for preference than that of the students from rural background. Percentage of students belonging to other communities protected by the constitution find power point as attractive lowest among the category is accorded by the students from general community background.

12. The question on content-wise variation in the teaching methods is almost unanimously agreed upon by the students it scores the overall agreement of 93.3%. This reveals that the students are aware about the pedagogy changes, essential for the delivery of certain modules in the class. The teachers need not expect the same technique of teaching will justify students' preference for the subject content at higher educational institutions. The Null hypothesis of no difference is set forth:

    Students' opinion about "content-wise variation in teaching method is essential" does not vary across the "demographic attributes," this stands not rejected and unanimity among the heterogeneous sample is indicative of their vigilant nature. Though statistically insignificant, community-wise variation to the tune of 11% between tribal and general category students reveals the higher expectations of students belonging to the general community.

13. The question on "relevance of content and ICT medium shall be synchronized" is preferred by 85.8% students in all. The opinions significantly vary across the program and the community of the students. MCom students prefer more than MBA students by 14%. It means there is a need to synchronized and is recognized by the students, the response varies significantly at 5% LoS across program. When observed community to which they belong, students belonging to tribal community accord higher preference than that of other community by 19%.

The Null hypothesis of no difference is set forth:
Students' opinion about "relevance of content and ICT medium shall be synchronized" does not vary across the demographic attributes," this stands rejected at 5% LoS as far as demographic attributes of program and community are concerned.
Across other demographic attributes, the variation of 9% among wards of business persons and farmers is found where wards of farmers accord higher preference. It is statistically insignificant but noteworthy that wards of farmers are vigilant about teaching at institutions of higher learning in Sikkim.

14. Question on "recollection is convenient when taught following participative interactive methods," it is observed that 86.7% of students admit it is an essential component in teaching; however, it varies across the seniority level at 10% LoS. The younger batch (juniors) exceeds the seniors by 11% and realizes its importance.
The Null Hypothesis of no difference is set forth:
Students' opinion on "recollection is convenient when taught following participative interactive methods" does not vary across the demographic attributes," this stands rejected at 10% LoS as far as demographic attributes of seniority is concerned.
There is an observed difference of 8% among business persons' wards and that of farmers though it is not significant, indicative need for focus on farmers' wards to include them in the learning process.
15. The statement on "power point used in class enables faster recognition and comprehension" scored 85% in all. The split across demographic attributes reveals that 79% of MBA and 91% of MCom students agree to it, and it varies significantly at 5% LoS across the "program."
"Students' opinion about 'Power point used in class enables faster recognition and comprehension' does not vary across the program" stands rejected at 5% LoS and alternatively one can state that "the program to which student belongs to have a significant bearing on their opinion about the 'power point used in class enable faster recognition and comprehension'." The pedagogy of teaching using power point slides is significant for MCom students, and teachers of MBA need additional efforts with a better blend.
16. To capture the notion on the motivation this question "inspiration is drawn among learners from the teachers for application of ICT" was asked. Around 82.5% of students in all agree to the notion. Further,

when read across demographic attributes, variation is observed in the "program" and "parents' occupation" at 5% LoS. Percentage of wards of farmers has been the lowest (66%) to agree, and surprisingly equal percent (88%) of wards of service persons and business persons have agreed to the motivational aspect.

The Null hypothesis of no difference:

Students' opinion about "inspiration is drawn among learners from the teachers for application of ICT" is independent of their demographic attributes. This stands rejected at 5% LoS with a given sample studied as regards the "program" and "parents' occupation." Alternatively one can infer that "students' opinion about "inspiration is drawn among learners from the teachers for application of ICT" is dependent on the program they are in and their parents' occupation. Program wise more percent of MCom students (91%) agree to the notion exceed 17% from their counter MBA part.
Similarly the wards of business persons and service persons accord 22% more than that of farmers. The wards of farmers draw relatively less inspiration from teachers' application of ICT in class.
The lacuna of teachers' for the interest in motivating students is revealed as on an average only 84% of students seem to take an interest.

17. Question on the link between ICT and career orientation has been agreed upon by 93% of students and it does not vary significantly across the demographic attributes of students. Most of the students irrespective of their demographic fabric accord unanimously toward an association between ICT and career prospects.
18. When the observed response of students on "identical method of ICT makes learning monotonous," it is observed that only 56.7% of students agree to it. This roughly confirms their response to the statement discussed in point 4 above. The response is independent of the demographic attributes of students. MCom students exceed the MBA by 11%. Wards of business persons are the greater supporter of the statement, with 63% and same is the case of urban students—63%, the general community students accord highest—67%. This warrants that pedagogical skills with ICT blend shall vary from subject to subject. It is a strong message from students.

    The Null hypothesis set forth:

Students' opinion about "identical method of ICT makes learning monotonous" is independent of their demographic attributes. Stands not rejected at 5% LoS within the given sample of students.

19. When sought information on the purchase of study materials through websites it is observed that in all 45% of students only support. Further, parent's occupation has a bearing on the response at 10% LoS. Relatively more percentage of the farmers' wards avail online purchase of study material (59%) followed by business persons' ward (50%) and service persons' wards being the lowest (35%). The reasons are multifold: (1) the material provided by the institution may be adequate, (2) students do not have adequate interest to know more, (3) free online material may suffice the needs and demotivates from buying study material.

The Null hypothesis set forth:

Students' opinion about "online study material purchases" is independent of their demographic attributes. Stands rejected at 10% LoS within the given sample of students.

Alternatively:

Parents' occupation wise there exists a difference in the notion of students as to the online purchase of study materials holds true. And one can infer that farmers' wards prefer online purchase of books and service person's wards prefer less of online purchase of study materials, parenting have a significant bearing on the students' preference.

Other demographic attributes have negligible bearing the difference between preferences for online buying of study materials. The difference is 13% between MCom and MBA students, more MBA students prefer. Wards of the tribal community prefer least among the community, and the general community prefers more of online buying.

20. Notion on students enquired on the "cross-checking of online material with required content" fetches 90.8% of students agree to notion. Relatively even percentage of response from students' with different demographic attributes and differences are statistically insignificant. However, the variation in percentage is not more than 5% across the groups within each of the demographic attributes around—percentage of students cross-check the online content with the required contents.

The Null hypothesis set forth:

Students' opinion about "cross-checking of online material with required content" is independent of their demographic attributes. Stands not rejected at 5% LoS within the given sample of students. So the teachers need to be vigilant about the directions given to students regarding the online sources. This will enable pinpoint achievement in the students' reference, to enhance enriched learning.

21. In order to follow the lead of the earlier question, students' notion on the reliability of the online content is checked. A total of 66.7%, almost two-third of students opine on the reliability of e-resources. Their strength of dependence on online material can be felt. It is noted that parents' occupation has a significant bearing at 5% LoS. The response is independent of other demographic attributes. Percentage (46%) of wards of service persons doubts more on the reliability of online content, followed by the farmers' wards (24%) and the percentage of wards of (22%) business persons. The percentages are not completely in favor of ICT as a medium of buying even if the educational level is high an element of doubt exists. Also one can infer that the motivation is low when read in context with the earlier question on buying habits for academics.

    The Null hypothesis is set forth:

    Students' opinion about "free to air e-resources are reliable" is independent of their demographic attributes.

    Stands rejected as far as parents occupation is concerned at 5% LoS. Alternatively, it can be stated that

    parents occupation have bearing on the reliability notion of free to air e-resources. The percentages do confirm with the online purchases.

22. Further enquiry into reliability and cross-reference aspect reveals that the need is felt by 84.2% of the students. However, the response is dependent on the community of the students at 5% LoS. The percentage of students belonging to tribal community is more to support the notion (92%), whereas only 78% of the general community students feel it essential. Across other demographic attributes the variation in response is not significant.

    The Null hypothesis set forth:

    Students' opinion about "reliability needs to be established by cross reference" is independent of their demographic attributes.

    Stands rejected as far as location background is concerned at 5% LoS. Alternatively, it can be stated that location to which students belong have bearing on the reliability notion of "reliability needs to be established by cross reference." Urban students practice more cross-reference than rural students.

23. When studied notion on "web sources are used for study materials," it is discovered that demographic attributes do not impact the percentage of students' response. Around 89.2% of students access the internet for study material. This shows that technology application by students and their attitude is unanimous despite of heterogeneity.
24. The response on the attraction quotient of the online animated content as to courses, it is observed that the demographic attributes do not have bearing on the students' response. A total of 90% of students agree that online animated content as to course is attractive.
25. Regarding the knowledge of students as to availability of free e-resources, 57.5% of students are aware of it. Efforts are needed by the teachers to improve awareness. It is discovered that locality of students to which they belong have implications at 5% LoS. A total of 69% of the urban students have the knowledge and 48% of rural students have the knowledge of free to air study material. As compared to other questions response to this question is not welcoming the percentage of students knowing the free e-resources is less. This needs to be addressed by the teachers in the class such that students can browse through much of the free information to improve their comprehension.

    The Null hypothesis set forth:

    Students' knowledge about "free e-resources" is independent of their demographic attributes. It stands rejected as far as location background is concerned at 5% LoS. Alternatively, it can be stated that location to which students belong have bearing on the knowledge of "free e-resources." Urban students know more than the rural students.

14.4 CONCLUSION

To sum up, it is broadly revealed in the study that students of B-School are well aware of the implications of ICT diffusion in the teaching and learning process. They have rational expectations from the teachers to apply ICT in the pedagogy of theoretical, applied, numerical modules, and case studies. The responses of students on the questionnaire regarding the activities of ICT and its implication on overall personality development have been encouraging. It is still at a stage where the applied nature of program and theoretical nature of program have bearing on four aspects of use of ICT, namely, quantum of material justifies or not; attention due to an attractive presentation; matching of ICT intervention with content, use of power point presentation helps in the recollection, recognition, and comprehension of contents delivered; and motivation by teachers to use ICT.

In above context based on the sample of 120 students of B-Schools in Sikkim, the Null hypothesis is rejected regarding the independence of students' notion, from the program (MBA/MCom). It seems though the contents of the modules are identical there exists a difference in delivery. Hence, it stands rejected at 10% LoS.

The seniority of students as an attribute has a significant bearing at 10% LoS on the following:

1. quantum of online material justified or not,
2. attention due to an attractive presentation, and
3. recollection is convenient when taught following participative interactive methods.

It is understood from the above that exposure to students for an year or above will have impact on their perception of course and the mode of delivery of information. This enables juniors to adapt to the pedagogy at later levels.

When observed across parents' occupation it is observed that

Null Hypothesis set forth:

Parental occupation has no significant impact on students' notion about ICT diffusion.

This stands rejected and alternatively, it can be said that parents' occupations have implications on the students' opinion about ICT diffusion in pedagogy for the following statements:

1. ICT medium for accessing Study material at 10% LoS.
2. Motivation drawn by students from teachers' use of ICT at 5% LoS.
3. Purchase of online study material. At 10% LoS.
4. And reliability of free to air e-resources at 3% LoS.

Student development and growth is a function of the greater environment. In addition to the type of program, seniority level, home environment have a bearing on their perception.

On observing the students' location (residential) background, in terms of rural or urban, Null hypothesis set forth:

> "Students' response to the opinion on ICT diffusion is independent of the location to which they belong; urban or rural."

This stands rejected in the following contexts at 5% LoS:

1. Reliability needs to be established by cross-reference.
2. Knowledge of free e-resources.

This confirms that the location disadvantage prevails among the perception and use of ICT among the B-School students. It is natural as the urban students have more accessibility and exposure to the application of technology and the rural students with poor connectivity and orientation more toward meeting the basic minimal services.

When observed across the community, if there exists the difference in opinion of students on ICT diffusion it is observed the Null hypothesis of no difference:

> "Students' response to notions on the ICT diffusion in teaching and learning process is independent of their community (constitutionally categorized for extending privileges)."

This stands rejected in the context of relevance of subject content and ICT medium shall be synchronized at 5% LoS. Tribal students are comparatively alert and general students are least.

To sum up there is a need to implement models to diffuse ICT in pedagogy in a systematic fashion. It is apparent that the academic administration is implementing the ICT; however, teachers are to diffuse the technology cautiously suitable for students and the content. On reviewing past works it is observed that models suggested by Nehru and Chakraborty (2019) and Nehru and Ton (2018) have relevance in developing scenario.

KEYWORDS

- **ICT**
- **students' perception**
- **B-schools**
- **higher education**
- **demographic attributes**

REFERENCES

Aubusson, Peter, Pau Burke, Sandy Schuck, Matthew Kearney, and Bart Frischknecht. Teachers Choosing Rich Tasks: The Moderating Impact of Technology on Student Learning, Enjoyment, and Preparation. *Educational Researcher* 43, no. 5 (June 2014): 219–229.

Blewett, Craig. From Traditional Pedagogy To Digital Pedagogy. Chap. 16 in *Disrupting Higher Education Curriculum. Constructing Knowledge: Curriculum Studies in Action.*, by Samuel M.A, Dhunpath R, and Amin N, 265–285. Rotterdam: Sense Publishers, 2016.

Bretag, Tracey and John Hannon. Negotiating Contested Discourses of Learning Technologies in Higher Education. *Journal of Educational Technology & Society* 13, no. 1 (January 2010): 106–120.

Debande, Olivier. ICTs and the Development of eLearning in Europe: The Role of the Public and Private Sectors. *European Journal of Education* 39, no. 2 (June 2004): 191–208.

Dooley, K.E. Towards a Holistic model for the Diffusion of Educational Technologies: An Integrative Review of Educational Innovation Studies. *Educational Technology & Society* 2, no. 4 (October 1999): 35–45.

Heider, Kelly, DeAnna Laverick, and Bethany Bennett. Digital Textbooks: The Next Paradigm Shift in Higher Education? *Association for the Advancement of Computing in Education Journal* 17, no. 2 (April 2009): 103–112.

Irani, T. Communication Potential, Information Richness and Attitude: A Study in Computer Mediated Communication in the ALN Classroom. *ALN Magazine*, March 1998: 1–12.

Lambert, N. M. and B. J. McCombs. Introduction: Learner-Centered Schools and Classrooms as a Direction for School Reform. In *How Students Learn: Reforming Schools Through Learner Centred Education*, by N.M. Lambert and B. L McCombs, 1–22. Washington DC: American Psychological Association, 1998.

Lamera, P., Levy, P, Paraskakis, I, and Webber, S. Blended University Teaching Using Virtual Learning Environments: Conceptions and Approaches. *Instructional Science* 40, no. 1 (May 2011): 141–157.

Law, N. Teacher Learning beyond Knowledge for Pedagogical Innovations with ICT. In *International Handbook of Information Technology in Primary and Secondary Education*, by J. Voogt and G. Knezek, 425–434. New York: Springer, 2008.

Lee, M. H., and C. C. Tsai. Exploring High School Students' and Teachers' Preferences toward the Constructivist Internet-Based Learning Environments in Taiwan. *Educational Studies* 31, no. 2 (October 2010): 149–167.

Liang, T.-H., Y.-M. Huang, and C.-C. Tsai. An Investigation of Teaching and Learning Interaction Factors for the Use of the Interactive Whiteboard Technology. *Educational Technology & Society* 15, no. 4 (October 2012): 356–367.

Lindblom-Ylanne, S., K. Trigwell, A. Nevgi, and Ashwin. P. How Approaches to Teaching Are Affected by Discipline and Teaching Context. *Studies in Higher Education* 31, no. 3 (January 2007): 285–298.

Mishra, P. and M Koehler. Technological Pedagogical Content Knowledge: A Framework for Teacher Knowledge. *Teachers College Record* (Elsevier) 108, no. 6 (June 2006): 1017–1054.

Nehru, R S, and S Chakraborty. The Education of Things (EoT) for Smart Learning Through IoT Intervention: A Case Study Based Analysis. *International Conference on Intelligent Computing and Communication Technologies -System Reliability, Quality Control, Safety, Maintenance and Management.* 2019. 529–536.

Nehru, R S, and Ton. Nehru - Ton Model Technologies for Blended Learning between India and Vietnam Teacher Education Programmes. *International Journal for Research in Engineering Application and Management* 4, no. 3 (June 2018): 322–335.

Neo. Mai., and Neo. Tse-Kian. Engaging students in multimedia-mediated Constructivist learning—Students' Perceptions. *Journal of Educational Technology & Society* 12, no. 2 (April 2009): 254–266.

Parrish, P, and J Linder-VanBerschot. Cultural Dimensions of Learning: Addressing the Challenges of Multicultural Instruction. *The International Review of Research in Open and Distributed Learning* 11, no. 2 (May 2010): 1–19.

Redpath, Lindsay. Confronting the Bias against On-Line Learning in Management Education. *Academy of Management Learning & Education* (Academy of Management) 11, no. 1 (March 2012): 125–140.

Renate, Motschnig Pitrik, and Mallich Katharina. Effects of Person-Centered Attitudes on Professional and Social Competence in a Blended Learning Paradigm. *Journal of Educational Technology & Society* (International Forum of Educational Technology & Society) 7, no. 4 (October 2004): 176–192.

Roberts, G. Teaching Using the Web: Conceptions and Approaches from a Phenomenographic Perspective. *Instructional Science* 5, no. 1 (January 2003): 127–150.

Shulman, L. Knowledge and Teaching: Foundations of the New Reform. *Harvard Educational Review* 57, no. 1 (April 1987): 1–20.

Venkatesh, V, M. G Morris, and G, B Davis. User Acceptance of Information Technology: Toward a Unified View. *MIS* 27, no. 3 (September 2003): 425–478.

Winke, Paula, and Senta Goertler. Did We Forget Someone? Students' Computer Access and Literacy for CALL? *CALICO Journal* 25, no. 3 (October 2008): 482–509.

Zayim, N., Yildirim, S. & Saka, O. Technology Adoption of Medical Faculty in Teaching: Differentiating Factors in Adopter Categories. *Educational Technology and Society* 9, no. 2 (2006): 213–222.

CHAPTER 15

MOOCs: The Paradigm Shift from Traditional to New Learning Spaces in Indian Higher Education

R.S.S NEHRU

Department of Education, Sikkim University, Gangtok, Sikkim, India

**E-mail: dr.rssnehru@gmail.com*

ABSTRACT

The future of higher education promotes active learning through new intensives and interactions. The massive open online courses (MOOCs) are powerful with diversified learning spaces and flexible time support, the interest of the student, different engagement, and motivation to the learners to design and implement new pedagogies. This chapter opens a possible collaborative and new learning spaces within MOOCs and suggests robust framework support. The new initiative on collaborative approaches finally discusses how to adopt new learning spaces to be adapted to situations reflected.

15.1 INTRODUCTION

Over the past few years, observers of higher education have speculated about dramatic changes that must occur to accommodate more learners at lower costs and year 2012 was declared as the year of MOOCs, the program entered in the picture and the creation impacted to such a power, The advancement steady growth in current years shows that this improvement is not embarking to the magnetic moment.

These inspirations for contributing MOOCs in India Higher Education Institutions (HEIs) and colleges have generally presented. The recommendations to be that educational advancement are essential instead of a plan of action. It delegated instructive change as the inspiration for colleges to offer MOOCs, are particular. It delegated instructive change as the inspiration for colleges to offer MOOCs, are particular. The huge factor in the development of instructive advancements is fundamentally straightforwardly engaged with creation or conveyance.

The impressions of learning advancements analysts and studies regularly routed to educators and chefs who are not constantly engaged with the creation of these courses offering institutions are less work done on the individual encounters inside the everyday improvement of MOOCs, that is, learning architects, teachers, teachers, postgraduate understudies, executives, lawful groups, and so on.

This exploration paper intends to delimit the change in perspective and impact of MOOCs in the Indians from the point of view of partners. That has to do with hierarchical change and information gathered from various MOOCs offering establishments sites like Edx, CWODL, Course Cara, UGC-SWYAM, and research papers distinguished with MOOCs.

15.1.1 A PARADIGM SHIFT FROM TRADITIONAL PEDAGOGY TO DIGITAL

In the course of the most recent two decades, there have been critical moves in the manner of instructing, and learning is seen and directed inside the training segment. One is a move from instructor focused on understudying focused training; conventional study hall innovation encourages information built perspectives innovation instruction the training segment. Before PCs, various types of change—TV in study hall shifting progress. The appearance of PCs, in any case, the instructional innovation, made its mark in a vast region of study. Innovation enhanced supported instructive cooperative program. By creating an educational plan, the program expects to nurture building training by, and large mutually did subsidize model crucial is current will deal with all the whole learning network's necessities. Incorporating the understudies took a crack at different instructive foundations, long-lasting students, by widely using ideas and procedure. Begun endeavor is upheld specialized assistance office encouraging understudies universities entrance addresses recognized resources and in this manner making a foundation of information sharing. PC Assisted Instruction-Some of the primary PC helped guidance (CAI), created, set principles ensuing programing. They were deliberately breaking down number-crunching different planned, profoundly organized PC frameworks, including student criticism, exercise fanning, and understudy vast persuasive wellspring of PC helped guidance framework many instructional exercise different frameworks assets accessible centralized PC.

Through PC gushing quickness more extensive scope substance brought together. Tapes just standard study hall innovation—a TV/VCR. Utilized related to each other, these two innovations can develop understudy commitment with exercises, reinforce comprehension of the associations among history and recent developments, and offer a down to earth path for understudies audit canvassed. Augmented cutting edge innovation provides for clients the fantasy drenched created, capacity associate. Computer-generated, associated with the make add a functioning advancements PC innovations carry perpetually ground-breaking illustrations to understudies' work areas, which will diminish a portion of the present recreation limitations and enlivened exhibits progressively utilized for introducing the usefulness of different PC.

Venture at the made an educator material science worried about participation whittling down area of expertise's power and attraction course. Teacher construct finding out power attraction—an intense, taking into account how hypothetical this topic will, in general, rearranges decreases issue order obscured instructor's more straightforward allowing confronting students, exercise prepared close by the overhead Projector accommodates—sparing, a bright arrangement unequivocally for all intents and purposes economic. It moderates the instructor's vitality student consideration kills need a partner.

15.2 REVIEW OF RELATED LITERATURE

Deryn (2001) has distinguished three obvious connections between proficiency and innovation. Usually and foundationally education through mechanical skill development. Value-based innovation proficiency cooperates advances produced execution starts varieties innovation; a steady condition in innovative education showed through human contact with new advances; established rules of quick social change execution vision.

There are three stages of information technology appropriation (1997):

(1) Individual profitability helps
(2) Enhancement, and
(3) A paradigm shift

Personal efficiency helps permit educators and students commonplace undertakings quicker all the viably (e.g., word, and spreadsheet). Enhancement includes ins infuses, educating an essential method of guidelines. Models incorporate e-mail, site pages, the utilization of video, mixed media, and reproduction to upgrade study hall introductions, and schoolwork assignments. The change in scope is the place of the workforce, and their institutions reconfigure training and learning activities to innovation—the outcome blended learning considering Nehru-ton model (Nehru et al., 2018; Nehru and Chakraborty, 2020) the New Learning Paradigm Shift Model which is more exceptional, and the model demonstrated in Table 15.1.

TABLE 15.1 New Learning Paradigm Shift Model

Traditional Model	Digital Model	Advantages
Statistic	Dynamic	Wide range of learning applications
Impassive	Supportive	Greater scope of remedial teaching and mentoring
Multimedia	Blended	Wide range of multimedia and technology
Synchronous	Hypersynchronous	Self-learning and new learning spaces with the education of things
Passive	Hyperactive	The effective level of personal learning
Uni-diretional	Multidirectional	Self-Pace learning and immediate feedback with education of things
Local	Global network	Global supportive learning environments
Teacher centered	Learner centered	New learning spaces for learning
Face to face	Mixed reality	Integrating mixed reality education of things like artificial intelligence (AI), virtual reality (VR), augmented reality (AR), Internet of Things (IoT), and other technologies

Ruler states that innovation plainly can befuddle, scare, and baffle students and clients. Figuring out how to adapt to these essential new information bases and aptitudes is an intricate procedure for all. Furthermore, teachers have other needs right now as they are encouraged to quickly and capably carry to figure out how to make a critical and instructive study hall.

Watson (2001) and Sellen (2002) portray because of idea advancements, as an impetus for change, yet additionally showing data altered how to improve training. Watson (2001) and Sellen (2002) portray that because of idea advancements, as an impetus for change, showing data altered how to improve training. Students could not be an influential reality if they prepared abilities.

Shinde et al. (2013) advance with the ramifications of innovations, which were introduced in Table 15.2. They additionally speaks to the training situation in India, by expressing that because of the effect of innovations on instruction, the status of the schools and universities and different establishments giving guidance has likewise changed. Presently these information houses are well furnished with all cutting edge, in fact, sound gadgets that encourage the learning condition. The utilization of projectors, DVDs, shrewd sheets, and other modernized devices means likewise creating equivalent entire diverse global showing up currently training gives the nation of the world.

Molebash (2013) clarifies that it is an undeniable fact that instruction is delayed to change, particularly in consolidating new advancements is portrayed as worldview loss of motion, the deferral or cutoff in our capacity to comprehend and utilize innovation because of past encounters. Contacts supplant essentially requires significant investment. Tragically, training can no longer take the time it needs. Patterns innovation making is showing up quicker instruction getting ready should, in this manner, patterns the direction adjust.

Mumtaz (2000) explains contextual investigation study halls that commitment guidance, including PC use. Commitment estimated instructors broadness collaborations different expansiveness inclusion explicit administration, tutoring, and the meeting usually partakes communications and homeroom manners insignificant widely almost certain ways of thinking perfect styles regularly use more modes. Constrained capability reached association intellectually, testing errands devices to imparting, thinking, creating.

Geraldine Blattner (2009) and Reginald Nnazor (2009) communicate that PC has gotten progressively installed in regular day-to-day existence. The present understudies depend on innovation for data gathering, to remain refreshed on social concerns and national issues, for between close to home correspondence, and as an approach to learning. It is not astonishing to locate a similar pattern literary most recent inquire about represented, and different correspondence advances bolster significant instructive encounters. Advances in data and correspondence innovation (ICT) have given excellent chances to innovation encouraged synchronous and offbeat instructing and learning nearby and in separation training conditions.

Rogers (2000) features adequacy advances expressing analysts infer that 60% of American understudies are prevailing visual students, 37% are sound-related predominant students, and 3% are dominant sensation students. Additionally, students complete courses with sight and sound in 33% of the time as those customary accepting directions and arrive at competency levels up to half higher.

TABLE 15.2 Integration of Techno-Pedagogies

Traditional Mode	Integrated Mode (Technology and Pedagogy)	Influence on Learner and Learning
Classroom teaching and Activities	Self-paced instruction	Networked instruction
Face-to-face interaction to teacher and student	Flexible and technology assistive	Different model of learning management systems
Self-learning	Collaborative learning	Virtual collaborative mentoring
Formal content knowledge systems	Dynamic knowledge and data systems	Sharing and analyzing databases for learning

15.3 A PARADIGM SHIFT FROM DIGITAL IN NEW LEARNING SPACES

Globally, the instruction part is experiencing a change in perspective that includes both imaginative manufactured learning situations and a huge change of the educational center, to more readily get ready understudies overall educational program regions and learning stages to prevail in a quickly changing and interconnected world. A variety of learning conditions are rising across instructive foundations as instructors endeavor to adjust their encouraging practices and upgrade learning results. These "adaptable learning spaces" contain an assortment of furniture alternatives in a generally open space, which can be arranged in different approaches to encourage a scope of instructing and learning encounters. They remain as a distinct difference to customary homerooms, which are described by lines of work areas and seats, confronting an educator at the front who utilizes transcendently educational instructing approaches. These customary conditions are currently viewed as insufficient to convey 21st-century abilities for students. The move by and by is reflected by a rising exploration center around the spatiality of training.

The vision for future-centered learning conditions is that learning will be improved through progressively utilizing understudy focused instructional methods. With regard to the schools partaking right now, umbrella term fuses venture-based and customized learning encounters that help further examination concerning zones of individual enthusiasm past what is conveyed to the entire class. Furthermore, these methodologies empower understudies to be locked in as co-makers of the learning experience, both freely and cooperatively. This is a lot of like the optional instruction changes in the Netherlands, where a move is happening from learning situations dependent on information transmission to those intended for information development. Besides, the spaces and how they are utilized, encourage sufficient chances to improve understudy innovativeness, advancement, correspondence, and critical thinking abilities, which are considered progressively urgent for the working environments of things to come that understudies are being set up for by schools. These learning situations in a perfect world help understudy decision of where and how to learn and empower simple access to a scope of instructive innovations intended to encourage learning. The joining of virtual space into learning conditions requires extra changes to both the assembled condition and the academic way to deal with benefit from the affordances of innovation.

It is implied that adaptable learning spaces characteristically bolster instructors to utilize understudy focused showing draws near and that these spaces can suit and encourage learning modes, for example, joint effort, express guidance, free work, criticism, and reflection just as experiential realization, which are accepted to prompt upgrades in understudies' commitment and inspiration. Thus the psychological, social, and conduct areas of understudy commitment are all in all related to improved learning results, for example, maintenance of information, test scores, and grades. It is comprehensively accepted that the educating and learning approach utilized in adaptable learning spaces will, at last, lead to enhancements in scholarly results.

15.3.1 EDUCATIONAL ADMINISTRATIVE PERSPECTIVE IN INDIA

Advanced innovation in India has been developing in the course of the most recent couple of years, changing the manner in which understudies learn ideas in school. The customary chalk and talk strategy have made ready for progressively intelligent training techniques as schools are progressively receiving computerized answers to keep themselves side by side with the innovative changes. The present age of education system is knowledge based with gadgets like PCs, tablets, and cell phones, these inventive techniques ensure more support. To oblige the school understudies' needs, instruction suppliers, for example, Educomp, Tata Class Edge, Pearson, and TeachNext, have been thinking of intelligent programing to help educators in study hall education. Notwithstanding, utilization of advanced innovations in organizations of advanced education is still in its incipient stages, and endeavors are being made to adjust these advances to adjust to the necessities of university understudies. To put it plainly, content advancement is yet to develop in schools and colleges across India.

India's dynamic economy with a blossoming working class and over 200 million Internet clients has made the nation the third biggest online market after China and the United States. With this sort of extensive effect, there is a promise for an expansion in the utilization of advanced advances in the training field. In any case, the absence of framework, poor power access, and low Internet infiltration in the semiurban and provincial regions have kept down the advancement of computerized benefits in schools. Obviously, this calls for government cooperation to address these difficulties and build up a sound biological system that can support the utilization of computerized advances.

Regardless of the previously mentioned difficulties, India's blasting urban regions give a great chance to the digitalization of instructive administrations. To build the nature of training with the most recent computerized innovative expertise, the lion's share of the schools and colleges is attempting to stay up with the advanced changes by executing them. Along these lines, by enabling instructors, computerized innovation holds the way into India's instructive difficulties.

15.3.2 DIGITAL INDIA VERSUS DIGITAL EDUCATION PROGRAM

All through the keep on multiplying mind-blowing speed essential time of change computerized a leader program—Digital Education Program—propelled a goal interfacing province or territory's fast web organizes and improving advanced proficiency. The vision of this program is to provide access to quality content, expand the learning opportunities, and stimulate higher engagement in students and teachers. Thus information economy is one of the most significant advances by propelling resident nation and associate to the astute information world.

15.3.2.1 PROPELLED UTILITY TO EVERY SUBJECT

This movement joins pass on correspondence developments electronic organizations constant. Open organizations confirmations significantly available.

15.3.2.2 ADMINISTRATION AND SERVICES

Driven portable advanced education universal access and testaments: advanced mobile phones to lightning-fast gadgets, difficult envision existence innovation. Mean across adroit educated. From advanced mobile phones to lightning-fast gadgets, difficult envision existence innovation. Mean across adroit educated. Tick, so computerized stage administration rouse associate a sagacious information wellbeing electronic storage.

15.3.3 INTEGRATION OF DIGITAL LEARNING IN INDIAN EDUCATION SYSTEM

Activities held received executed by activity way breaking movements gave an immense chance to improving instructive measures; their latent capacity is yet to use entirely. The advancement in the execution of this computerized action plan, particularly with regard to embracing mixed principles of educating/learning process, Ministry of Human Resources Development took the initiative in different ways:

- *Swayam:* utilization of data correspondence and innovation intended to give incorporated stage entryway advanced education subjects and expertise are more than students tried outgo. Outside colleges likewise, seminars stage assessments should possible rules.
- *Swayam Prabha:* activities administration intended to give great instructive TV channels over the expansiveness nation on-premise.
- *National Digital Library:* activity legislature builds up a system of vault assets solitary office. Up until this point, digital reports accessible substance patrons more than clients from instructive foundations enlisted.
- *e-Shodh Sindhu:* venture administration targets giving assets, including content authentic scholarly establishments pace structured membership task blending activities.

- *Free and Open Source Software for Education:* venture targets advancing the utilization programing in instructive organizations nature training, lessening reliance exclusive programing.
- *Virtual Lab:* activity targets giving labs in different controls understudies look into task likewise build up a total understudies profit different instruments picking up other assets, addresses, energized exhibits assessment.

15.4 ABOUT MASSIVE OPEN ONLINE COURSES

Allude created by Stephen D and George S is titled Connectivism and Connectivity Knowledge (2008). They expected opportunities for collaborations of an assortment of members conceivable apparatuses give more extreme condition customary devices permit. Twenty-five understudies went seminar grounds University Manitoba; further 2300 globe took interest web. Twenty-five understudies went seminar grounds University Manitoba; further 2300 globe took interest web presently MOOCs. Nothing underlying enlistment more than understudies finished secure communication among understudies abusing potential outcomes contacting large crowd. Established an organization started to create nothing teachers and began an organization called Coursera, which joined forces with colleges to get ready. It built up stage renamed edX when an organization framed, which creates gives accomplices, accessible from platform utilized produced different foundations additionally completes examination into getting the hang of using new advances by dissecting information it gets from understudies. Undoubtedly, a previous venture occupied.

Understudies enlisted selected understudies yet minimal effort mixed help nearby educators instructing associates. Achievement had to propose maybe 10 organizations advanced. Try not precisely useful, in specific radically mixed classes customary. Moreover, the pace of many expressed that pull together professional training leader consortium, understudies colleges profiting by arrangement.

15.5 HOW CAN MASSIVE OPEN ONLINE COURSES EDUCATION BE HELPFUL FOR THE MASSES?

Online education has demonstrated to be useful on ongoing occasions. It very well may be taken by both full-time understudies and by the regular workers' individuals. The use of advanced education online courses by the understudies will create and expand the present atmosphere of the online instruction framework and will develop its interest in producing more income.

Obviously, it will likewise refine the previously working experts with further advancement of their aptitudes and streamlining of the information. This will help the business the experts are related to generating enormous benefits and better arrangements.

15.5.1 CURRENT SCENARIO

The current diagram of the advanced education framework in India is very grievous, where unplanned over-expansion is often criticized as one of the biggest downfalls of Indian higher education. Many institutions suffer from the subpar quality and a lack of funding. As a result, entry into the top institutions is highly competitive and translates into a contest for higher entrance test scores and better private coaching institutes. What is more, the state of ladies advanced education is much more terrible where guardians are not slanted toward subsidizing their advanced education for marriage reasons. At present, there are practically 30 million understudies, 779 colleges, 39,701 universities, and 11,923 instructive establishments, which are not adequate to get the job done.

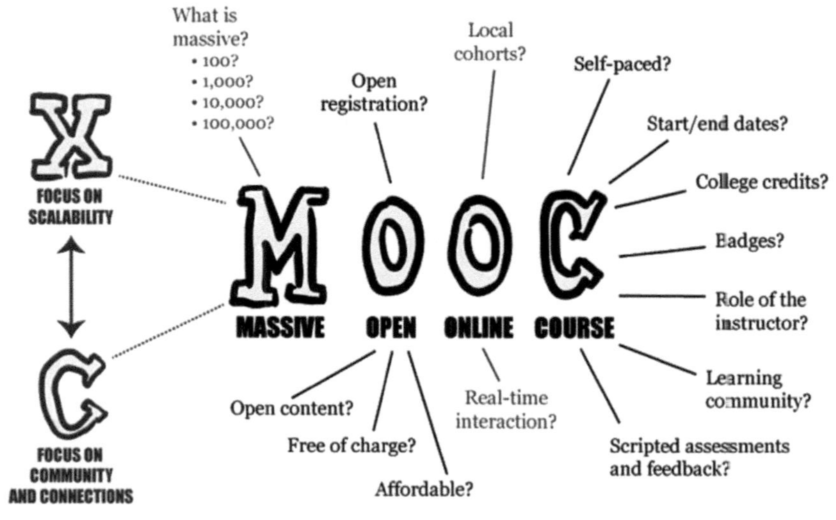

FIGURE 15.1 The meaning of massive open online courses.
Source: By Mathieu Plourde (Flickr, 2013). https://creativecommons.org/licenses/by/2.0/deed.en

There is a trust that with the headway of computerized instruction in India, an ever-increasing number of understudies will have the option to get to quality training without trading off the accounts. The administration is concentrating acutely on the advancement of online instructive arrangements for the simple entry to the understudies (Figure 15.1). MOOCS will be later incorporated with other instructive approaches in the nation planned by K. Kasturirangan Committee.

15.3.6. UGC INCEPTIONS ON ONLINE EDUCATION (MASSIVE OPEN ONLINE COURSES AND SWYAM)

Welcoming instructive organizations allow power confer training using online mode. From 2017, the administration permitted the colleges to offer 20% of their courses using the online technique plan.

Online Education Regulation Act 2018

- UGC spent an ago, permitting advanced organizations directing on the web exhorted.
- In standard separation, college likewise permitted led using the online technique at any rate for the main bunch.
- Recognitions, endorsements perceived establishments will fill in as comparing foundations.
- It permitted direct will in any event of the age.

Give empowering arrangements to keep up the holiness of confirmations, educating learning, assessment, the validness of the student, and required divulgence of Program-wise data, for example, span, start and expense, understudies, understudies an open area.

- Programs that will require down to earth/lab necessities won't grant for online courses, which prompts designing, medication, dental, drug store, nursing, engineering, and physiotherapy programs cannot do using.
- They directed and administered way of UGC standards.
- An educational plan is constrained addresses, appraisal uniquely, conversation gathering explain questions and the program head entrance on the web.

15.4 HIGHLIGHTS OF THE ACT 2018

(1) The quality will additionally improve because the colleges cognizant showing crowd, guarantee teaching method utilized correctly. The quality will also improve because the colleges are aware of showing crowd, guarantee teaching method utilized correctly and comprehensively functioned admirably.
(2) It is wrong to state that quality may bargain in a shortage value tutors; however, successful utilization innovation empowers tech organizations to give observing students urban communities training.
(3) Many presumed foundations in India and in the world are utilizing the web courses as a method for conferring quality instruction through a quality workforce and connecting the information hole at a lot quicker rate than the physical establishments that gets conceivable.

Issues absence leading standards of colleges and the answers include:

- There is some measure of interest as far as assets, yet with the cloud-based advancements embraced nowadays, there is no critical framework speculation required, and the staff can likewise prepare without any problem Huge and so on. pervasive, occur savvy configuration making likewise considerably adequately,"
- Most colleges in India are asset compelled regarding staff scope of specializations give utilizing gets conceivable top-notch instruction minimal expense. Top-notch instruction opens moderate understudies.
- The presentation of online courses does not mean a generous increment in workforce and assets. Includes instructors getting innovation—innovation ventures furthermore all around if there should arise an occurrence of online training—colleges can take the administrations of tech organizations who bolster colleges in giving the stage, content just as conveyance.
- The NAAC appraisal criteria had set up to fathom issues, for example, substance and reach. Guarantees that these colleges are in a situation to institute the essential framework or venture required or tie-up with reasonable Edu-tech accomplices to complete online courses.
- Different measures taken to forestall cheating during the time spent leading assessments through online mode.
- Delegated test kinds—administering utilizing advanced facial acknowledgment, checks, and so on. With the assistance of these,

one can screen understudies, permit them to chip away at frameworks where they cannot flip; on the off chance that they do, the screens get logged out.

- The leaps forward in innovation have become so propelled that if attempt considerably virtually rapidly homeroom. Hence, the test can look at present utilizing design comprehensively various instructive foundations are using.
- Administered whole individual online-test personal computer (PC)/ device, delegated personal computer PC /device camcorders, and, consequently, preventing instances.
- Propelled frameworks have programmed web-based delegating caught using iris development and some other information catching procedures integrated IoT, and they are equivalent to if worse than typical physical administering, include.
- The individual takes their whole test PC, delegates PC camcorder along these lines forestalling instances of cheating.
- Tests can use the online mode.

Innovation has made directing both emotional and target tests conceivable. Target questions can be appraised and assessed through frameworks, though for moving inquiries, they get prompted on to particular staff, who can additionally evaluate and rate them on the web.

15.4.1 THE BENEFITS OF ONLINE MODE

- Credits are transferable across colleges.
- The means currently transferable a creative advance that can change the essence of training.
- Better skilling openings.
- Customarily, advanced education in India has viewed as identification to a great job. Be that as it may, the situation is evolving quick innovation requirements embrace a progressively adaptable way deal with learning.
- Reskilling and upskilling is required nearly by working experts, and it is challenging to recover every single working proficient to class. Online conveyance experts fulfill adapting needs permit to stay pertinent quick advancing business situations.
- Future colleges may have an altogether different structure.

- The UGC choice is that colleges, as we were probably aware, may stop to have the sort of impact they appreciate today. Rather, the best colleges of things to come might be communitarian endeavors between scholastics, policymakers, business people, and speculators.
- This implies the measurements for estimating the effect and execution of advanced education organizations will change. Ideally, to improve things!.
- It could make instruction more effective in open zones.

The choice guarantee training comes to open pieces nation. A significant consideration and a chance to ensure advancement are accessible and could bring better scholastic industry.

15.4.2 THE INFLUENCE OF MASSIVE OPEN ONLINE COURSES ON HIGHER EDUCATION

15.4.2.1 THE EVENTUAL FATE OF ADVANCED TECHNOLOGIES

Surprisingly, the general across the board subject of HEIs is progressively regarding, examining, and dreaming about the job of MOOCs—and computerized instruction further comprehensively—the future models. Throughout the previous two decades, numerous teachers, professors, and administrations drew in with the advanced period. Today, these transfers between alternates and workforce day by day; The fascination for the calling of instructing inquires about involves a position of right at colleges and universities; however, MOOCs have focused consideration on the educating and learning process. In the scholarly community, MOOCs intend to try different things with mixed understanding. Many inquired about procedures for improving dynamic education.

15.4.2.2 GROUP-BASED COURSE DESIGN

Designing MOOCs requires individuals over the foundation to work together with instructional planners, programing engineers, learning scientists, curators, videographers, and experts group. The accomplice's methodology and situations engage singular experts to share their aptitude—the difference in educating as a push to building numerous approaches to apply group approaches and settings.

15.4.2.3 LIMIT WORKING OVER OUTSOURCING

The individuals from the consortium center around every establishment to gather encounters with the capacity to envision and support the accompanying improvement. In India, the HEIs to molding than being framed by the advanced period by SWYAM moreover.

(1) *Making of new space for experimentation:* The educators and teacher bunches generally recognized institutional difficulties, for example, shared and expanded dynamic cycles, which, while having numerous fundamental advantages, regularly render the foundations slower to analyze and change. Numerous members found that the authoritative structures planned around MOOC creation gave safe spaces to experimentation and advancement in educating and learning. Isolated from customary corporate procedures and structures and combined with the group based methodology, groups at our foundations have started to examine acquired suppositions concerning advanced education. The MOOCs that license us to envision what is conceivable along every vertical of the abbreviation—the monstrous, the open, and the on the web (or computerized). The MOOCs publicity might be passing on, yet its energy has prepared for expanded experimentation with mindful and intense thoughts for advanced education in a computerized time.

In the information time, HEIs have become more diverse in type, ownership, and educational offerings, have also diversified their missions, targeting specific groups of students (women, minorities, disadvantaged or students with special needs, adults and lifelong learners, international students, etc.), serving specific local or regional needs. The particular instruments, administrations, and encounters of customary advanced education will keep on being unbundled by a scope of organizations and new companies. It is up to schools and colleges to package and reincorporate these new components such that encapsulates the firm beliefs of instruction with the pragmatic deep-rooted adapting necessities of people.

The early effects of MOOCs on advanced education are an indication that this progress is precarious yet altogether conceivable. As the following rushes of hype impact colleges, likely as versatile learning and competency-based instruction, the effective frameworks will be those that address the full range of information: human sciences, professional and deep-rooted.

(2) *Difficulties for students:* there are provokes intrinsic to creating instructive substances in developing organizations, for example, MOOCs, for example, repetitional hazard and absence of conviction about future headings.
(3) *Openings:* Involvement in MOOCs can bring an assortment of events, for example, the democratization of instruction.
(4) *Computerized change:* MOOCs can have a job in the advanced change of colleges from an assortment of points of view.
(5) *Needs/demands:* those associated with MOOC advancement have explicit requests for an actualizing of MOOCs effectively in the establishment.
(6) *Tending to target crowds:* Tending to target crowds: The specialists express the trouble of managing obscure objective groups' necessities likewise now and again. For instance, the member underneath battled when endeavoring to inspire interest from the learning network of the MOOC, because of the understudies; take part. Isn't include a great deal, yet in any event, What's more, a few members were not intrigued by the profundity of the discussions.
(7) *The outstanding task at hand:* Many of the teachers express about the remaining task at hand and feel that, albeit compensating somehow or another, the endeavors associated with making MOOCs were huge, given that they made on their statutory obligations.
(8) *Openings:* Those who are members associated with MOOCs and their everyday exercises somehow. Along these lines, it was a positive sign in the institutional commitment in MOOCs.
(9) *Social strategic free training:* MOOCs can give profoundly moderate access to information to all residents because of mechanical advancement, in an exceptionally adaptable path regarding commitment, timing, and accessibility instruction principal, indeed, fundamentally engaged.
(10) *Computerized change and Institutional advanced culture:* The MOOCs have a huge impact and, for the greater part of them, chance to get computerized instruction experience, basically using virtual learning environments like Moodle and Blackboard too.
(11) *Advanced material creation:* The MOOCs give involvement with utilizing computerized instructive apparatuses for overseeing and conveying their substance; the utilization of media, for example, sounds a large number. Simultaneously use of these new arrangements a test a few scholastics like talking before a camera was testing

and scripted the accounts since progression talk characteristic. Some of them think they are misusing the understanding of showing experience. However, they feel perfect in a talk room loaded with understudies, yet the camera was overwhelming! The assortment of arrangements was a primary point inside the foundation, as the instructional fashioner implies peer audit, countless advanced organizations. We were extremely enthusiastic about creating instructive natural assets.

(12) *Requests for students:* The understudies perceived and got with energy and excitement the proposition to take an interest observed latent capacity invocations, establishment, fate instruction. In any case, after their cooperation, they conceded that MOOCs underpins and satisfy their objectives.

(13) *Motivating forces:* To the creating MOOCs includes a critical interest in exertion and time, and in return, there were a couple of requests that teachers might want to make on the off chance that they got the opportunity. There is a requirement for more help from the establishment. This course seriously for a while; this errand ought to be perceived and remunerated with the lightening of showing hours at any rate. HEIs, assess endeavors in MOOCs that are like the battles in our up close and personal talks.

(14) *Interest in assets:* Another solicitation from instructors engaged with MOOCs was an expansion in asset speculation, both in people and advances, and these days gave numerous apparatuses for creating varying media content. In any case, there are numerous new instruments out there that can be of incredible assistance to us for growing better assets for subsidizing to build up these advances with a generous venture since this would upgrade the situation of HEIs.

15.5 DIFFICULTIES

As talked about creation, it conveys differed difficulties to an education scholarly network collaborating obscure crowds regular test instructors focus on connected systems. A portion of the everyday encounters, for example, getting students to take an interest in the gatherings and deal with this support, is fundamentally testing additional includes a substantial outstanding burden. An overview uncovered this reality—the difficulties related to joining general academic obligations with the recently presented MOOCs on the instructive workplace.

15.6 OPENINGS

The MOOCs are an essential wellspring of chances at a level the social advantages, giving training appears inspiration putting resources into simultaneously, recommends appearance is a critical sign change in the digitization of specialized devices assumed a job training. The advanced innovations examine the core utilization of digital and stretch out of information worldwide.

15.7 REQUESTS

Present days MOOCs are exceptionally requesting and they require inventiveness, development, time, and a lot of exertion. Institutional help in different structures should, hence, from the establishment to those scholastics that consent to be associated with the advancement of these sorts of courses. One of these types of help to be a lot of motivating forces, particularly the acknowledgment of their exertion in the method for an easing of different obligations, for example, addressing time, checking, look into, and official duties. Another regular interest expansion specialized assets help scholastics creation advanced for making new learning spaces. It brings up the issue of whether institutional methodologies distribute assets size needs of instructive development and innovation, an issue hard to adjust (Salmon, 2005).

15.8 END

This paper has talked about the impact of Indian HEIs on MOOCs at the college level. From the exploration, audits expressed that the individuals who engaged with MOOCs hold an uplifting mentality to this open training approach. They are enthusiastic about making confronting exhibit equivalent give training likewise consider to be developed a chance to change, sorting them out toward ideal utilization.

In any case, there is an expense related to being associated with these sorts of instruction sources. Instructors think that it is difficult to convey their substance in new configurations, for example, imparting through video and composing for enormous, different, and obscure crowds. Regularly prompts scholastics giving an increasingly critical measure of time and exertion that remunerated. It is in this manner boundless that they request that their endeavors perceived in different structures, for example, the decrease of instructing obligations. Another regular interest advancement both inside organizations, to stay manageable. Greater interest in assets for MOOCs from the establishments is likewise profoundly mentioned, as members accept that they would become markers of greatness.

KEYWORDS

- **cooperative learning**
- **new pedagogies**
- **MOOC**

REFERENCES

Mumtaz, S. Factors affecting teachers' use of information and communications technology: a review of the literature. Journal of Information Technology for Teacher Education 9 (3) (2000):69–84.

Nehru, R.S.S. E-Learning—Theory and Practice. APH Co. New Delhi, 2013.

Nehru, Ravipalli & Ton, Quang Cuong. Nehru-ton model technologies for blended learning between India & Vietnam Teacher Education Programmes. International Journal of Engineering Application & Management (IJREAM), 3 (4) (2018):322–335. DoI:10.18231/2454–9150.2018.0344.

Nehru, R.S.S., Chakraborty S. The Education of Things (EoT) for smart learning through IoT intervention: a case study based analysis. In: Gunjan V., Garcia Diaz V.,

Reginald Nnazor. A conceptual framework for understanding use of information and communication technology in teaching in universities. International Journal of Instructional Technology & Distance Learning, 3 (1) (2009):47–57.

Rogers, Donna. A paradigm shift: technology integration for higher education in the new millennium. Educational Technology Review. 1(2000):19–28.

Shinde, S. P. and V. P. Deshmukh. *Web–based education in schools: a paradigm shift in India.* International Journal of Computer and Informatics (IJCS) Inter-science Open Access Journals, 2 (1) (2002): 90–94.

CHAPTER 16

Application of ArcGIS StoryMaps in Teaching Chemistry

VU MINH TRANG,[1,*] BUI THI THANH HUONG[1], LAI PHUONG LIEN[1], and NGUYEN QUANG NGOC[2]

[1]*University of Education, Vietnam National University, Hanoi, Vietnam*

[2]*Cao Ba Quat High School, Hanoi, Vietnam*

[*]*Corresponding author. E-mail: trangvm84@gmail.com*

ABSTRACT

Chemistry, a subject with theory and practice section, is a difficult science requiring students to have thinking abilities, abundant imagination, sensitivity, intelligence through observation, synthesis, analysis, evaluation, reasoning skills, and skills applying chemistry into reality. By the outstanding features of ESRI story map—one of the web applications created by cloud platform of ESRI, ArcGIS Online can design topics and lesson plans for teaching chemistry following three different groups of students (low, medium, high level) to evaluate their improvement of chemical awareness, their interest in learning, and their abilities to connect to realities by location-based thought. In addition, the chapter not only indicated the results of the investigation in applying ArcGIS on Chemistry lessons at high schools in Hanoi capital but also gave the rules and process to guide teachers on how to use ArcGIS in designing contents as well as lesson plans in chemistry.

16.1 INTRODUCTION

Today, over 80% of the world's spatial-related information requires a powerful transformation in life and education. The digital transformation in education, which is taking place strongly, needs to be associated with the transformation of spatial data to suit the current trend. In the learning materials system, teachers' lectures and students' lessons need to integrate spatial data into all subjects and fields. Geographic information system (GIS) is a technology recommended by managers to use in education, currently integrated into the subjects of Natural Sciences, Geography, History, and school management in many countries worldwide, especially the United States. However, the integration of GIS in teaching and educational management is still at a low level, so this field is quite new in education in Vietnam.

ArcGIS StoryMaps, an application of ArcGIS Online created by Esri company (United States), is a tool for building stories from content blocks such as narrative text, multimedia, maps, and embedded web content. Story viewers can read stories on any device, share social links, and embed stories on their own websites. Applying story maps to teaching contributes to increasing visualization to stimulate attention, arouse affection, attract students into lessons, thus mobilizing multiple senses of students to participate in the process of acquiring knowledge. Therefore students are able to develop the ability of observation, imagination, and scientific thinking.

The application of ArcGIS StoryMaps in teaching Chemistry is a new approach in Vietnam, has not been implemented in teaching at high schools and universities level. Using ArcGIS StoryMaps combined with simulation features of Chemistry utility software in teaching will help students access and master knowledge quickly and effectively.

In this chapter, the teaching content is developed in the metal section of the Chemistry subject. This chapter also outlines the principles and procedures for designing teaching topics and organizing learning content with the help of the ArcGIS StoryMaps integrated into the lesson. The tool kit for assessing Chemistry awareness, informatics competency, and learning interest of students is also designed for quantitative assessment before and after conducting at-school pedagogical experiments to evaluate the effectiveness of lessons.

16.2 OVERVIEW ARCGIS—STORYMAPS

16.2.1 APPLICATION OF GEOGRAPHIC INFORMATION SYSTEM *IN TEACHING*

According to the author Mohammad Robiul Hussain, a GIS is a useful information technology application tool that helps bridge the gap between requirements and reality. In particular, it is worth mentioning that the GIS can help answer questions and solve problems based on the database provided quickly and easily shared because the GIS can link individual data based on common geographic Location, creating new information from existing data sources (Mohammad Robiul Hussain, 2016), An Overview of Geographic Information System (GIS).

Although there are many definitions of GIS currently, in this study, we use the definition of GIS from Esri, "GIS is a computer-based tool for mapping and analyzing existing data and events happening on the earth. GIS technology integrates common database operations such as questions and statistical analysis with unique imagination and benefits of geographic analysis provided by maps."

Based on the benefits of GIS, Nihal Ugurul has studied the use of GIS for educational purposes in Turkey when it is recognized that the use of information technology in the learning environment has an extremely important role in meeting the integration of teaching and learning activities with the technology of the curriculum in modern society in order to bring education and pedagogical processes in schools attached to the conditions of the living. In the study, the author pointed out that the use of GIS for educational purposes has brought positive results, that is, GIS has supported the structure of a positive learning environment and is suitable for a student-centered learning environment based on interactive maps. In addition, the study also demonstrates the positive impact of GIS maps on the different expressions of spatial data on learning thinking skills and positive attitudes with technological tools of students (Nihal Ugurul, 2011)

Following the announcement of Nihal Ugurul, in 2012, Soon Singh Bikar Singh and colleagues continued to conduct a survey on the use of GIS in teaching Geography at six smart schools in Sabah, Malaysia. The research results show that nearly 90% of Geography teachers are aware of the importance of GIS in teaching Geography and especially make recommendations to the Board of Program Building and the Ministry of Education of Malaysia recognizes the need to develop new Geography teaching topics in the direction of integrating GIS technology (Soon Singh Bikar Singh et al., 2012)

In 2018, Yavuz Degirmenci conducted a qualitative study gathering opinions of teachers from different schools in a city of Turkey through semi-structured interviews developed by the researcher and data series obtained by analysis through a content analysis method. And the results also show that the teachers here agree on the need to use GIS in Geography lessons because

GIS in Geography lessons has helped develop reading and questioning skills in maps for students (Yavuz Degirmenci, 2018)

The above bases show that Vietnamese educators need to take advantage of and exploit the strengths of GIS in education to help students be not only aware of integrating technology in studying lessons but also be able to connect theory with reality and vice versa to observe, discover, and solve problems in reality to discover features and operation principles of things and phenomena.

16.2.2 OVERVIEW OF ARCGIS—STORYMAPS

ArcGIS StoryMaps is an extraordinary educational tool with infinite possibilities in classrooms. With the story map, students can describe places, explore changes over time, retell history, investigate environmental issues, or write down any topic of their own. Story maps promote the development of powerful communication skills through the ability to combine spatial visualization of maps with text, images, audio, and video data (Kathryn et al., 2018).

At a high-level conference on Geography at the University of South Carolina, thanks to the combination of Esri's story maps, students can communicate results from long projects throughout the semester. The authors discussed the successes and challenges that point to the advantages and minimize the limitations of using story maps in their classrooms. The outstanding advantage of the story map is that it allows students at different levels of mapping experience and GIS to create professional dynamic web mapping applications to support and present spatial stories tied to their study projects. In particular, the story map also creates opportunities for the majority of students to participate in projects. Besides, the story map also allows teachers to introduce students to the concepts and techniques involved in creating a vivid, geocentric web application (Sarah and Remington, 2013).

A study of Caitlin and Mitchell (2014) conducted by 42 teachers confirms the effectiveness of story maps in teaching that story maps can be used to present the materials in line with learning standards. In addition, story maps also help convey the enthusiasm of collaboration among teachers to create interdisciplinary, integrated story maps used as a teaching tool. Therefore the story map has a huge role in promoting spatial thinking skills and geographic knowledge in K-12 and informal classes because educators have opened up the story map of a budding future as a successful and effective educational technology (Caitlin and Mitchell, 2014).

In Vietnam, educators have also begun to be aware of the outstanding features of GIS when applied to the teaching process of the subjects. Typically, the study of Phan Van Trung has proved that GIS, when integrated into teaching and studying History, will "promote abstract and practical thinking for learners through turning word channels into picture channels, creating excitement for learners through visual channels, promoting logical thinking through frame maps, adding more means of teaching following the teaching method innovation trend" and is an "effective support tool for intensive history studies" (Phan, 2017)

16.2.3 ARCGIS—STORYMAPS IN TEACHING CHEMISTRY

Chemistry is a science in the field of Natural Sciences. It closely combines theory and experiment, bridging other Natural Sciences. Therefore advances in Chemistry are associated with the development of new discoveries in the fields of Biology, Medicine, and Physics. Besides, Chemistry plays an important role in life and production, contributing to socioeconomic development.

In the education program of high schools in Vietnam, Chemistry is a subject belonging to the group of Natural Sciences and has relationships with many other fields of education. The content of Chemistry is designed into topics that both reinforces the content flow, develop knowledge and practical skills, and help students gain a deeper understanding of the general foundations of Chemistry as a basis for studying, working, researching, and especially solving practical problems.

The above bases show that Vietnamese educators need to take advantage and exploit the strengths of GIS, especially "story map" in education, to help students not only be aware of technology integration in the study but also have the ability to connect theory with reality and vice versa to observe, discover, solve problems in practice so that the properties and operating principles of things, phenomenons can be discovered.

Therefore the content of the book presents the process and principles of integrating the advantages of story map in the design and organization of teaching integrated topics in the metal part of Chemistry in order to achieve the goals:

- It creates interest in learning for students in Chemistry.
- It improves the awareness capacity of Chemistry for students.

- It improves the ability of practical connection based on spatial thinking through improving computer literacy for students.

16.3 DESIGN INTEGRATED TOPICS IN CHEMISTRY TEACHING BY ARCGIS—STORYMAPS

16.3.1 ANALYSIS OF CHEMISTRY PROGRAMS IN HIGH SCHOOLS IN VIETNAM

The draft of the general education program in high schools issued in 2017 clarified six essential qualities that need to be formed and developed in students: loving the country, loving people, studying hard, working hard, being honest, responsible along with 10 core abilities, including common abilities: self-reliance and self-study, communication and cooperation ability, problem-solving and creative ability; professional abilities: linguistic ability, computational ability, natural and social discovery ability, technological ability, informatics ability, aesthetic ability, physical ability.

Chemistry has both a theoretical and an experimental and abstract character with the process of Chemistry performed on a microscopic level that is not visible to the naked eye, so that, it requires students to think, have a rich imagination, sharpness, and intelligence expressed through skills of observation, synthesis, analysis, evaluation, and reasoning. Thereby Chemistry helps students develop their chemical competencies, including chemical cognition ability, ability to understand the natural world, the ability to apply knowledge.

In the newly issued national high school education program system, Chemistry is introduced at the high school level as an elective subject selected by students in accordance with appropriate career orientation with abilities, interests, conditions, and circumstances. Therefore Chemistry, along with other topics, helps students form and develop core qualities and competencies to become a person who is the creative, autonomous, flexible ability, and can be adaptable in the context of global integration.

With these goals, the new Chemistry program of the high school level is built on a scientific basis inheriting and developing the knowledge of Natural Science at the Secondary School level according to both concentric and linear structures to equip students with basic Chemistry knowledge from basic to intensive and advanced levels, which focuses on training, developing experimental skills and applying Chemistry knowledge in practice.

Along with that, students are interested in learning Chemistry, have faith in the transformation of matter, people's cognitive ability, the role of Chemistry in life, and propagate the application of Chemistry in life.

The Chemistry program in high schools is developed systematically, scientifically, and logically to ensure the inheritance and development, practicality, career orientation, and promotion of students' learning activeness with three core content streams including knowledge of common Chemistry, knowledge of inorganic Chemistry, knowledge of organic Chemistry, and learning topics. Specifically:

- The Chemistry program in high schools is built in a logical way, the knowledge and concepts are formed and developed continuously and completely.
- The Chemistry program in high schools is built from two systems of knowledge about substances and chemical reactions. These two concepts are developed in parallel and supported by each other on the basis of the main theory of the program.
- The Chemistry program in high schools is built mainly on the principle of straight lines, some knowledge and concepts are built on the principle of concentricity.
- In the whole program, the knowledge of key substances, concepts, and theories is arranged alternately, ensuring the leading role of the theory and the effectiveness of the process for cognition and development of concepts.
- The Chemistry program in high schools also focuses on practical skills through experiential activities to help students understand and apply the main theory and, at the same time, expand concepts to continue solving problems in practice, meet the requirements of life.

16.3.2 ANALYSIS OF THE CONTENT ABOUT METALS IN CHEMISTRY PROGRAM IN HIGH SCHOOLS IN VIETNAM

The Metals Section of the 12th-grade Chemistry textbook program deals with metals of groups IA, IIA, IIIA, and metals of group B along with their important compounds and alloys. It is worth noting that the metals introduced in the program are important metals that have many practical applications in life. When studying metals, students need to learn about the location of metals in the periodic table, the configuration of atomic electrons, the formation of oxidized digital states of metals, and the understanding of physical, chemical, and application properties as well as methods to prepare them.

16.3.3 RULES FOR DESIGNING INTEGRATED TOPICS IN CHEMISTRY BY ARCGIS—STORYMAPS

Rule 1: Ensuring the goals of education, forming and developing necessary abilities for learners.

The goal is to help students develop completely of ethical, intellectual, physical, aesthetic matters and basic skills, develop personal abilities, dynamism and creation, and shape human character, build citizenship and responsibility, prepare students to continue their education or enter working life, participate in building and defending the Fatherland.

According to the orientation of developing the ability for learners of high school education programs, education not only equips knowledge and skills for students but also needs to pay attention to develop qualities and competencies, odd learners. Therefore the selection of contents for integrated teaching topics must aim to develop the necessary competencies of citizens to meet the development requirements of the country. That is cognitive ability, computer ability, language ability, cooperation ability, problem-solving ability.

Rule 2: Meeting the development requirements of society, with practical and meaningful quality to learners

The education sector needs to train human resources for the society with a broad educational background that is able to perform a wide range of tasks and specializations in order to ensure the quality of work with high efficiency. Therefore each trained person needs to be dynamic, creative, have professional knowledge and skills, ready to shoulder the responsibilities. Thus the selection of content for integrated topics should streamline academic knowledge, select practical, meaningful and life-related knowledge, meet the social innovation requirements, and create conditions for learners to adapt to life while being able to be proactive and flexible in acquiring knowledge and lifelong learning on the basis of high school education.

Rule 3: Ensuring education quality for sustainable development

The content of selected integrated topics should contribute to the formation and fostering of students not only about the world but also to express their attitudes toward the world and foster the quality of citizens in the new era:

love of homeland, country, responsibility for family and society, cooperation, solidarity and equality, study and respect for cultures.

In addition, the era of globalization and sustainable development not only creates opportunities but also poses great challenges for education, namely the challenges of teaching technology and sustainable development challenges. Therefore in each integrated teaching subject, it is necessary to integrate technology suitable to each content unit of knowledge and lesson purposes in order to create learners' confidence in science through the feature of connecting learners with the real world, thereby helping learners to be passionate about discovering and creating in science.

Rule 4: Ensuring science and access to the achievements of science and technology, and at the same time to fit students.

The development of integrated topics must both ensure science and access to new achievements of science and technology but must be suitable for the cognitive abilities of students as well as teaching plans. In order to do this, integrated topics must enhance practical knowledge enabling students to experience and explore knowledge through the support of technology tools and software in teaching and learning.

Rule 5: Increasing usefulness, practicality, attention to social issues in the local area

All science is the result of human perception in the process of practical activities. Therefore the content of integrated topics should enhance the usefulness and practicality in order to train students to apply knowledge to learn and solve practical problems, contributing to meet demands of life. Therefore through integrated teaching topics, students need to be directed to social issues in the local area in order to help students have an understanding of where they are living. Thus this helps students to prepare themselves to participate in local socioeconomic activities. These are also the strengths that GIS–Story Map helps learners to learn about nature and society based on spatial thinking because each lesson is no longer limited within the classroom, but practical connectivity features will bring learners to experience knowledge on geographical spaces in a real and vivid way.

Rule 6: Ensuring integrated topics built around knowledge based on the current curriculum of subjects

Integrated topics are determined based on the intersection content of current topics and the issues that need international, national, and meaningful education for students' lives. Integrated topics are not only done between topics, between contents that have similarities but also are done between topics, between different content but complementary to each other.

***Rule** 7*: Ensuring the specificity of the subject

Building an integrated teaching topic needs to ensure the specificity of the subject, but it is also necessary to balance theory and practice, considering subject theory to be the basic foundation, practice is the goal, toward the educational goals. Therefore the content chosen to build a topic should ensure basic, practical, instrumental, and as a basis for learning other topics as well as lifelong learning of students.

16.3.4 STEPS FOR DESIGNING INTEGRATED TOPICS IN CHEMISTRY BY ARCGIS—STORYMAPS

***Step** 1*: Defining the goals and focus of the topics

When setting goals, it is necessary to comply with the standards of knowledge and required skills promulgated in the program by the Ministry of Education and Training and orientation of education renovation in the direction of developing capacity. The objectives of the topics include elements like knowledge, skills, attitudes, and competencies, especially when defining integrated contents, need to pay attention to the skills and attitudes hidden in the topics.

***Step** 2*: Identifying problems that need to be solved in the integrated topics

Identifying the problem to be solved is one of the important steps having orientability in topics building. Teachers rely on teaching objectives and program delivery of the current subject to identify relevant content to build into a common problem of the teaching topics.

***Step** 3*: Identifying integration basis and topics contents

Researching contents of topics regarding contents of Chemistry in the topics, identifying knowledge used to explain practical issues. When determining the teaching contents integrated into the topics, teachers need to identify the students' existing knowledge, thinking ability, and educational level so that the lesson is effective and affordable for students. The selected contents for integration must be suitable for topics and practical, highly applicable, meeting the objectives of the topics, creating learning interests for students.

***Step** 4*: Building the detailed contents of the topics on GIS–StoryMap

Teachers need to be based on the spatial connectivity features of GIS–StoryMap to structure teaching content in a logical, scientific, and compatible manner with the features of this technology and especially fully exploit advantages of GIS–StoryMap in lesson simulation, practical connection.

Step 5: Identifying teaching methods

Methods are a way for teachers to achieve their teaching goals. Choosing the right methods will help teachers succeed in the lesson. The methods chosen should be appropriate to the goals and contents of the lesson to promote students' activeness, initiative, and creativity.

Step 6: Designing teaching process according to integrated topics

The teaching process is designed into activities through learning tasks corresponding to the contents of knowledge. Teachers can organize students to perform tasks at home or in class.

In the process of organizing teaching activities, teachers need to put students in situations that come from real life to make students ready to participate in solving those situations. In the process of acquiring knowledge, students need to argue in order to defend their opinions before the collective and vice versa know to raise discussion issues for the class to generalize the general conclusions for the lesson. These learning activities need to be guided, organized by teachers according to the pedagogical process of the selected teaching method. The acquisition of knowledge through learning activities will help students to be more initiative, more active, creative thinker, and a better learner. Thus the development of the teaching process according to integrated topics starts from building learning situations, then proposing solutions to solve problems, implementing measures, reporting, discussing, and concluding.

Step 7: Consulting the educational experts

Step 8: Teaching according to integrated topics

Step 9: Editing and improving the teaching topics.

After every lesson, teachers need to get feedback from students about the lesson to improve the lesson and complete the integrated teaching topics.

16.3.5 STEPS FOR TEACHING CHEMISTRY BY ARCGIS—STORYMAPS

The process of teaching Chemistry by ArcGIS—StoryMaps is shown in Figure 16.1.

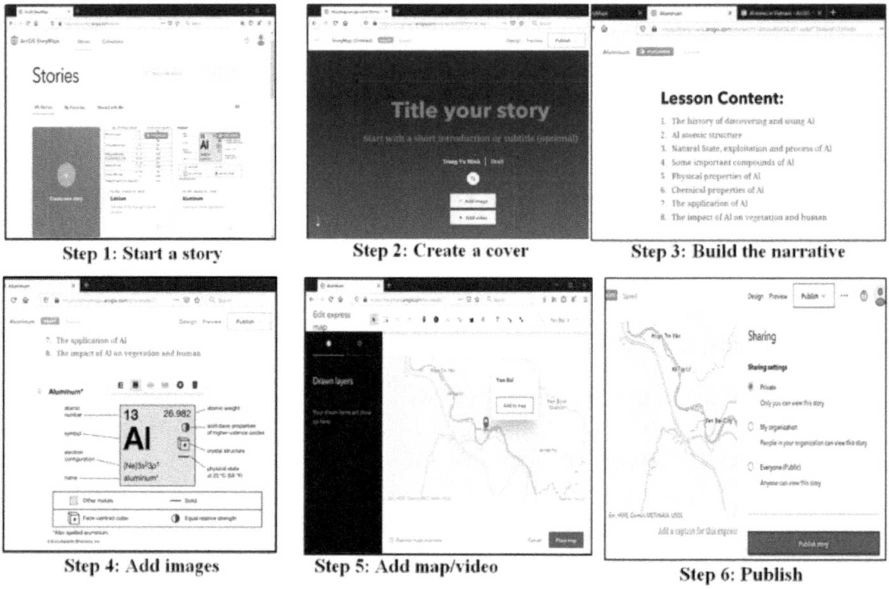

FIGURE 16.1 The steps building Aluminum StoryMap in ArcGIS Online.

Step 1: Teachers need to understand clearly contents and determine the meaning of each teaching activity.

Step 2: Teachers should plan to use teaching methods that appropriate to each content of knowledge and practical conditions.

Step 3: Introduce students to technology software that can support the learning of students and especially guide students to use unique technology features to support knowledge learning.

Step 4: Teachers assign learning tasks to students, such as learning materials, equipment, requirements for learning products.

Step 5: Anticipate supportive measures when students encounter difficulties in the process of performing learning tasks.

Step 6: Organize for students to report learning products and discuss in class.

Step 7: Strengthen lessons to inculcate knowledge and guide students to apply knowledge at an advanced level.

Step 8: Set up a plan to assess student learning results.

16.4 ARCGIS—STORYMAPS TECHNIQUE DESIGNING IN METAL LESSON

16.4.1 TECHNIQUES DESIGNING ARCGIS—STORYMAPS FOR ALUMINUM LESSONS

The steps building ArcGIS—StoryMaps for Aluminum lessons are shown in Figure 16.2.

Step 1: Start a story

Logging into ArcGIS StoryMap/Navigate to: https://storymaps.ArcGIS.com/ Click: Get started/Type username and password/Verify login

By completing logging in, you will be navigated to your story page. If you have no stories created yet, there will only be a "Create New Story" button on the page. To create a new story, click on the "Create new story" button. Clicking on that button, a brief notification appears—this notification will disappear momentarily. After the launch is done, we proceed to add multiple elements for your story map. The elements that make up our story map are as follows:

- Cover
- Narrative
- Image
- Map

Step 2: Create a cover

Demo Story Text [TITLE]/Aluminum/[Subtitle]/Aluminum and its application/[Byline] /By Vu Minh Trang. Next, you'd want to add an image or video to the cover. Click "Add image" and look for the image files you want to add. That image will fill the title page background. If you choose the Add video option, you can upload an MP4 from your files. The video will fill the screen, and will automatically loop.

Step 3: Build the narrative

Now the cover is done, there is a little plus sign below the cover. Hovering over it makes it green. Click on it, and a menu of options appears. It is common to start with a Paragraph option. Choose Paragraph, then type in the text.

Step 4: Add an image

Next, we will add an image to our story. Here, we would like to put an image of Aluminum with its brief physical properties to give the learners a quick image of how Aluminum looks like.

To do this, open the block palette, then pick Image from the menu. Just like you did for the cover image, navigate your folders to find an image you would like to include. Click Open, and it will appear in your story. When you hover over the image, you will see a little set of options in a panel at the top of the image. Note that there is a place for a caption under each image that you add. Captions are optional; the placeholder Add a caption text will not appear when you publish your story.

Step 5: Make a map

Here you can either choose a map that you have created in ArcGIS Online (tab My Maps) or Start an express map. Look for the Aluminum Factory point that you would want to place on the map. Here we have three factories. We can add points with the fixed symbol, or we can add points with numbered symbols. Also, we can draw a line or a polygon to show some areal elements on the map. Other useful options provided are as follows:

- Add annotation to give more explanation;
- Add an arrow to show the direction;
- Undo/Redo what we just did.

Because there are several videos associated with this factory, therefore, we will put this map on the left side of the page. On the right side, we put the links associated with the factories. To put it on the left side, we choose the Float option. After that, we put the link on the right side of the map. For doing that, we first type the text we want to display. Then we turn the text into a hyperlink by selecting the entire text, then click on the hyperlink icon, and type in the address we want to navigate to when we click on the link. Note that the line under the text means that text is a hyperlink.

For each factory, since there are also multiple pictures, go with it to give a closer look at the factory. Therefore an album underneath the map could be useful and will provide the audience more detail about the map.

When the slideshow widget appears, we press "New slide" to add a new slide. For each slide, we can add an image as its background. Along with the image, a slide title could be utilized to explain the image.

For the factory's point, an express map was used. For aluminum mines, we also use an express map. In this case, there are 18 aluminum mines, if you add each point separately on the story, the story will become too long to keep track of. Therefore we will use one map for all 18 mine points.

In order to do this, we first add the express map and add point just like we did with the factories above. After we finish adding the first point, we remain on the same map and look for other points.

At this phase, we will have a set of points, each point has its own information displayed when we hover them. At this phase, we can choose the base map for our map. Esri provided us with several types of base map. Upon finishing adding points, we press the "Place map." The newly created map will be placed on our story. Again, we can either choose to place it on the left, at the center with a small size, or occupy the entire screen.

Step 6: Publish

You might have noticed that there is no save button, but that the top bar is telling you that your story is saved, and is a draft.

You can preview your story at any point (the Preview function is next to Publish) to see what it will look like in published form on PCs, laptops, tablets, and cell phones.

When you are ready to publish, go ahead and click Publish. Not to worry, you can continue to edit your story after you publish it. The first time you click this button, you see these sharing options. Private means it visible only to you; My organization makes it accessible to other people in your ArcGIS organization; Everyone makes your story public. You can also share your story with ArcGIS groups you are a part of by using the search box.

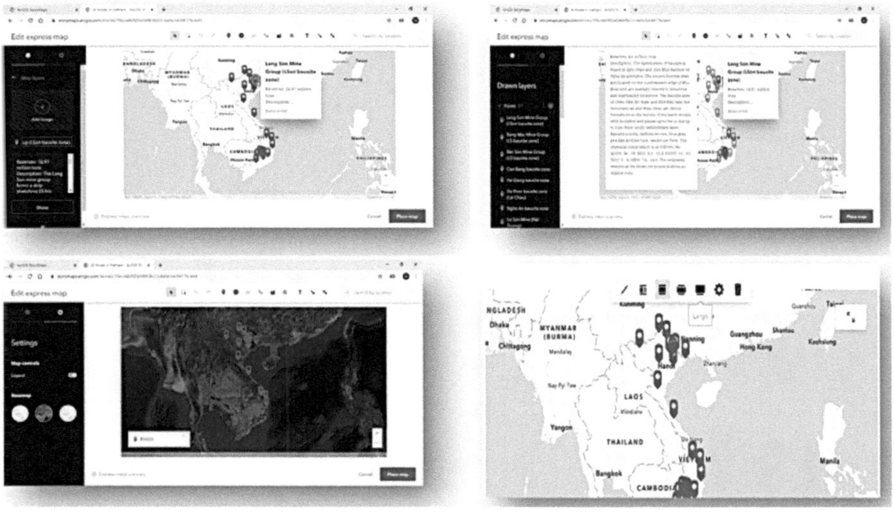

https://storymaps.ArcGIS.com/stories/f91ddabc89434301ae8ef726dee4123f

FIGURE 16.2 Aluminum StoryMaps published on ArcGIS Online.

16.4.2 TECHNIQUES DESIGNING ARCGIS—STORYMAPS FOR IRON LESSONS

With the same steps described earlier, the research group built Iron lessons by ArcGIS—StoryMaps following the steps shown in Figure 16.3.

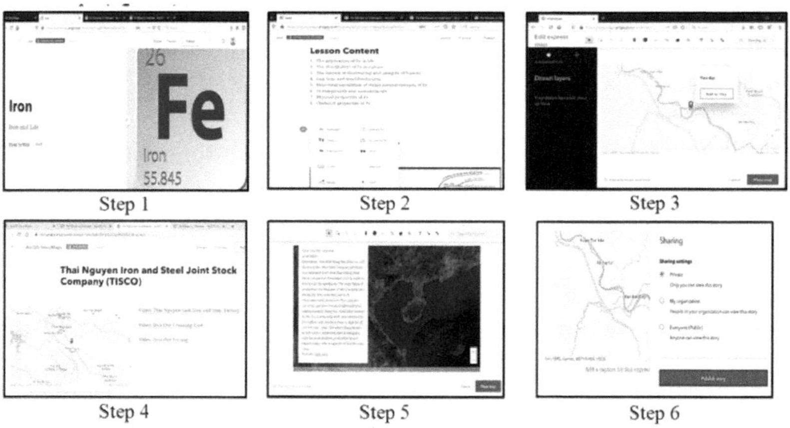

FIGURE 16.3 The steps building Iron StoryMaps in ArcGIS Online.

The results of building ArcGIS—StoryMaps follow the steps present in Figure 16.4.

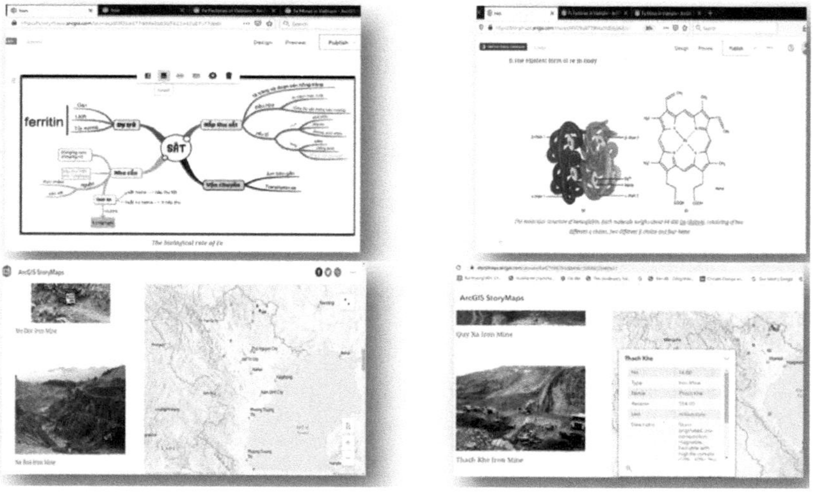

https://storymaps.ArcGIS.com/stories/df3f2ba8779d4a2b83b5622e42c87a77
FIGURE 16.4 Iron StoryMaps published on ArcGIS Online.

16.4.3 TECHNIQUES DESIGNING ARCGIS—STORYMAPS FOR CALCIUM LESSONS

With the same earlier described steps, the research group built Calcium lessons by ArcGIS—StoryMaps following the below steps shown in Figure 16.5. Figure 16.6 shows the Calcium StoryMaps published on ArcGIS Online.

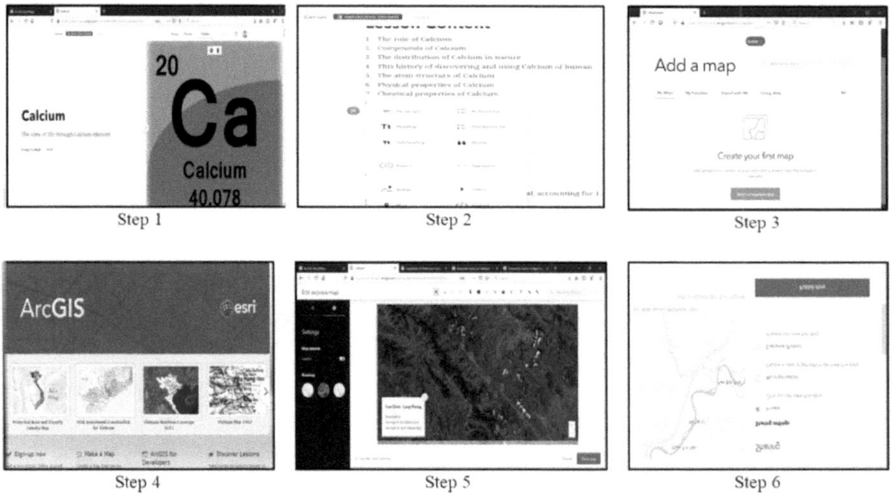

FIGURE 16.5 The steps of building Calcium StoryMaps in ArcGIS Online.

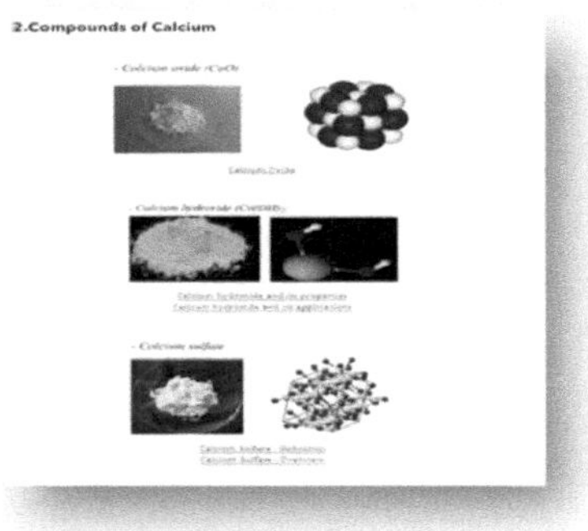

Information about Calcium

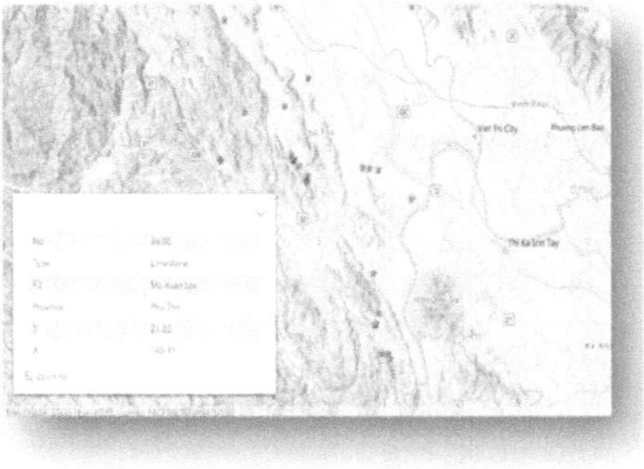

Limestone mines in Phu Tho Province

FIGURE 16.6 *(Continued)*

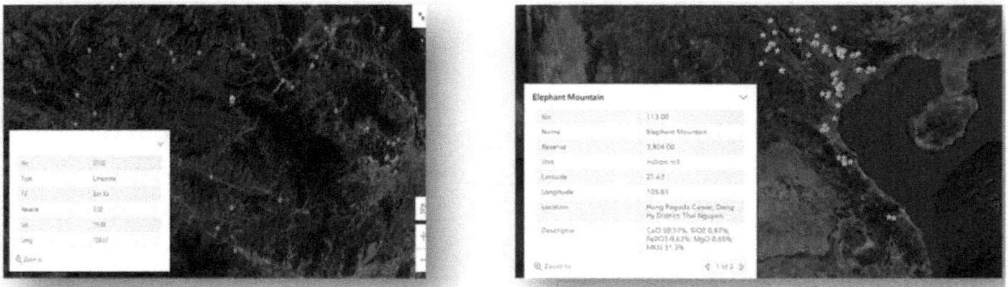

Limestone mines in NgheAn province Limestone mines in Vietnam

https://storymaps.ArcGIS.com/stories/9b190d76d3474a0dab87b5ea7b23327a

FIGURE 16.6 Calcium StoryMaps published on ArcGIS Online.

16.5 EVALUATION TOOLS THE EFFECTIVENESS OF TEACHING CHEMISTRY BY ARCGIS-STORYMAPS

16.5.1 EVALUATION OF LEARNING INTEREST

Evaluation of learning interest was built by 4 levels achieved through 10 criteria, detailed in Tables 16.1 and 16.2.

Table 16.1 is standards for teachers to assess students' interest in learning, include 10 criteria (from 1 to 10). Each criterion is assessed at 4 levels: not passing (level 1), passing (level 2), good (level 3), and very good (level 4). Each level has a corresponding expression to assess students' level of interest in learning.

TABLE 16.1 Criteria for Evaluation of Students' Learning Interest *(For Teachers)*

Criteria	Evaluation of the Levels Achieved			
	Level 1 (Not Passed)	**Level 2 (Passed)**	**Level 3 (Good)**	**Level 4 (Very Good)**
Criterion 1. For the positivity in lessons.	Not volunteer to speak at any time.	Volunteering to speak 1 time.	Volunteer speak 2–3 times.	Volunteer to speak more than 3 times.
Criterion 2. Creativity in learning and practical activities, reading skills, books and materials related to the problems posed in learning.	Not active in learning and practical activities, not reading more books and papers related to problems raised in learning.	Active but without creativity in learning and practical activities, seldom reading more books and papers related to issues raised in learning.	Active and creative in learning and practical activities, but rarely reading more books and papers related to the issues posed in learning.	Active and creative in learning and practical activities, regularly reading more books and papers related to the issues posed in learning.

TABLE 16.1 *(Continued)*

Criteria	Evaluation of the Levels Achieved			
	Level 1 (Not Passed)	**Level 2 (Passed)**	**Level 3 (Good)**	**Level 4 (Very Good)**
Criterion 3. Ability to concentrate on learning	Not focused and attentive in class.	Relatively focused on the lesson.	Concentrate, pay attention to the lesson.	Completely focus and pay attention to the lesson.
Criterion 4. An attitude of interest in performing tasks, learning tasks.	Not interested in and not perform work, learn tasks.	Not interested but still work, learn tasks.	Excited but do work, the task of learning through speakers.	Very interested and well do work, study tasks
Criterion 5. Students' excitement in the learning process, learning new issues, creativity, active activities.	Not interested in the learning process, explore new issues, not coming up with initiatives to improve activities more actively.	Excited, excited in the learning process, passive in learning new issues, not coming up with initiatives to perfect activities.	Excited in the learning process, or explore new issues, not coming up with initiatives to perfect the activity.	Always interested in the learning process, or explore new issues, be creative, think of initiatives to improve activities more actively.
Criterion 6. Student spirit throughout the lesson.	Not feel interested in the activities in lessons; the atmosphere of the classroom is mysterious and uncomfortable.	I feel excited about the activities in lessons but the atmosphere of the classroom is not comfortable.	Feel interested in activities in the lessons, the atmosphere of the classroom is normal.	Feel very interested in activities in the classroom, the atmosphere of the classroom is very comfortable.
Criterion 7. Feeling of students if they are absent from this lesson.	Feel very happy if can skip this lesson away.	Feel normal if skip this lesson.	Feel a bit sorry if skip this lesson.	Feel sorry if skip this lesson.
Criterion 8. Student feeling about the duration of the lesson.	Lessons passed a long time.	Lessons passed as usual.	Lessons passed relatively quickly.	Lessons passed very fast.

TABLE 16.1 *(Continued)*

Criteria	Evaluation of the Levels Achieved			
	Level 1 (Not Passed)	**Level 2 (Passed)**	**Level 3 (Good)**	**Level 4 (Very Good)**
Criterion 9. Apply the learned knowledge to solve problems in study and in life.	Awarded, embarrassed, not know how to apply possible ability, not solving the problem until the end.	Not afraid of hardship, not know how to apply all abilities to solve the problem to the end.	Not afraid of hardship, and use all possibilities but not solve the problem to the end.	Not afraid of hardship, and use all abilities to solve the problem to the end.
Criterion 10. Taking notes, accumulate necessary knowledge for yourself.	Not taking notes, not know how to accumulate knowledge.	Taking notes when required, not accumulate the necessary knowledge.	Voluntarily take notes but don't know how to accumulate necessary knowledge.	Voluntarily take notes, accumulating the necessary knowledge.

Table 16.2 is a checklist for teachers to assess students' interest in learning based on the performance of each of the criteria in Table 16.1. List of students is listed by the teacher in column 2. Each student will have 10 points correspond to the 10 criteria to assess students' interest in learning. The score of each criterion will be assigned with a score from 1 (not passed) to 4 (very good).

Class: .... Evaluated teacher: ....

TABLE 16.2 Survey Students' Learning Interest *(For Teachers)*

No.	**Criteria / Full name**	**1**	**2**	**3**	**4**	**5**	**6**	**7**	**8**	**9**	**10**
1											
2											
3											
4											
5											
6											
.....											

Table 16.3 is the content of the questionnaire evaluating students' interest in learning. This content is self-assessed by students.

Full name:......................................Class:...................Group:.........

Lesson:...

Please evaluate the following possibilities by circling the answer that best suits you according to the levels: 1. Disagree, 2. Confuse, 3. Partially agree, 4. Totally agree

TABLE 16.3 Evaluation Form of Learning Interest (For Students)

1. I completely focus and pay attention to the lesson.	1	2	3	4
2. I feel interested in the assignments and tasks assigned by the teacher and I always complete it well.	1	2	3	4
3. I am interested in learning new knowledge and I always think of ideas and creativity to solve problems in class.	1	2	3	4
4. I am interested in the activities in class and I feel the atmosphere in the classroom is very comfortable during the lesson.	1	2	3	4
5. I feel sorry if I miss this lesson.	1	2	3	4
6. I feel the lesson has passed quickly.	1	2	3	4
7. I know how to apply my existing knowledge to solve the same problems I encounter in my study and life.	1	2	3	4
8. I voluntarily write down the full content of the lesson and feel that I have accumulated a lot of knowledge after the lesson.	1	2	3	4
9. I feel that I am very active and creative in both study and practical activities; I often find more reading materials and books to improve my knowledge.	1	2	3	4

16.5.2 EVALUATION OF CHEMISTRY COGNITIVE COMPETENCE

Table 16.4 is the criteria for teachers to assess students' chemical cognitive competence, include 7 criteria (from 1 to 7) which are component skills. Each criterion is assessed at 4 levels: not pass (level 1), pass (level 2), good (level 3), and very good (level 4). Each level has a corresponding expression to assess students' chemical cognitive competence.

TABLE 16.4 Evaluation Criteria of Chemical Cognitive Competence (For Teachers)

Criteria	Evaluation of the Levels Achieved			
	Level 1 (Not Passed)	**Level 2 (Passed)**	**Level 3 (Good)**	**Level 4 (Very Good)**
Criterion 1. Ability to identify and name objects, events, concepts, or chemical processes	Not identify names of objects, events, concepts, or chemical processes have not been identified and identified	Identify but not name the objects, events, concepts, or chemical processes	Identify and identify names of objects, events, concepts, or chemical processes but are not accurate	Identify and name objects, events, concepts, or chemical processes correctly
Criterion 2. Ability to present facts, characteristics, roles of objects, concepts, or chemical processes	Failure to present facts, characteristics, the role of objects, concepts, or chemical processes	Present facts and characteristics but not yet presenting the role of objects, concepts, or chemical processes	Present facts, characteristics, roles of objects, concepts, or chemical processes but not exactly	Describe facts, characteristics, roles of objects, concepts, or chemical processes correctly
Criterion 3. Ability to describe objects by means of speaking, writing, formulas, diagrams, charts, tables	Not describe objects in forms of speaking, writing formulas, diagrams, charts, tables	Describe the object in forms of speaking, writing formulas but not yet described by diagrams, charts and tables	Describe the objects in forms of speaking, writing formulas, diagrams, charts, tables but not accurate	Describe the object in forms of speaking, writing formulas, diagrams, charts, tables
Criterion 4. Ability to compare, classify, select objects, concepts or chemical processes according to certain logic	Not compare, classify, select objects, concepts, or chemical processes according to a certain logic	Compare but not yet classify, selected objects, skills, or chemical process according to a certain logic	Compare, classify, select objects, concepts, or chemical processes but not following a certain logic	Compare, classify, select objects, concepts, or chemical processes according to a certain logic

TABLE 16.4 *(Continued)*

Criterion 5. Ability to explain and reason about the relationship between objects, concepts, or chemical processes	Not explain and reason about the relationship between objects, concepts, or chemical processes (structure - properties, causes - results ...)	Explain but not reason about the relationship between objects, concepts, or chemical processes (structure - properties, causes - results ...)	Explain and reason about the relationship between objects, concepts, or chemical processes (structure - properties, causes - results ...) but not accurate	Explain and reason correctly about the relationship between objects, concepts, or chemical processes (structure - properties, causes - results, etc.)
Criterion 6. The ability to connect information logically makes sense	Not find keywords, use scientific jargon, connect information with meaningful logic, make an outline when reading and presenting scientific documents	Find keywords, use scientific terminology but not connect information with meaningful logic, make an outline when reading and presenting scientific documents	Find keywords, use scientific terminology, connect logical information with meaning but have not been outlined when reading and presenting scientific documents	Find keywords, use scientific terminology, connect logical information with meaning, make an outline when reading and presenting scientific documents
Criterion 7. The ability to make critical comments that are relevant to the topic	Do not discuss, make critical comments related to the topic	Discuss but not yet made critical comments related to the topic	Discuss and make critical comments but not yet closely related to the topic	Discuss and make critical comments that are closely related to the topic

Table 16.5 is a checklist for teachers to assess students' chemical cognitive competence based on the performance of each of the criteria in Table 16.4. List of students is listed by the teacher in column 2. Each student will have the top seven points correspond to the seven criteria to assess students' chemical cognitive competence. The score of each criterion will be assigned with a score of 1 (not passing) to 4 (very good).

Class: .... Evaluated teacher: ....

TABLE 16.5 Evaluation Table of Students' Chemical Cognitive Competence (For Teachers)

No.	Criteria / Full name	C1	C2	C3	C4	C5	C6	C7
1								
2								
3								
4								
5								
6								
.....								

16.5.3 EVALUATION OF COMPUTING COMPETENCE

Table 16.6 is the criteria for teachers to assess students' ICT competence, include 5 criteria (from 1 to 5) which are component skills. Each criterion is assessed at 4 levels: not pass (level 1), pass (level 2), good (level 3) and very good (level 4). Each level has a corresponding expression to assess students' ICT competence.

TABLE 16.6 Criteria for Evaluating Students' Computing Competence (For Teachers)

Criteria	Evaluation of the Levels Achieved			
	Level 1 (Not Passed)	**Level 2 (Passed)**	**Level 3 (Good)**	**Level 4 (Very Good)**
Criterion 1. Ability to use and manage ICT tools	Not use and manage ICT tools	Use and manage ICT tools but excluding the ability to exploit common applications on the Internet and other digital services	Use and manage ICT tools, including the ability to exploit common applications on the Internet and other digital services but not for learning and life	Use and manage ICT tools, including the ability to exploit common applications on the Internet and other digital services to support learning and life

TABLE 16.6 *(Continued)*

Criteria	Evaluation of the Levels Achieved			
	Level 1 (Not Passed)	**Level 2 (Passed)**	**Level 3 (Good)**	**Level 4 (Very Good)**
Criterion 2. The ability to identify and behave in accordance with the law in the use of ICT	Not identify and behave in accordance with the law in the use of ICT	Identify and behave in accordance with the law in the use of ICT but not in accordance with the ethical and cultural standards of Vietnamese society and not ensuring information security for yourself and the community.	Identify and behave in accordance with the law in the use of ICT and ensure information security for yourself and the community but not in accordance with the ethical and cultural standards of Vietnamese society.	Identify and behave in accordance with the law in the use of ICT, conform to the ethical and cultural standards of Vietnamese society and ensure information security for yourself and the community.
Criterion 3. The ability to detect and solve problems with the help of ICT tools	Not detect and solve problems with the help of ICT tools	Identify and solve problems with the support of ICT tools but not innovative	Identify and solve problems creatively with the help of ICT tools, but do not include the full range of capabilities: automation and control thinking, planning, management capabilities, find, evaluate, use flexibly, optimize resources, and design systematically and problem-solving design	Identify and solve problems creatively with the help of ICT tools, including automation and control thinking capabilities, the ability to plan, manage, search, and fight price, flexible use, optimization of resources and the ability to systematically solve problems

TABLE 16.6 *(Continued)*

Criteria	Evaluation of the Levels Achieved			
	Level 1 (Not Passed)	Level 2 (Passed)	Level 3 (Good)	Level 4 (Very Good)
Criterion 4. Ability to learn and self-study with the support of ICT	Not learn and self-study with the support of ICT	Learn and self-study with the support of ICT but do not include the ability to exploit applications and services of digital technology of the ICT environment	Learn and self-study with the support of ICT, including the ability to exploit applications and services of digital technologies of the ICT environment but learning ineffectively in different fields	Learning and self-learning with the support of ICT, including the ability to exploit applications and services of digital technologies of the ICT environment to effectively study in different fields
Criterion 5. The ability to use ICT tools and environment to share information and collaborate	It is impossible to use ICT tools and environment for information sharing and cooperation	Use ICT tools and environment to share information but not cooperate with all members of the school, in the community and in society to improve work efficiency, improve quality of life	Use ICT tools and environment to share information and cooperate with all members of the school, in the community and in society but have not improved work efficiency, improved quality of life	Use ICT tools and environment to share information and collaborate with all members of the school, in the community and in society to improve work efficiency, improve quality of life

Table 16.7 is a checklist for teachers to assess students' ICT competence based on the performance of each of the criteria in Table 16.6. The list of students is listed by the teacher in column 2. Each student will have The top 5 points correspond to the 5 criteria to assess students' ICT competence. The score of each criterion will be assigned with a score of 1 (not passing) to 4 (very good).

Class: .... Evaluated teacher: ....

TABLE 16.7 Evaluation Table of Students' Computing Competence (For Teachers)

No.	Full name \ Criteria	1	2	3	4	5
1						
2						
3						
4						
5						
6						
...						

Table 16.8 is the contents of the assessment sheet of students' chemical cognitive and ICT competence. The criteria (from 1 to 12) are the component skills of ICT competence (from criteria 1 to 6) and the chemical cognitive competence (from 7 to 12). After each lesson, students are given this form to self-assess their level of ICT competence and chemical cognitive competence.

Full name:..Class:...........Group:...........

Lesson:...

Please evaluate the following possibilities by circling the answer that best suits you according to the levels: 1. Disagree, 2. Confuse, 3. Partially agree, 4. Totally agree

TABLE 16.8 Form of Informatics Cognitive Competence and Chemical Cognitive Competence (For Students)

1	You can use and manage the tools of ICT (*Information and Communication Technology*), including the ability to exploit common applications on the Internet and digital services. other to serve learning and life.	1	2	3	4
2	I identify and behave in accordance with the law in the use of ICT, in accordance with the ethical and cultural standards of Vietnamese society and ensure information security for myself and the community.	1	2	3	4
3	I identify and solve problems creatively with the assistance of ICT tools, including automation and control thinking skills, the ability to plan, manage, search, evaluate, flexibly use, optimize resources, and the ability to design and systematically solve problems.	1	2	3	4

TABLE 16.8 *(Continued)*

4	I can learn and self-study with the support of ICT, including the ability to exploit the applications and services of digital technologies of the ICT environment to study effectively in many fields.	1	2	3	4
5	I use ICT tools and the environment to share information, cooperate with all members of the school, in the community, and in society to improve work efficiency, improve quality of life.	1	2	3	4
6	I identify and name the objects, events, concepts, or chemical processes	1	2	3	4
7	I present facts, characteristics, roles of objects, concepts, or chemical processes.	1	2	3	4
8	I can describe objects by forms of speaking, writing, formulas, diagrams, charts, tables.	1	2	3	4
9	I compare, classify, select objects, concepts, or chemical processes according to a certain logic	1	2	3	4
10	I explain and reason about the relationship between objects, concepts, or chemical processes (structure, properties, causes, results ...)	1	2	3	4
11	I find keywords, use scientific terminology, connect information with logical meaning, make an outline when reading and presenting scientific documents.	1	2	3	4
12	I discuss, make critical comments related to the topic	1	2	3	4

16.6 RECOMMENDATION

Some recommendations for educational managers are as follows:

1. Educational administrators must see the role of GIS in visualizing Chemistry lessons to improve the quality of teaching and learning Chemistry through a direct pedagogical interaction between teachers and students. Hence, giving priority to appropriate investment in technology equipment for education.
2. The Ministry of Education and Training needs to disseminate and devise policies both to support and to create needs and motivation for teachers to realize the need to integrate GIS–StoryMap in teaching.
3. Schools should regularly open refresher courses, skills training in the application of information technology in general, and Geographic Information Systems in particular in teaching.

4. Leaders at schools need to come up with measures to encourage teachers to be active and proactive in integrating GIS–StoryMap in the teaching process.
5. Regularly organizing exchanges of experiences and initiatives on GIS–StoryMap integration in developing lectures between teachers in professional groups and between domestic and foreign experts in the field of GIS with teachers.
6. Connecting copyright providers to provide ArcGIS accounts for teachers to build a foundation of teaching and education management for integrating spatial data into the school's educational process.

Some recommendations for teachers are as follows:

1. The process and system of the syllabus form, research results of the topics are references for Chemistry teachers to apply GIS in education.
2. Combining positive teaching methods to improve the effectiveness of the teaching quality.
3. Teachers need to cultivate GIS knowledge as well as professional knowledge of Chemistry, passion for research, and learning.
4. However, teachers need to know how to select appropriate Chemistry knowledge to integrate GIS, which helps to increase the simulation and practical connection of lessons to help students interested in learning and raising awareness of Chemistry.

16.7 CONCLUSION

Integrate special contents in teaching Chemistry in K12 schools by using ArcGIS—StoryMaps is a modern teaching method adapting to digital transformation in education in Vietnam. When using this application, Chemistry teachers need to be attentive to identify the integrated topics, contents, teaching scenario by STORY MAP, adjust teaching process of integrated topics in Chemistry with 8 steps, and tools for evaluating the effectiveness of teaching Chemistry by ArcGIS—StoryMaps, recommended ones. The criteria for teachers to assess students' chemical cognitive competence include 7 criteria (from 1 to 7), which are component skills. The standards for teachers to assess students' interest in learning include 10 criteria (from 1 to 10). Each level has a corresponding expression to assess students' level of interest in learning. The criteria for teachers to assess students' ICT competence include 5 criteria (from 1 to 5), which are component skills. Each criterion is assessed at 4 levels: not pass (level 1), pass (level 2), good (level 3), and very good (level 4). Each level has a corresponding expression to assess students' ICT competence. Some recommendations were given for educational managers and Chemistry teachers.

ACKNOWLEDGMENT

This research was funded by the University of Education, Vietnam National University, Hanoi under Research Project Code: QS.19.08.

KEYWORDS

- **ArcGIS StoryMap**
- **location-based thought**
- **chemical teaching**

REFERENCES

Caitlin Strachan, Jerry Mitchell. Teachers' Perceptions of Esri Story Maps as Effective Teaching Tools. *Review of International Geographical Education Online* (2014): 4 (3), pp. 195—220.

Kathryn Keranen, Lyn Malone, and Michael Wagner. Teach with GIS implementation guide for the classroom. *The Science of Where, ESRI* (2018): pp. 6—8.

Mohammad Robiul Hussain. An Overview of Geographic Information System (GIS). Estimation of Floods Using Geographic Information System, India, 2016, pp.1—14.

Nihal Ugurul. Use of Geographical Information Systems (GIS) for educational purposes, *social studies review. Journal of the California Council for the Social Studies.* 50 (1) (2011): pp. 79—88.

Nguyen Phuong Chi, Nguyen Thi Hong Phuong. Process of building and intergrated teaching topics in Mathematic–Chemistry–Biology in high school. *Vietnam Journal of Education,* 398(2) (2017): pp. 53–57.

Pham Hong Thai, Tong Thi Hanh. Designing topics intergrated with natural science in teaching in secondary school for students of Dien Bien teacher training College. *Journal of Education* (2017): pp. 189–191.

Phan Van Trung. Applying geographic information systems in studying and teaching history. *Science Journal of Thu Dau Mot University,* 2(33) (2017): pp. 135–141.

Sarah E. Battersby and Kevin C. Remington. Story maps in The Classroom. *Computer Lab Activities Develop Spatial Reasoning and Analysis Skills, Education,* (2013): pp. 62—65.

Soon Singh Bikar Singh, Grant Kleeman, Penny Van Bergen. Opportunities to implement GIS in teaching and learning geography: a survey among smart schools in Sabah. *Procedia—Social and Behavioral Sciences, Malaysia, 2012.* 69, pp. 884—889.

Stephan Forster. *Methods of Teaching Chemistry*, Global Media, 2009.

Yavuz Degirmenci. Use of geographic information systems (GIS) in geography lessons according to teachers' opinion. *World Journal on Educational Technology: Current Issues*, 10 (3) (2018): pp. 186—196.

CHAPTER 17

E-Tools: A Need for Enhancing Teaching and Learning English Language Skills

SHABREEN SULTANA SHAIK

Department of English, Bapatla Engineering College, Bapatla, Andhra Pradesh, India

Corresponding author. E-mail: shabreen.sulthana@gmail.com

ABSTRACT

Technology has been advancing rapidly for the last few decades. It has become an integral part of personal and professional lives. Language learning has become a continuous process. It is not confined to conventional classroom learning. The educational technology helps to improve the learning. It is mostly rote memory in school education, and "learning" does not naturally take place. The teaching of English is needed to be more practical oriented. Children are preparing for the next hundred years. The information and communication technology tools offer a powerful platform for better education in Indian schools and colleges. Electronic gadgets have become an integral part of personal and professional lives. People from all walks of life use technology. Students across the world are using technology more. It is a fact that students are more knowledgeable and skilled when it comes to using technology than teachers. Teachers are needed to embrace technological tools for teaching. They facilitate models of the language used to develop the students' language skills. The integration of electronic tools in the classroom creates a powerful environment, encourages learning, and nurtures a desire among students to have success in the future. E-tools are more powerful than any other teaching aid used so far. They boost students' self-confidence, facilitates independent and encourage self-paced learning, and hence, student-centered learning. The objective of this chapter is to exhibit a variety result in of E-tools to enhance English language skills of the students. Technology has changed both the role of a student and teacher. Hence, technology enables multimodal language activities for the students and let them grow as independent learners and act as the master of their own time.

17.1 INTRODUCTION

The teaching of English is needed to be more practical oriented. The students should be proficient in all four basic skills. There is no single, accepted method of teaching. Teachers are needed to think innovatively to design new methodologies to enhance teaching and learning. Learning is supposed to take place all the time because that is what education is about. But, more often, there is mostly rote memory in school education, and "learning" does not easily take place.

Teachers are no longer educating the children for the 20th century. They are educating children in the 21st century. Children are preparing for the next hundred years. Teachers are needed to embrace the dizzying pace requires tools that help them to understand the future. They need to seek the help of technology for teaching. The integration of technology in the classroom creates a powerful environment, encourages learning, and nurtures a desire in students to succeed in the future.

Technology has been advancing rapidly for the last few years. There are hundreds of studies on educational effects that were being investigated. It is difficult to measure the learning that technology enhances. The majority of the findings have indicated that English teachers understood to make the most benefit to learners through a variety of technology-rich learning environments.

Language learning has become a continuous process. It is not confined to conventional classroom learning. Indian institutions, organizations, and individuals are forced to contend in the globalized world, and success is possible only if our approach to education is changed or modified to suit the present-day need. The ICT tools present a powerful platform for better education in India. People from all walks of life use technology. Students across the world are using technology more. It is a fact that students have more knowledgeable and skilled when it comes to using technology than teachers.

The question regarding the relevance of technology in teaching, the majority of the reviews of current literature on "a variety of technologies and language learning" are proved positive. There is the potential to have dominant works in accomplishing educational visions. In addition to this, it is also revealed that students feel comfortable with the technological tools (Peck and Dorricott, 1994), and they would learn and master the skills required for the jobs effectively than ever before (Alavi, 1994).

There is some disagreement among experts as to what is meant by educational technology. Sometimes it is impounded to hardware like tele media, computers, and the teaching machines. It is an application of techniques for the systematic design. In this context, the prominence is more on a problem-oriented approach rather than on the use of individual media (Atkins and Clift, 2006). Tools can be selected based on their appropriateness for the specific educational objective of the particular course and their achievement. So, the word "technology" refers to any practical art using scientific knowledge.

17.1.1 RECENT TRENDS AND LANGUAGE LEARNING

Students can develop language and literacy skills in the following ways:

1. To incorporate viewing and representing English language competency.
2. To recognize children begin to develop learning ability before joining in the schools and their experiences help to influence language learning.
3. To recognize that children construct knowledge actively based on their knowledge and experiences (Vassiliki and Hamel, 2005).
4. To encourage an interactive process that facilitates social communication.
5. To acknowledge the importance of students' responses in learning through situations, fostering interactions, and connecting learning to their interests and experiences.
6. To encourage strategic thinking among students to solve the problems in the tasks effectively.
7. To promote interrogative and critical thinking skills among the students to evaluate and reflect their learning skills.
8. To emphasize planning, monitor progress, and evaluate their learning.
9. To promote active-based learning beyond the school environment.
10. To emphasize the potentiality of technology to teaching language and encourage learning.

11. To recognize the consequence of electronic gadgets on students to extend their learning.

17.1.1.1 MOTIVATION: A PART OF TEACHING

Motivation is an important factor in learning. Very little learning takes place when there is no motivation. The colorful pictures, animated action, audio, and other stimuli, which appeal to the senses, will catch the students' attention and motivate them to learn. However, the greatest motivational source is an achievement. Learning a new skill, being able to see progress, and being able to demonstrate a skill are all more highly motivating than much of the entertainment (J. Res. Computer Education 1990).

17.2 THE EDUCATIONAL TECHNOLOGY

The educational technology helps to improve the learning practice. It involves people, procedures, ideas, evaluations, implementations, and managing solutions to the problems related to learning (A.E.C.T. Task Force, 1977. p.164). It is the design of instructional materials and then interactions for maximum assistance. Hence, educational technology recommends structures of learning interactions for getting the greatest benefit (Figure 17.1).

FIGURE 17.1

https://www.indiatoday.in/education-today/featurephilia/story/digital-education-1027965–2017-08–08/

The shifting of research focuses on teachers to learners in second language acquisition. The learning process depends on a series of factors such as attitudes and motivation, which are accounted for language learning outcomes. Language learning is a scientific process. Learners develop the ability to generate language in their day-to-day life for specific communication purposes. The teacher facilitates models of the language used to develop the students' language skills. It has observed widely that a few students experience anxiety in the English classroom. As a result, they stutter, and their heartbeat increases (Cui, 2011). Anxiety is one of the important factors that affect L2 acquisition.

The student-centered model demands the equal involvement of both students and teachers. Teachers design lessons and learning tasks for the students' learning. They identify how students work themselves in a meaningful interaction. They plan the modules to generate appropriate language, and then students involve themselves to practice activities, thus simulate real communicative situations. Commitment by both of them leads to a dynamic classroom setting. Learning becomes enjoyable. In the learner-centered classroom, the instructor plans the activities and content, sets the time limits for each activity, and models the language competency. Hence, students' learning activities are still in control of the teacher.

With the convergence of computers and communication, there is the emergence of a technology that is more powerful than any other teaching aid used so far. One of the justifications for using the computer in language learning is that it offers a powerful self-access facility. Lerner-centered, self-pacing activities can be easily generated on computers. Multimedia presentations can be designed to learn a language. It acts as a powerful motivating factor for the students, thus reducing anxiety.

17.2.1 MULTIMEDIA TOOL FOR TEACHING

The multimedia tool is a combination of many media. Multimedia is the presentation of video, audio, animation as well as computer graphics and text, all on the same computer monitor. In a multimedia system, a variety of learning experiences are presented to the learner through the media available to the teacher. Different medium serves different educational functions. So several media used as multimedia serve the purpose better than separate ones. Certain instructional situations may require more than one medium to achieve maximum results. Each medium may serve a different purpose. So, when planned systematically, all media together can be effectively used for optimum learning to take place. The different medium serves to stimulate the different senses. Learning can be accelerated because the sensory experience is the foundation of any intellectual activity. Learners differ in their sense of reception. So, multimedia has the advantage of appealing to the individual, to the learner's pace, interest, and readiness.

17.3 THE INFORMATION AND COMMUNICATION TECHNOLOGY

The ICT experts in the field of English language teaching (ELT) predict that ICT will become a normal tool in the English language classrooms in the coming years. Gavin Dudeney and Nicky Hockly have listed the benefits of ICT that are mentioned below:

- Internet access is available to all learners in the 21st century.
- Technology has become an integrated part of the digital-born generation. The integration of technology as a tool brings the outside world into the classroom.
- Internet helps the English teachers to access e-books and a variety of ready-made ELT materials for designing authentic tasks and materials.
- The Internet offers an excellent connection for collaboration and communication between learners around the world who are geographically dispersed.
- Young children are fond of learning through the integration of technology into teaching.
- Technology offers a variety of exercises to practice language items and assessing their performance.
- Technology and ICT tools provide learners with exposure to practice, develop, and improve their L2 receptive and productive skills.

But even now in India the majority of the schools are without imaginative and creative language classrooms. Thus both the teachers and the students are felt bored entering the language classroom without any motivation or enthusiasm.

Technological devices offer a variety of grammar exercises that motivate students and facilitate their efforts to understand grammatical concepts and use them effectively in their daily conversation. The Internet allows learners more opportunities to practice online grammar quizzes at their own pace. This type of teaching has proven to be more effective than textbooks traditionally.

Technology-based authentic grammar-activities and web-based videos enable English teachers to design their context-based grammar exercises. Hence, technology helps the teachers to make their teaching interactive and effective.

17.3.1 LANGUAGE LABORATORIES

There are certain learning experiences unique to a language laboratory. The most striking feature has a variety of voices to the learner to expand the listening experience.

The language laboratory provides the following features that are as follows:

- The language laboratory reinforces the classroom work and accelerates the students' progress.
- It provides an opportunity and environment for practicing English pronunciation.
- It is the correct model for imitation.
- Isolation by headphones prevents self-consciousness.
- Increase motivation.
- It saves classroom time.
- It saves the teacher's voice and energy. But there is a lack of immediate feedback.

17.3.1.1 Advantages of the Language Laboratory

According to Hayes (1968), some of the main advantages of the language laboratory are as follows:

- In a language laboratory, all learners can practice aloud simultaneously, yet individually. This is not possible in a regular classroom situation.
- The teacher is free to focus attention on the individual learner's performance without interrupting the rest of the group.
- The wearing of headphones isolates each student to concentrate on the language items effectively and equal sound to the learners. This phenomenon provides complete concentration.

- The language laboratory facilitates the testing of each learner for listening comprehension, which otherwise becomes almost impracticable.
- The language laboratory is used as a self-access study center that can very well be used to conduct machine-directed activities as well. This factor enables the learner to self-instruct, self-monitor, and self-evaluate his performance.
- Finally, the language laboratory helps the low-level learner to get over his inhibition to speak in English, which in turn creates self-confidence.

17.4 WEB-BASED TECHNOLOGY AND TEACHING

New educational theories lay emphasis on student-centered classrooms and learners' autonomy. The web-based activities have mentioned a few qualities of meaningful learning that are as follows:

- active
- practical and collaborative
- deliberate
- interactive
- contextual and contemplative

1. The web-based activities support active learning where learners are engaged in "meaningful context" (Jonassen, 1995, p. 60).
2. Learners generate new information with the existed knowledge and team up with their peers in learning.
3. Learners deliberately work toward an objective.
4. The learners engaged in L2 conversations with their classmates through sharing the information that they have found on the Web and role-playing.
5. Learners are involved in a meaningful context, for instance, using web content.

The modern technological devices play a significant role in teaching grammar communicatively and excitingly. While teaching present continuous tense, the teacher can show numerous online visual prompts containing different activities such as swimming, running, cooking, playing, and so forth and ask the students what they are doing. By answering the questions, the students can practice present continuous tense. Instead of telling the students to memorize the rules of grammar, the teacher should design a variety of situations and ask them to frame different sentences that are suitable for the given situation. Likewise, the teacher can make students participate in the tasks actively, which makes them communicatively competent in producing grammatically correct and situation based sentences.

17.4.1 LEARNING THEORY AND TECHNOLOGY-ASSISTED LANGUAGE LEARNING

Technology-based language learning strongly supports the audio-lingual method that is based on B.F Skinner's behaviorist theory of learning popularly known in the 1950s and 1960s, with special emphasis on drills to teach structural patterns, memorization of set phrases, pronunciation, intonation, the teaching of vocabulary in context and reinforcement of correct responses. The basic assumption of this theory was that a learner responds to the environmental stimuli, and his repeated response reinforces language learning. Some of the electronic learning resources are based on the foundation of behaviorism. Numerous language software have immense potential for immediate feedback and encourage learners to follow the trial and error concept of learning. If learners give incorrect responses, they receive feedback and attempt the question once again to improve their mistakes.

Pitler et al. (2007) have pointed out that the behavioral theory of learning could be better taught with technological devices through drill activities leading to the acquisition of accurate pronunciation of difficult English words. Research data also show that EFL learners easily understand and retain new words with technology. For example, Nakata (2006) investigated that repeated exposures to media technology have far-reaching effects on learning new vocabulary. Hulstijn and Laufer (2001) advocated that the retention of unfamiliar vocabulary depends on learners' involvement in the use of technology.

17.5 TECHNOLOGY AND COMMUNICATIVE LANGUAGE TEACHING

Some software has simple drill and skill programs. For example, the website, Prentice-Hall, gives students an adequate opportunity to practice language a great deal. Electronic devices substantially facilitate communicative language teaching prominent between the 1970s and 1980s. Technology offers real-world language situations and motivates students to communicate with the correct pronunciation. It boosts students' self-confidence, facilitates independent and self-paced learning and student-centered learning. It turns students more active and engaged. Technological tools provide realistic language experiences through the blended learning environment. A teacher is needed to find a variety of activities for authentic communication.

They are language websites contain English as a second language (ESL) learning materials in abundance to teach listening, speaking, reading, writing, vocabulary, grammar, and pronunciation. LeLoup and Ponterio (2006) have considered a huge online video as a “priceless opportunity” to practice listening skills in an “entertaining environment.” According to them, television, news, documentaries, music videos, and any videos have inherent potentials to enhance learners’ primary skills. According to Beckett and Miller (2006), technology provides all items of language activities such as oral, reading, and writing skills. Zhanghongling (2006) strongly supports the technology in teaching language skills.

17.5.1 TECHNOLOGY-MEDIATED TEACHING AND THE INTEGRATED LANGUAGE SKILLS

The ESL students should be proficient in all the four skills for effective communication. There is no single accepted method of teaching and learning. Teachers are always looking for a new methodology to improve teaching and learning. Learning is supposed to take place all the time. Skyes (2013) believes that technology gives a very enjoyable experience for both learners and teachers. And integration of educational technology into ELT classes instill in learners substantial motivation for learning English (Figure 17.2).

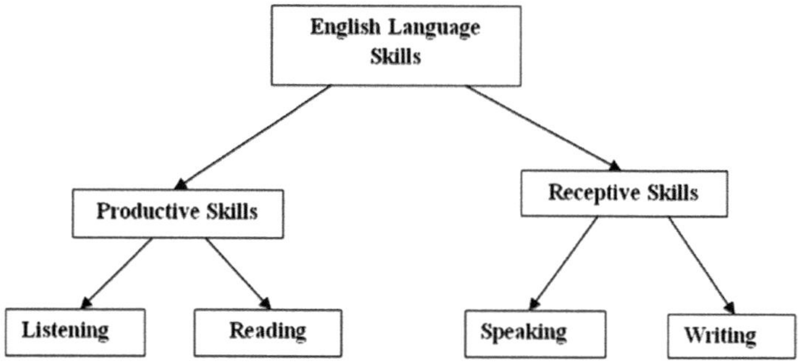

FIGURE 17.2
(*Source*: Original by Dr Shabreen Sultana Shaik)

17.5.1.1 Listening Skill

Listening is the first language skill that children acquire. It is an essential skill for all other skills of the language. It plays a significant role in acquiring language, whether it is in a first or second language. It is not just an act of recording the speaker's utterances and repeating them. Listening has the following components:

1. Listening is to recognize the systematic sounds of the language.
2. It recognizes the construction of sounds into words.
3. It recognizes the meaning of the words in sentences.
4. Listening helps comprehend the speaker's utterances.

To improve listening skills, ESL learners have to make the best use of YouTube, simple audio materials, and audiobooks. Extensive listening is a prerequisite to improve both listening comprehension and speaking skills. Listening to the native speakers of English through the online and reading the text are the two effective ways to understand, learn, and improve listening comprehension and accentual patterns of English words. Podcasts are also important tools to improve listening skills. Some of the popular podcasts the EFL learners can try are "Listen to English of Peter Carter," "Elementary Podcasts of the British Council," and others. There is always a misconception that modern technological devices can replace teachers and make them less necessary for the students. Teacher's role in the modern technology-led pedagogical setup is more challenging than before.

17.5.1.2 Speaking Skill

Speaking is an intricate skill. It is an art of generating language. The important components of speaking are mentioned below:

- The conscious usage of grammar, lexis, and enriching context of the language
- The context-based speaking ability
- Appropriate pronunciation
- Ability to present the ideas in a meaningful manner

Hosam (2012) considered TALL a better tool for learners of English because conventional teaching is very dreary and uninspiring. The traditional method of teaching makes learners passive recipient, whereas TALL increase learners' enthusiasm for learning a great deal. Many researchers strongly supported technology in the ESL classroom. For example, Munir, Amelia, Isham, and Siti Nur Afiqah (2012) have strongly supported the teaching of grammar via SMS because he considers SMS a significant pedagogical tool for teaching grammar. Issham Ismail and Rozhan M. Idrus (2009) are also of the view that learning grammar via technology, such as a computer, mobile/ SMS, offers more fun to the learners and gives them a unique excitement to learn grammar communicatively.

According to Solanki (2012), the use of authentic technological resources is effective in teaching English. It has the edge over traditional methodologies and revolutionizes the learner's dynamism for learning. Madhavaiah et al. (2013) also validate that technology propels the entire learning process and invigorates ESL learners to learn language skills faster than before because technology has multifaceted features to meet all the requirements of learners.

Al-Subary (2013) has revealed that the web-based teaching tool helps to enhance learners' performance in acquiring language skills. He has believed that technology equips learners to flourish 21st-century professional skills (2).

17.5.1.3 Reading Skill

Reading is one of the four basic skills. The level of reading skill varies from person to person. They learn according to their linguistic competence and background knowledge. However, the following are important components of reading skills:

1. To recognize the graphemes
2. To recognize the correlation between graphemes and words
3. To recognize word limit and sentence limit
4. To recognize the meaning of words and their relationship in a sentence
5. To recognize the relationship among sentences in a discourse
6. To infer implicit and explicit information and ideas of a text

The technology has changed both perspectives: the role of a student and the role of a teacher. Instead of just being a passive recipient in the class, the students have to engage themselves actively in problem-solving, communication, collaboration, and to share knowledge. They are knowledge managers. They have to grow as autonomous learners. They are supposed to improve their skills by using authentic online resources such as YouTube, e-mail, Twitter, Discussion Forums, Skype, WhatsApp, iPods, tablets, and e-readers. These technological resources with Internet connectivity can provide authentic communication opportunities to ESL learners beyond the ESL classrooms because they can use them anytime and anywhere to improve their interactive reading experience. This interactive reading can expose ESL learners to different syntactical structures and can expand a wide range of vocabulary.

17.5.1.4 Writing Skill

Writing skill is dynamic. The writing skill has several subskills. They are as follows:

- Mechanics—such as handwriting, spelling, and punctuation.
- Appropriate words—vocabulary.
- Paragraph organization— such as support and cohesion.
- Syntax— such as sentence structure and style.
- Grammar—such as rules of verbs, nouns, adjectives, adverbs, prepositions, and articles.
- Content—such as tone, clarity, originality and logic.
- The writing process—such as writing drafts and revising.
- Purpose of writing, justification, and others.

Cetto (2010) has stated that the technology expands the horizon of teaching and empowers communicative competency among learners. Houcine (2011) has also affirmed the effectiveness of the Internet and hypermedia in English language teaching. He considered the Internet as a powerful resource for both ESL learners and teachers. Technology offers several advantages that make it easier to understand and to improve communication skills. These are as follows:

- Comprehension exercises (oral or written)
- Dialogues, role-play, tasks
- Dictation
- Gap-filing tasks (suffixes, inflectional endings, word phrase in a sentence)
- Grammatical manipulations
- Message production such as e-mail, letters, memos, telegrams.

- Monologs, telling or retelling a story, present argument
- Multiple-choice activities
- Appropriate pronunciation like sounds, stress, intonation
- Vocabulary work

In addition to the more traditional exercises listed above, there are also communicative based activities and real items like message taking or giving, asking permission, requesting, handling telephone calls to which language is put into the context.

17.6 LATEST TECHNOLOGY

The modern ESL teachers are not just supposed to mesmerize ESL learners with their lecturing potential, but they have to facilitate learners and engage them in the whole learning process. Their entire focus should be on how to make learning more effective. In the modern technological era, the concept of pedagogy has completely changed. They have to evolve their teaching to substitute with the latest gadgets. They have to be competent in using the following technologies:

- WhatsApp
- Twitter
- Blogs
- E-mail
- Discussion Forums
- Wikis
- Skype

Chapelle (2010: 66) considered TALL as an effective tool for language learning. He realized the role of interactive CD-ROMs, e-mail, blogs, and wikis in ESL contexts. Baturay et al. (2010) have also endorsed that technology-based instructions enrich multimedia language content.

All the above can provide authentic communication opportunities to the learners. They have to cope with the paradigm shift, so-called, from traditional learning dubbed as a teacher-centered approach to student-centered learning. The ESL teachers have to encourage learners' full potential to embark on authentic online resources to stay independent learners rather than passive receptors relying on teachers. The ESL teachers should introduce the communicative approach, where students can be actively engaged in blended learning. They have to facilitate students to tackle learning problems based on their capabilities, which empowers them with a wide range of integrated language skills. They are expected to play the role of an instructional designer, facilitator, trainer, mentoring, and assessment specialist where they have to customize teaching materials, prepare effective summative assessment policies, and integrate technology in ESL classrooms to accomplish unique language learning tasks.

Collaborative teaching focuses on informal sharing teaching experience with colleagues, and team teaching is becoming very common in the modern pedagogical scenario. Here teachers are supposed to share their innovative and practical pedagogical experiences with their colleagues to make the whole learning process more informative and effective. There is an immense challenge for the teachers in this digital era. They have to plan and make the best use of electronic gadgets in the classroom. Several researchers have pointed out that it becomes worst when students become addicted to smartphones. They feel homophobia when they are prohibited from using mobiles in the classrooms. Learners' frenzied addiction to smartphones entertains their concentration and engagement in learning. The teacher and parents can obliterate the nasty side effects of smartphones on learners if they deal with students tactfully.

According to Odera and Ogott (2014:4), Internet-capable technology assists in collaboration and social networking, which can provide a better link between teachers and parents. Teachers can customize and upload their teaching materials in conformity with the level of the students and track the language activities and their progress online. The studies manifest that the ELT teachers and applied linguists strongly believed that TALL has enough and multifaceted potentials to improve learners' comprehensive language skills.

17.6.1 OVERVIEW OF MODERN ENGLISH AS A SECOND LANGUAGE TECHNOLOGY

The following points have an overview of technology for enhancing language skills (Figure 17.3):

FIGURE 17.3

https://zipzap.in/learning-and-smart-schools/

- Technology reduces the learner's anxiety and increases their motivation and participation, such as game-based activities and digital storytelling.
- It enables learners to collaborate, learn new vocabulary words, and build communities.
- It allows learners to construct a social interaction online, which gives them the confidence to interact with native speakers of the target language.
- Technology facilitates individual learning experiences where the learner monitors and tracks his progress
- It enables access to large corpora (authentic learning activities) that helps teachers to design a variety of tasks. (Godwin-Jones, 2017a; Kessler, 2018).
- It enables authentic contextual interaction through tools like virtual reality, online games, and video conferencing.

- Technology improves learner autonomy and allows informal learning experiences to empower learners.
- It allows learners to engage in meaningful interaction with native speakers through e-mail, online discussions, blogs, wikis, and web-based word processing, Google Docs, and others.
- It enables computer-based testing, which improves test security.
- It evaluates automated feedback on written tasks via chat boards, which is designed by teachers for text chat practice. It allows real-time feedback on assessments.
- Technology situated and personal learning via mobile devices (Godwin-Jones, 2016).

17.6.2 CHALLENGES IN ENGLISH AS A SECOND LANGUAGE CLASSROOMS

Computers in the learning and teaching of languages challenge the teachers throughout the world. Are computers compatible with English language teaching is the question asked by thousands of language teachers around the world. Teachers have to use them and view them from an innovative and creative angle. Second language acquisition theory has played a tremendous role in the development and the usage of E-tools in the ESL classroom.

17.7 CONCLUSION

Technology enables multimodal language activities (Blake, 2016; Felix, 2008). It is inferred that listening to native speakers on TV, radio, BBC, chat forums, and on the Internet is consequential to learners' high performance in pronunciation. The technology-based teaching of vocabulary produced a better result than traditional learners. The learners not only learn the words, but they also find competent enough to use the words in their day-to-day conversation. Technology-based grammar learning navigates various websites to have more grammar exercises contextually. Thus they harness well with the potential of e-learning and learn language skills more competently and effectively. Technology-based language learning seems vital for the teaching of integrated English language skills.

KEYWORDS

- **education technology**
- **E-tools**
- **language skills**
- **student-centered learning**

REFERENCES

Ahmad, Jameel. English Language Teaching (E.L.T.) and Integration of Media Technology, *Procedia—Social and Behavioural Sciences*, Vol. 47, 2012:924–929.

Ahmad, Jameel. Traditional and Socio-Cultural Barriers to EFL Learning: A Case Study, *English Language Teaching*, 2015.

Ambridge, B., and Lieven, E.V.M. *Language Acquisition: Contrasting Theoretical Approaches*. Cambridge: Cambridge University Press, 2011.

Atkins, K. J., Clift, J. C. Application of Educational Technology in the Development of a Learning Experience for a First Year Engineering Course, *Programmed Learning and Educational Technology*, 2006. Accessed May 20 April https://doi.org/10.1080/1355800750120103

Blake, Robert, J. *Brave New Digital Classroom.* Washington, D.C: Georgetown University Press, 2008.

Chinnery, G. M. Going to the MALL: Mobile Assisted Language Learning. *Language Learning and Technology*, Vol. 10, No. 1, 2006: 9–16.

Cui, Jingjing. Research on High School Students' English Learning Anxiety, *Journal of Language Teaching and Research*, Vol. 2, No. 4, 2011.

Ellis, Rod. *The Study of Second Language Acquisition.* UK: Oxford University Press, 2008.

Hackbarth, S. *The Educational Technology Handbook: A Comprehensive Guide: Process and Products for Learning.* Englewood Cliffs, NJ: Educational Technology Publications, 1996.

Hirsch, R. *Audio-Visual Aids in Language Teaching.* Washington, DC: Georgetown University Press, 1954.

Hlynka, D., and Mason, R. PowerPoint in the Classroom: What Is the Point?, *Educational Technology*, No. 38, 1998: 45–48.

Hocking, Elton. *Language Laboratory and Language Learning.* Washington, DC: Oxford University Press, 1964.

Simich-Dudgeon, J.C. Providing Support for Non-native Learners of English in the Social Studies Classroom, *The Social Studies*, Vol. 92, No. 1, 2001: 22–25.

Jhonston, J. Electronic *Learning: From Audiotape to Video Disk.* Mahwah, NJ: Lawrence Erlbaum Associates, 1987.

Singh, P.P., and Sandhir Sharma. *E-Learning: New Trends and Innovations*. New Delhi: Deep and Deep Publications, 2005.

Vassiliki Simina, Marie-Josée Hamel. C.A.S.L.A. through a Social Constructivist Perspective: Web-Quest in Project-Driven Language Learning, ReCALL, 2005.

https://www.researchgate.net/publication/231952049_CASLA_through_a_social_constructivist_perspective_WebQuest_in_project-driven_language_learning

Web Link. Educational Technology. Accessed April 5, 2020 https://www.indiatoday.in/education-today/featurephilia/story/digital-education-1027965–2017-08–08/

Web Link. C.L.I.L.: Teaching English. Accessed April 7, 2020 https://www.teachingenglish.org.uk/article/clil-a-lesson-framework/

Web Link. Smart Schools. Accessed April 10, 2020

https://zipzap.in/learning-and-smart-schools/

Williams, Eiko Torii. Incorporating the Use of E-Mail into a Language Program, *Computer Assisted Language Learning,* Vol. 17, No. 1, 2004: 109–122.

CHAPTER 18

Digital Pedagogy: Technology for Enhancing Learning Process

ADAM PAUL PATTETI

Maulana Azad National Urdu University, College of Teacher Education, NUH, Mewat, Haryana, India

Corresponding author. E-mail: pattety06@gmail.com

ABSTRACT

The 21st-century students are digital natives. Their thinking process is varying from their predecessors. The education system has been rapidly changing with innovative pedagogy. Students are fond of using the Internet, video games, mobile phones, and other digital tools. Teachers are needed to design lessons to make learning interactive and meaningful. The conventional methods of teaching are rigid and laborious. Education facilitates learning, knowledge, skills, and moral values. The Web 2.0 is being shifted from content-based pedagogies to communicative based pedagogies. The methodology of novel teaching is reproduced in the lesson plan of the teacher. Online resources such as e-tutoring, e-journals, e-books, and e-library facilitate learning without time constraints. This chapter explores the digital pedagogy for effective teaching and learning. The web pages, multimedia, podcast, lessons, assignments, multiple-choice questions, quiz, discussion group, and so many others are possible methods to deliver the content. E-learning has also been supplementing in the Traditional Classroom. The enthusiastic students use e-learning to progress their skills. Synchronous and asynchronous learning is self-paced learning. Students are allowed to swap over their data to the peer. Teachers design the text, methods, strategies, and tools of technology in education. The educational reforms depend on both the teachers and their learners. The state should encourage innovative pedagogy and conduct training and the Continuous Professional Development programs for the teachers.

18.1 INTRODUCTION

The modern students are digital natives. They are tech savvy. Their thinking process is varying from their predecessors. They are fond of by means of the Internet, video games, mobile phones, and other digital tools. New "digital pedagogy" is innovative teaching tools and abstract approaches. The education system has been rapidly changing with the latest tools and innovative pedagogy. Students would like to employ e-tools to learn new things and skills, solve problems, communicate, innovate, and collaborate. Thus teachers are needed to use appropriate digital pedagogy. The methods and approaches of teaching should be shifted from the teacher-centered approach to the learner-centered approach.

18.1.1 DIGITAL PEDAGOGY WITH WEB 2.0

Digital pedagogy is the study of digital technology in the classroom. It is an online and face-to-face learning environment. "New pedagogy" demands innovative teaching tools and learner-centered approaches. It supports adaptive and personal learning. Hence, digital pedagogy contributes to the design of new modes of learning.

Digital citizens (students) have invariable relationships with electronic tools. Teachers should be responsible for engaging them in activities. They are needed to possess digital pedagogy. They have to update their pedagogy suitable for the latest technology. Teachers are needed to design lessons to make learning interactive and meaningful. The digital pedagogy, that is, Web 2.0, is an innovative pedagogical tool. Teachers are needed to think about the innovative way to integrate e-tools in the classroom to construct and connect knowledge in real situations. Thus teachers design their lessons to practice digital pedagogy in their classrooms.

18.1.2 DIGITAL PEDAGOGY VALUES

Digital pedagogy has the following values:

- Open instructional resources,
- Input syllabi,
- Contribution of teaching resources,

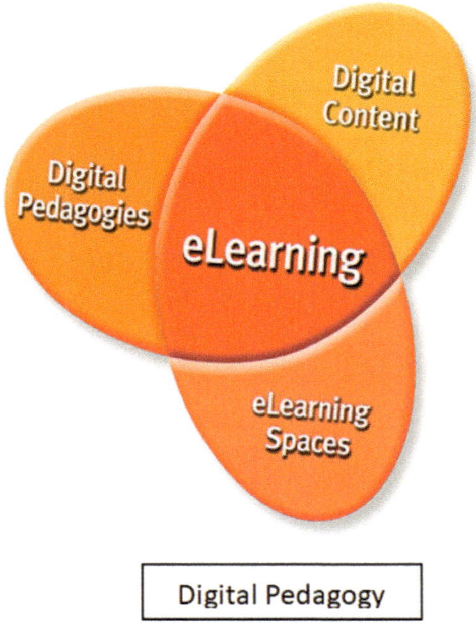

FIGURE 18.1 Digital pedagogy.

(Source: Reprinted from https://sites.google.com/site/andreasportfoliopedagogy/15-digital-pedagogy)

The long-established methods of teaching are found rigid and outdated. The new teaching–learning process steps are mentioned below:

1. Measurement
2. Preparation
3. Execution
4. Assessment and evaluation

18.1.3 SEVEN PRINCIPLES OF THE EFFECTIVE TEACHING–LEARNING PROCESS

The learning principles are needed to reflect from approaching to retrograde. It is enacted between exposure and justification in the interaction. Conventional learning environments are escorted normally through a sequence of the planned ladder to rational education consequences. It is classically displayed in a products-assisted method and approach where the activities guide in ordered comportment on the way to attain education. Hence, denotation is the potential where the procedure and content are learned in advance. It is followed to encourage learner–faculty contact; encourage coordination and cooperation; encourage dynamic learning; constructive feedback; emphasize the set-time for the task; expect the fluent communication; and respect the assorted way of learning.

The education is the course of assisting in learning, and the acquiring data, skills, and credence. Teaching and learning techniques are incorporated into teaching, training, and research. However, the learner can also edify himself. Education can be in a strict or relaxed manner. The methodologies and approaches of teaching are called pedagogy. Official education has been segregated into phases: nursery, primary school, high school, college, and university.

18.2 PEDAGOGICAL ELEMENTS

No definition of learning is found excellent, and it is agreed in worldwide. The following are one of the best definitions that are reliable with the perspective of learning:

1. (How) it has emerged through significance and association.
2. (What) it has emerged an unwrap system.
3. (Who) are self-directed, and self-organized managers.
4. (Where) information is stored inside individuals and superficially within the setup.
5. (Why) fabricates accomplished skills.

This connectivity-based explanation of learning is remarkably diverged from conventional based learning that is adopted more positive-based results. This definition emphasizes the role of network engagement among agents. Hence, the dispensed nature of learning is discussed.

Pedagogical elements are the units of learning material or content. These units have an independent format. The unit is delivered in various ways: the textbook, web page, multimedia, social media, podcast, lessons, assignments, multiple-choice questions, quiz, discussion group, and others. All of these are possible methods to deliver the content (Table 18.1).

TABLE 18.1 Five Aspects of Meaningful Learning

Aspects	Characteristics
Dynamic	(Calculating or attentive)
	Students are dynamic. They make an active function in activities. They enthusiastically manipulate objects and information, and curious to get feedback on their tasks.
Productive	(Articulate or contemplative)
	Learners create their comprehending information. They reproduce and communicate their individual kind. Their tasks are scrutinized.
Reliable	(Intricate or relative)
	Students are engaged in reliable tasks and difficulties rather than commit to memory concepts. They solve their real issues.
Planned	(Goal-directed or regulatory)
	Students set their goals and then plan their trails.
Supportive	(Mutual or informal)
	Learners are divided into groups and given the task to each group. They share their existed knowledge among the group and confer to enhance their learning and pertain to their understanding.

18.2.1 NEW PEDAGOGIES

The first educational type of pedagogy is the change from utilization to a construction. Conventional learning focuses on using up of content. Naturally, the learners are given a great portion of material that they must learn and understand. Then after, they have to replicate their memorized points in assessment and evaluation. Such sort of teaching centers on the transmission of evidence and perceptive and memorizing this information, as it is often found in conservative theories and definition of learning as conferred above.

The second style in teaching methodology is a change from accurate to accept. Conventional lessons focus on content that is accurate. This type of lecturing is centered lying on learners' knowledge procedures to rise with possible and accurate responses. Web 2.0 content improves and corrects the designed content.

The third style in teaching method is a change from text-assisted approach to a dialogues-based pedagogy. Formerly, education took place through conversation, which is named as the Socratic method. Afterward, learning has begun to center increasingly around the material. This method of education dominated the pedagogy for 500 years. However, technological innovations are being impacted on teaching pedagogy. Web 2.0 is being changed from corpora-based pedagogies to dialogue-based pedagogies. Hence, it is now called as "a Socratic 2.0 method."

The fourth change in teaching method is a change from fixed methodology to the methodology of turmoil. The conventional approach of learning focuses on sequence and organization. Here, the organization is educator-centric. An organized-based methodology is a content choice for content engagement. Typically, in this type, the learner is provided with reliable details and is directed to resolve the problem.

Teaching has been forged for centuries. It is a new paradigm shift. The pedagogy is where learning cannot be condensed to the preparations of (behaviorist theory), or the representation of (cognitive theory), or the tasks of (constructive theory). Education is a procedure of association between human and nonhuman.

Foreign studies have demonstrated new approaches to teaching (Table 18.2).

TABLE 18.2 Old Versus New Approaches to Teaching

Twentieth-Century Generation	New Generation
• Books and reading	• Display: Visual perception
• Single tasks	• Multitasking
• Linear approach	• Hypermedia
• Perception through reading	• Iconic perception
• Self-government	• Connection
• Vagueness	• Cooperation
• Passive school	• School as game
• Conversation and argument	• Warning
• Reality	• Fantasies
• External technology	• Internal technology
• Fact awareness	• Self-awareness

Innovation is a method, technique, technology, and new content. It is the introduction of the new and the novelty. Taubaeva and Laktionova have stated that there is a need to introduce innovative pedagogy to teach the digital natives in the education system.

18.2.1.1 INNOVATIONS IN PEDAGOGY

1. The individual power is boundless.
2. The realistic academic approach.
3. Stimulation of thinking.
4. Innovative pedagogy: the pleasure of learning, the delight of accomplishment, and the methodology of triumph.
5. Learning is a never-ending process.

The very pedagogy of novel education is student-centered learning. It is synthesized systemic, communicative competence, and so on. it is to establish the attitude of the innovative practice in teaching. The methodology of modern teaching is reflected in the lesson plan of the teacher. Hietajärvi et al. have named digital production as the "Sociodigital Generation" (Table 18.3).

TABLE 18.3 Sociodigital Generation Practices Versus Old Generation Practices

Sociodigital Generation Practices	Old Generation Practices
• Technological tools	• Conventional tools
• ICT tools	• Traditional aids
• Internet access	• Text
• Online	• Face-to-face
• Monitor and screen, notepad	• Blackboard, notebook, and pen
• Flexible and collaborative	• Individual performance
• Learning universally	• Fixed Traditional Classroom
• Task and creativity	• Laborious attainment

18.3 EDUCATIONAL TECHNOLOGY

Educational technology is facilitated learning and improved performance by providing appropriate technological resources. It is also called as EdTech or EduTech. It brings the pupils, teachers, and technology mutually in an efficient manner. It emerges education, psychology, sociology, and electronic gadgets. Technology in teaching includes several domains: learning theory, computer-assisted learning, Web-assisted pedagogy, and mobile-assisted pedagogy.

The educational technology allocates students to get learned from their own pace:

1. Educational technology enables part-time students, particularly to continue their education.
2. Educational materials are easy to get from e-resources.
3. Students can engage in online resources.
4. Web-based sources facilitate to expend time on tough portion.

Although learners miss some aspects of a classroom situation, these are helpful tools for the educational system.

18.3.1 PRESCHOOL

An assortment outline of electronic tools can be a future of nursery education. In the present era of technology, if a mobile phone is given to a kid, it starts operating it. It depends on technological resources to the candidate's level of maturity phase (labeled by Swiss psychologist, Jean Piaget). The consideration is such as appropriate age, consistency, ethical and educational features, which has recommended choosing the technology suitable for them.

At the preschool level, the rhymes on computers are introduced in classrooms. Age-appropriate educational videos or audio recordings, websites, Edu-apps, e-books assistive technology are introduced to them to participate with their peers. Educational websites and apps are available targeting the educational needs. Thus traditional learning and educational technology come together effectively. It allows children to experience screen concentration and motivation to complete learning tasks.

18.3.2 PRIMARY AND SECONDARY SCHOOLS

E-learning environments permit learners to be present from their homes. Several states are being used virtual learning platforms in the country, and they are kept on to enhance. Virtual classrooms in school facilitate them to log into synchronous or asynchronous everywhere. Online schools are heaven for students to obtain excellency.

E-learning has also been supplementing in the Traditional Classroom. The enthusiastic students get benefits to master their skills through electronic tools. Several online courses facilitate to perceive knowledge on the digital platform. Nationalized personal schools are available online. They allow students with great flexibility, and sometimes they are exempted from testing.

18.3.3 HIGHER EDUCATION

Higher level institutions present online classes. The majority of private institutions are involved in online presentations. Trained staff are also hired to teach the students online. They are needed to understand the content and also to get qualified to operate computers and surf the Internet. Online education is rapidly increasing at leading research universities.

18.4 DIGITAL REVOLUTION

Educators and students need not rely completely on conventional resources. Rather, they can access endless channels accessible through the Internet where students, teachers, experts, scholars, business leaders, and peers can access around the world. 21st-century learning requires a technological contribution to the rapidly transforming international environment. It has become essential for the teachers to employ electronic teaching aids: for statistical rationales, e-library, Internet, workplace tools, websites, and so on.

18.4.1 THE PARADIGM SHIFTS

The alteration in teaching is not just technological; it is rather renovated the entire setup of teaching. The ideology of globalization in the background of the global village is unbiased. It rather contains ideological reinforcement, which persuades the technoadapters not only the way they work but also their observation of methodology; education has gone through a transformation. Thus these concept changes from one point to another one as follows:

1. Teaching to production
2. Teacher-centered to learner-centered
3. The traditional teacher to the facilitator
4. Rote memorization to navigate and to learn
5. Learning is torture to learning is as fun.
6. Limited learning to lifetime learning.

The present education system is information overloaded. The role of the teachers is shifted from transferring content to facilitating and instructing learning procedure.

18.4.1.1 STUDENT-CENTERED LEARNING BACKGROUND

The learner-centered approach is originated from the constructive theory. It is argued that information is neither dependent nor inactively takes delivery of it. It is rather formed through information. The use of digital technology in teaching recommends supplementary student-centered settings, which are based on constructivist in nature. It enables learning related to the context. The Web has met the need for innovative ways to people disseminated across the lane. Experts have cultured an enormous contract the Web to raise and facilitate learner fraternity.

The National Research Council, United States has defined learner-centered situations to focus carefully on the knowledge, skills, behavior of the students convey with them to the classroom. The method is shifted from a contented-centered course to a proficiency-based syllabus. It is a shift from instructor-centered learning to student-centered learning. Technology-facilitated approaches encourage students to obtain their own learning significance. Instructional blogs and virtual classroom learning put forward additional opportunities to engage students (Table 18.4).

TABLE 18.4 Traditional Pedagogy Versus Digital Pedagogy

Aspect	Traditional Pedagogy	Digital Pedagogy
Active	• Activities approved by the teacher • Class instructions • Distinction in activities • Determined by the teachers	• Activities are determined for the learners • Group instructions • Several activities • Determined by learners
Collaborative	• Individual • Homogenous group • all and sundry	• Team Work • Heterogeneous group • Collaboration
Creative	• Traditional and fixed • Existed solutions	• Creative • Innovative solutions
Integrative	• No connection between assumption and outcome • Detach subjects • Personage teachers	• Integrating theory and practice • Connections between subjects • Group of teachers
Evaluative	• Teachers fixed • Cumulative	• Students engaged • Diagnostic

18.4.2 ONLINE COURSES

Online courses are available on the Internet. They are free or payable. Students enroll themselves in these courses and get training easily. Examples of the online courses are massive open online course, Future Learn, National Program on Technology Enhanced Learning, IITs, and so on. Students can access online and offline learning.

Online interactions ⇒ Blog posts, sharing artifacts online

Real encounter ⇒ Skype, Webinars, Online chat

Online learning ⇒ MOOCs

18.4.2.1 Massive Open Online Course

MOOCs are programs akin to edX founded by Massachusetts Institute of Technology, and Harvard University offers free and paid courses with an ample variety of disciplines. They are used to report themselves with extremely specific online courses for the interested candidates by the repudiated universities. They have been observed and found that the majority of them discontinue the course.

Several facilities are available on the Internet for the growth of learning. Online interactions would facilitate learning without time constraints.

Some of them are as follows:

1. E-training
2. E-lessons
3. E-journals
4. E-books
5. E-library
6. Virtual classes

Technology-assisted learning is both synchronous and asynchronous. Synchronous learning is concurrent learning and to act together simultaneously. It swaps over information with more than one participant simultaneously. They are as follows:

1. face-to-face
2. online instruction
3. feedback
4. Skype interaction
5. chatrooms
6. virtual classroom, and others

Students are facilitated to work collaboratively. Synchronize learning helps them to be open-minded. They can actively listen to their peers. It improves their writing skills.

Asynchronous is self-paced learning. It allocates applicants to exchange their ideas depending on the other participants' involvement. Streaming audio on the Internet and podcasts are examples of asynchronous. The following are the listed technologies:

1. E-mail
2. Blogs
3. Wikis
4. Web content and corpora
5. Hypertext documents
6. Multimedia courses
7. Web 2.0

Asynchronous collaboration provides helpful guidance to complete the tasks and assignments. Tools use for online courses are multimedia, class discussions, projects, and text material. Such students can complete their diploma courses and get certificates online or discontinue courses. Enthusiastic students can get registered with a variety of online courses to improve their knowledge and skills.

18.5 ELECTRONIC TOOLS: FEATURE FILMS IN TEACHING

Films based on documentaries, education, and entertainment enhance learning. All of these can create a high level of motivation. Consequently, the learning becomes satisfied.

Don Morrison (2003) has stated that the "educational constitution has to switch from 'push form' to 'pull form technology'." Previously, the teacher has stayed as the storehouse of data. Now students can surf their relevant information from the Internet.

18.5.1 GAMES IN THE CLASSROOM

Students can be engaged in many ways in the lesson. One of the effective ways is to use games in the activities. Games are highly motivating for students, and they also provide priceless opportunities to practice speaking. A set of rules is there to track and to reach their goals. It also helps students to interact with each other, which is difficult for the teacher to achieve, especially with self-conscious teenagers. Games create a friendly and enjoyable learning environment, linking "learning" with "fun" in the students' minds.

Another benefit is the motivation to work on mistakes. The particularly difficult aspect of learning new things is, in fact, students make mistakes at the beginning level. It is a normal human response to become embarrassed when we make a mistake, and this makes it difficult for the teacher to convince them to practice something new. The teacher can give feedback immediately. Thus the game motivates students to correct themselves to win the game. Games motivate themselves for repetition. Students always try to repeat a game to have more chances of winning without realizing that they are perfecting their learning at a similar time.

18.5.1.1 MS Office

The importance of using MS Word, MS PowerPoint, XP version, and Visual Basic in the current programmed leaning is as follows:

1. The learning material can be presented in a sophisticated manner.
2. The content to be presented in different types of learners.
3. To build up complex ideas animations are produced.
4. Hyperlinks and other programs are built .

18.5.1.2 Flipped Classroom

The flipped classroom is an instruction-based teaching approach. It is computer-based teaching in the classroom. Students are given basic instruction before assigning the task or activity. Sometimes, instructions are delivered online. Videos, texts, online, and other gadgets are used in this type of classroom.

18.5.1.3 Audio and Video

Video technology is DVDs, digital video through the server. Interactive electronic video games are being used at higher level students. Screen-casting shares the video available online viewers. Learners can pause, rewind, and review them. Webcams have enabled to create a virtual learning environment.

"Tandem language learning" is a technique of language learning based on language exchange between native and nonnative speakers who both want to learn each other's language. The condition is you must have a good grip on your mother tongue. At the beginning of the 19th century, Joseph Lancaster and Andrew Bell started as "Mutual System." It is a kind of peer teaching—exchange between Frenchman and Englishman of both languages. Tandem learners can exchange knowledge through e-mail or multimedia. These two people can also have a counsellor, or they can learn in groups. Tandem learning first appeared in 1971 with the influence of audio–visual method. German–French language learners were first to study through Tandem learning.

18.5.1.4 Disadvantages

The technology does not result in a fundamental enhancement in educational practice. It is needed to focus on the learner's communication with electronic tools but not the tools alone. It is recognized as "tool" rather than "addictive" or "subtractive." Branford et al. have stated that "technology does not guarantee effective learning." He also believed that the inappropriate use of these tools can cause obstacles. Technology supports individuals rather than collaborations.

18.6 CONCLUSION

Teachers design the text, methods, strategies, and tools of technology in education. The educational reforms depend on both the teachers and their learners. The state should encourage the teachers for innovative pedagogy and provide them with professional training programs, and thus Continuous Professional Development of the teachers. Active learning approaches, higher and the facilitating technologies affords a series of emerging innovative educational approaches. This need has led many organizations to take hold of the tool to sustain this transformation. However, the force of traditional teaching for centuries and informative pedagogies are not basically a matter of swapping e-tools with smartboards and tablets to chalkboards and textbooks. The progress from a conventional methodology to digital methodology is mutually as nuanced and indeterminate in the digital planet.

KEYWORDS

- **Web 2.0**
- **interactive learning**
- **digital pedagogy**
- **online content**
- **digital natives**

REFERENCES

Adams, MA. *Beginning to Read: Thinking and Learning about Print*. Cambridge: MIT Press, 1990.

Anderson, T. Towards a Theory of Online Learning, *Theory and Practice of Online Learning*, No. 2, 2008: 15–44.

Atkins, M.J. Theories of Learning and Multimedia Applications: An Overview, *Research in Education*, Vol. 8, No. 2, 1993: 251–271.

Babalola, B., A Olubiyi, M Adarabioyo. Modeling Students' Attendance with Respect to Lecturers' Qualities/Attributes, *Journal of Education, Society and Behavioural Science*, 2018.

Best, JW and J.V. Khan. *Research in Education*, 7th ed. New Delhi: Practice Hall of India, 1999.

Blewett, C. (2016). Chapter 16 From Traditional Pedagogy to Digital Pedagogy: Paradoxes, Affordances, and Approaches. In *Disrupting Higher Education Curriculum* (pp. 265–287). Brill Sense.

Breunig, Karl Joachim. *Limitless Learning: Assessing Social Media Use for Global Workplace Learning*. The Learning Organization, 2016.

Chaudron, Craig. *Second Language Classroom, Research on Teaching and Learning*. Cambridge: Cambridge University Press, 1988.

Chiknaverova, K. (2020). Integrating Foreign Languages and Business Informatics Teaching in Russian Universities. The Problem of Teaching Content. In *Proceedings of the Conference Integrating Engineering Education and Humanities for Global Intercultural Perspectives*, March (pp. 47–56). Springer, Cham.

Crystal, D. *Language and The Internet*. New York: Cambridge University Press, 2001.

Dalgarno, B., and Lee, M. J. What are the Learning Affordances of 3-D Virtual Environments?, *British Journal of Educational Technology*, Vol. 41, No.1, 2010: 10–32.

Downes, S. Learning Networks in Practice, *Emerging Technologies for Learning*, No. 2, 2007: 19–27.
Heider, K., and Laverick, D. A., & Bennett, B. Digital Textbooks: The Next Paradigm Shift in Higher Education? *AACE Journal*, No. 17, 2009: 103–112.
Kasemsap, K. (2016). Chapter 12 Exploring the role of web-based learning in global education. In *Revolutionizing Education through Web-Based Instruction* (pp. 202–224). IGI Global.
Oliver, M. The Problem with Affordance, *E-Learning and Digital Media*, Vol. 2 No. 4, 2005: 402–413.
Robninson, K., and Aronica, L. *The Element: How Finding Your Passion Changes Everything*. London: Penguin Books, 2009.
Rogers, E. M. *Diffusion of Innovations* (3rd ed.). New York, NY: The Free Press, 2010.
Song, Donggil., and Lee, J. Has Web 2.0 Revitalized Informal Learning? The Relationship between Web 2.0 and Informal Learning, *Journal of Computer Assisted Learning,* Vol. 30, No. 6, 2014. DOI: 10.1111/jcal.12056.
Venkatesh, V., Morris, M. G., Davis, G. B., and Davis, F. D. User Acceptance of Information Technology: Toward a Unified View, *Management Information Systems (MIS) Quarterly*, No. 1, 2003: 425–478.
Warschauer, M. The Paradoxical Future of Digital Learning, *Learning Inquiry*, No. 1, 2007: 41–49.
Web Link. Digital Pedagogy. Accessed April 6, 2020.
https://sites.google.com/site/andreasportfoliopedagogy/15-digital-pedagogy
Web Link. Educational Technology. Accessed April 13, 2020
https://www.brookings.edu/opinions/five-ways-teachers-can-use-technology-to- help-students/
Web Link. Emerging Technology. Accessed April 13, 2020. https://en.wikibooks.org/wiki/Web_2.0_and_Emerging_Learning_Technologies /Learning_Theory.
Web Link. Enquiry Based Teaching and learning. Accessed April 13, 2020
https://www.researchgate.net/publication/327399259_What_Is_Inquiry- Based_Science_Teaching_and_Learning.

Index

D

E

F

G

T

V

W